Interferon and Nonviral Pathogens

IMMUNOLOGY SERIES

Editor-in-Chief
NOEL R. ROSE
Professor and Chairman
Department of Immunology and
*   Infectious Diseases*
The Johns Hopkins University
School of Hygiene and Public Health
Baltimore, Maryland

European Editor
ZDENEK TRNKA
Basel Institute for
*   Immunology*
Basel, Switzerland

Additional Volumes in Preparation

Interferon and Nonviral Pathogens

edited by

Gerald I. Byrne

University of Wisconsin Medical School
Madison, Wisconsin

Jenifer Turco

University of South Alabama College of Medicine
Mobile, Alabama

MARCEL DEKKER, INC.   New York and Basel

Library of Congress Cataloging-in-Publication Data

Interferon and nonviral pathogens / edited by Gerald I. Byrne, Jenifer
Turco.
         p. cm. – (Immunology series ; 42)
     Includes index.
     ISBN 0-8247-7973-8
     1. Interferon--Therapeutic use--Testing. 2. Infection-
-Immunotherapy. 3. Intracellular pathogens. I. Byrne, Gerald I.
II. Turco, Jenifer. III. Series: Immunology
series ; v. 42.
     [DNLM: 1. Immunity. 2. Interferons. W1 IM53K v. 42 / QW 800
I593]
RM666.I58I58 1988
616.07'9–dc19
DNLM/DLC   88-20318
for Library of Congress   CIP

MARCEL DEKKER, INC.
270 Madison Avenue, New York, New York 10016

Current printing (last digit):
10 9 8 7 6 5 4 3 2 1

PRINTED IN THE UNITED STATES OF AMERICA

To our families, friends, students, and colleagues

Series Introduction

The interferons represent a particularly fascinating chapter in the study of infection and immunity. First considered merely a laboratory curiosity, the interferon systems became the focus of enormous attention in the 1970s. They were touted as the panacea for cancer in its many manifestations as well as for infectious disease. This wave of overenthusiasm was followed by unwarranted discouragement and disillusion as the results of clinical trials came to publication.

The present volume corrects these wild swings in expectations. The interferons are neither a universal cure nor a hollow promise. With judicious attention to application, they represent important additions to the medical treatment of many infectious diseases and a few forms of cancer. They herald another step in the evolution of clinical immunology from a heuristic to a therapeutic specialty.

Noel R. Rose
Professor and Chairman
Department of Immunology and
Infectious Diseases
The Johns Hopkins University
School of Hygiene and Public Health
Baltimore, Maryland

Foreword

It is most fortunate that Dr. Gerald I. Byrne and Dr. Jenifer Turco have undertaken to edit a volume devoted to interferon, immunity, and nonviral intracellular pathogens. Although the inhibitory action of interferon on the multiplication of chlamidiae was demonstrated almost twenty-five years ago by Sueltenfuss and Pollard, this finding, and the somewhat later demonstrations of an inhibitory action by interferon on protozoan agents, had long been largely ignored. Until about ten years ago, most investigators viewed interferons as selective antiviral agents. The early findings of antichlamydial, antirickettsial, antitoxoplasma, and antiplasmodial activities did not fit the then-accepted clichés about interferon action. As frequently happens, valid and important experimental findings were either overlooked or dismissed because they could not be accommodated by existing theories.

By now a large body of experimental evidence, accompanied by a much better understanding of the molecular mechanisms of interferon action, clearly points to the important role of the interferons in infections due to nonviral intracellular pathogens. In general terms, these actions can be divided into two broad categories—those that affect primarily the development of the immune response and those that result in a direct inhibition of the multiplication of the pathogens in the target cells. Both of these general mechanisms of interferon action are thoroughly described and analyzed by the contributing authors in this volume.

Much of the information included in this volume until now was scattered in hundreds of publications or not available at all. The initial two chapters provide a summary of essential information about the nature and the actions of interferons. The balance of the text is devoted to a description of the role of interferons in infections caused by individual groups of intracellular pathogens. Many chapters reach far beyond their primary topic. For example, the chapter on leishmania covers in considerable detail the basic mechanisms of macrophage activation and the interactions of interferon-gamma with other cytokines in this process. The chapter on interferons, immunity, and chlamydiae provides an analysis of the molecular mechanisms of interferon action that are likely to operate also against other types of intracellular agents.

This volume is being published at a time of an emerging awareness of the intimate interplay of the two classes of interferons (alpha/beta and gamma) with other cytokines. It is already clear that in infections caused by intracellular pathogens, the interferons often act in concert with interleukin-1, tumor necrosis factor, and other cytokines, some yet to be identified. Much more experimental work remains to be done to gain a better understanding of these interactions. This book could play an important role as a catalyst of more exciting research in this area. Those investigating the role of interferons and other cytokines in intracellular infections today and those who will be stimulated to enter the field after reading this volume are not likely to encounter the same skepticism and reservations that faced the pioneering workers two decades ago. Much new information about the mechanisms discussed in this volume will emerge in the next few years. (Perhaps Dr. Byrne and Dr. Turco should start thinking about a second edition.) But until that time, this unique book fills a big gap in the literature on interferons and nonviral infectious agents.

Jan Vilček, M.D.
Professor of Microbiology
New York University Medical Center
New York, New York

Preface

The inspiration for this book arose as a result of a seminar on Interferon, Immunity and Intracellular Pathogens convened by us at the annual meeting of the American Society for Microbiology in 1986. We eventually expanded our group of contributors from the original six speakers at the seminar to include other international experts on the role of interferons in recovery (or lack thereof) from infections caused by a variety of nonviral intracellular pathogens.

The volume has been divided into three parts. The first part provides background information on the interferon system—how interferons are induced and how these important regulatory proteins exert their influence on cells and organisms. The remaining two parts deal with the induction of interferons by a variety of intracellular microorganisms and the effects of interferons on the host cells and the microorganisms. The pathogens have been divided into obligate intracellular parasites (Part II) and facultative intracellular microorganisms (Part III).

We asked the contributors not only to provide up-to-date accounts of research in their particular fields (with emphasis on their own contributions), but also to go beyond bare facts to give their insights and perspectives. The facts have been presented; in addition, the authors have included comments on what is relevant and important, what needs to be done, how each piece of research fits in with the work of others, and what the future may hold. This format provided contributors with freedom of expression that is not allowed in the primary literature: We hope this has provided the reader with a text that is informative and enjoyable.

We would like to thank Neil Schiller for suggesting that we organize the seminar, our numerous colleagues who contributed to this volume or read the contributions of others, our publisher, and especially our production editor, Carol S. Carol, for her patience.

Gerald I. Byrne
Jenifer Turco

Contents

Contributors

Peter W. Andrew, Ph.D. University of Leicester, Leicester, England

Fausto G. Araujo, D.V.M., Ph.D. Palo Alto Medical Foundation, Palo Alto, California

Miodrag Belosevic, Ph.D. Walter Reed Army Institute of Research, Washington, D.C.

*Carolyn M. Black, Ph.D.** Palo Alto Medical Foundation, Palo Alto, and Stanford University Medical Center, Stanford, California

Ernest C. Borden, M.D. University of Wisconsin Clinical Cancer Center, Madison, Wisconsin

Geir Bukholm, M.D., Ph.D. The Wilhemsens Institute of Bacteriology, University of Oslo, National Hospital, Oslo, Norway

Gerald I. Byrne, Ph.D. University of Wisconsin Medical School, Madison, Wisconsin

Joseph M. Carlin, Ph.D. University of Wisconsin Medical School, Madison, Wisconsin

Miklos Degré, M.D., Ph.D. The Wilhelmsens Institute of Bacteriology, University of Oslo, National Hospital, Oslo, Norway

Arturo Ferreira, D.V.M., Ph.D.† New York University School of Medicine, New York, New York

Edward A. Havell, Ph.D. Trudeau Institute, Inc., Saranac Lake, New York

David L. Hoover, M.D. Walter Reed Army Institute of Research, Washington, D.C.

---

*Present affiliation: Centers for Disease Control, Atlanta, Georgia
†Present affiliation: Faculty of Medicine, University of Chile, Santiago, Chile

Marcus A. Horwitz, M.D.   Center for the Health Sciences, UCLA School of Medicine, Los Angeles, California

*Thomas R. Jerrells, Ph.D.**   Walter Reed Army Institute of Research, Washington, D.C.

Felipe Kierszenbaum, Ph.D.   Michigan State University, East Lansing, Michigan

Douglas B. Lowrie, Ph.D.†   Hammersmith Hospital, London, England

Robert E. McCabe, M.D.‡   Martinez Veterans Administration Medical Center, Martinez, California

Monte S. Meltzer, M.D.   Walter Reed Army Institute of Research, Washington, D.C.

Carol A. Nacy, Ph.D.   Walter Reed Army Institute of Research, Washington, D.C.

Donna M. Paulnock, Ph.D.   University of Wisconsin Medical School, Madison, .
Wisconsin

E. R. Pfefferkorn, Ph.D.   Dartmouth Medical School, Hanover, New Hampshire

Jack S. Remington, M.D.   Palo Alto Medical Foundation, Palo Alto, and Stanford University Medical Center, Stanford, California

Israel Sarov, Ph.D.   Ben-Gurion University of the Negev, Beer Sheva, Israel

Louis Schofield, Ph.D.   New York University School of Medicine, New York, New York

Yonat Shemer   Ben-Gurion University of the Negev, Beer Sheva, Israel

Gerald Sonnenfeld, Ph.D.   University of Louisville Schools of Medicine and Dentistry, Louisville, Kentucky

Jenifer Turco, Ph.D.   University of South Alabama College of Medicine, Mobile, Alabama

Herbert H. Winkler, Ph.D.   University of South Alabama College of Medicine, Mobile, Alabama

---

*Present affiliation: University of Texas Medical Branch, Galveston, Texas
†Present affiliation: National Institute for Medical Research, London, England
‡Present affiliation: University of California, Davis, California

Part I

THE INTERFERON SYSTEM

1

Interferons and Their Induction

Joseph M. Carlin / University of Wisconsin Medical School, Madison, Wisconsin

Ernest C. Borden / University of Wisconsin Clinical Cancer Center, Madison, Wisconsin

I. INTRODUCTION

Since the original demonstration of interferon by Issacs and Lindenmann (1) 30 years ago, in which a factor released by chicken egg chorioallantoic membranes exposed to heat-inactivated influenza virus was shown to inhibit the replication of live virus, the complexity and impact of the interferon system has expanded substantially. Several excellent reviews (2-5) have been written concerning interferons and their physical characteristics, inducers and mechanisms of induction, mechanisms of regulation of production, genetics, and their biological activities. Since an all-inclusive review of the interferon system is beyond the scope of this chapter, its purpose will be to present a general overview of interferons and their induction processes.

II. PHYSICAL CHARACTERISTICS OF INTERFERONS

Interferons are defined as inducible proteins that inhibit viral replication. Additional characteristics which serve to distinguish interferons from antibodies and nonspecific inhibitors of viral growth include several biological and physicochemical criteria (Table 1). Interferons are divided into three main classes, α, β, and γ (Table 2). Although immunologically distinct, the type I interferons, interferon α and interferon β, share many characteristics and have been shown to be physicochemically related (7), with approximately 30% shared amino acid sequence and with 46% DNA sequence homology. It is speculated that these two interferon classes arose from a common ancestral gene $5-10 \times 10^5$ years ago.

Table 1 Definition of Interferon

Biological criteria

1. Inducible protein which inhibits replication of a broad range of viruses, both DNA and RNA.

2. Virus replication is inhibited intracellularly through a process that involves de novo cellular RNA and protein synthesis.

3. Antiviral effects of interferons from individual vertebrates each have characteristic and limited host range specificities.

Physicochemical criteria

1. Protein with antigenic cross-reactivity with known interferon.

2. Amino acid or DNA sequence similar to known interferon.

3. Antiviral effects mediated through activation of 2–5A synthetase.

Source: Adapted from Ref. 6.

Table 2 Human Interferon Classification

Characteristic	Class		
	α	β	γ
Synonym	leukocyte lymphoblastoid type 1	fibroblast type I	immune type II
Cell source	leukocytes lymphoblastoid cells	fibroblasts	T lymphocytes NK cells
Molecular weight	18,000–22,000	20,000–25,000	20,000–25,000
pH 2 stable	yes/no	yes	no
Heat stable (56°C for 30 min)	yes	yes	no
Glycosylation	no	yes	yes
Species number	20	1–5	1
Introns	no	no/yes	yes
Chromosomal location	9	9,7,5,2	12

Interferon γ (type II) is both antigenically and genetically distinct from type I interferons (8,9) although a common evolutionary ancestor for type I and type II interferons has been proposed (10).

A. Interferon α

Interferon α (IFNα) represents a family of related interferons produced by leukocytes and lymphoblastoid cell lines in response to virus infection (11), hence its former names "leukocyte interferon" and "lymphoblastoid interferon." The human interferon α species are physicochemically heterogeneous when analyzed by sodium dodecyl sulfate-polyacrylamide gel electrophoresis (SDS-PAGE), affinity chromatography, isoelectric focusing, or high performance liquid chromatography (HPLC) (12-20), and display a range of molecular weights, from 18,000 to 22,000. Since most IFNα species are not glycosylated (21,22), heterogeneity in molecular weight is due to differences in structure and amino acid sequence. Interferon α is characterized further by its acid stability at pH 2. One exception is a recently described species of interferon α found in the circulation of patients with the immune system disorders, systemic lupus erythematosus (SLE), acquired immunodeficiency syndrome (AIDS), generalized lymphadenopathy, and juvenile arthritis (23-27). Although interferon obtained from the serum of these patients is characterized as interferon α by neutralization assays with anti-IFNα antibody, it is acid labile. This species also can be obtained from normal donor blood, produced either by Sendai virus-induced polymorphonuclear cells or produced spontaneously in culture upon incubation of a subfraction of large granular lymphocytes (28,29). The exact relationship of acid-labile interferon α to other IFNα species awaits further characterization.

More than 20 interferon α genes and pseudogenes (α_1, α_2, α_3, etc.) have been identified (30-35) and classified into two large subfamilies based on amino acid replacement (36). Amino acid sequences of interferon α species derived from virus-induced leukocyte mRNAs show about 80% homology and nucleotide sequences show 90% homology (34,37,38). On the basis of this homology, it has been calculated that the human interferon α genes diverged $2-3 \times 10^5$ years ago. Mature interferon α polypeptides are 165 or 166 amino acids in length. Somatic cell hybridization studies have shown that all human interferon α genes reside on chromosome 9 (39,40).

B. Interferon β

Although interferon β also is secreted by virus-induced leukocytes (although to a lesser degree) (41), it is the major class of interferon generated by polyribonucleotide-induced fibroblasts. Purified natural interferon β_1 has a molecular weight of approximately 20,000-25,000 daltons (42-44). Its amino acid sequence has been determined from cloned cDNA (45-47) and IFNβ₁ shows homology with IFNα at both the amino acid and nucleotide level (7). Its similarity to interferon α does not end at the sequence level. Interferon β_1 is also 166 amino acids in length and also contains no introns in its genetic structure (48,49). Not only is it acid stable like interferon α, but it also binds to the same receptor (50). However, unlike IFNα, interferon β_1 produced by eukaryotic cells is glycosylated (51).

In addition to interferon β_1, up to six additional interferon-β mRNAs have been described (52). Only one of their products, interferon β_2, has been characterized sufficiently to compare to interferon β_1. The physicochemical relationship of interferon β_2 to interferon β_1 is still being elucidated. Although initial identification of interferon β_2 was based on its neutralization and precipitation by polyclonal anti-IFNβ (53,54), highly specific anti-IFNβ$_1$ antibody fails to precipitate IFNβ$_2$ (55). Furthermore, mRNAs specific to interferon β_1 and interferon β_2 fail to cross-hybridize, indicating a lack of homology between mRNA sequences (53,54). The interferon β_2-specific mRNA codes for a 26,000 dalton primary transcript (54). This polypeptide is glycosylated and cleaved to a 21,000–22,000 dalton product (55), approximately 180–184 amino acids in length (56), before secretion from the cell. The gene coding for the 26,000 molecular weight protein has been sequenced (56). Approximately 20% amino acid homology has been defined between interferon β_1, and interferon β_2 (57) with similarities especially in the region of interferon β_1 amino acids 42–52 (54), which contains conserved sequences between IFNα and IFNβ$_1$. Unlike IFNβ$_1$, the IFNβ$_2$ gene sequence is interrupted by at least three introns (56–58). The sequence of interferon β_2 has been shown to be identical to the reported sequence of human B-cell differentiation factor (BCDF or BSF-2) (59).

Some investigations have failed to induce antiviral activity with translation products of IFNβ$_2$ mRNA (55); others have induced antiviral activity with similar preparations (53,54,58). In addition, antiviral activity induced by tumor necrosis factor (TNF) has been shown to be neutralized by antisera to IFNβ. Blot hybridization analysis has shown that TNF induces IFNβ$_2$ mRNA, but not IFNβ$_1$ mRNA (60). It has been speculated that differences in biological interferon assay system sensitivities may account for the conflicting results obtained from various laboratories (56).

Characterizations of genomic clones and chromosomal mapping studies have demonstrated that the interferon β_1 gene is present on human chromosome 9 (39,40,61), the same chromosome which contains the interferon α gene family. Similarly, both the murine interferon β_1 and interferon α genes have been located on mouse chromosome 4 (62,63). However, somatic cell hybridization studies have indicated that human interferon β genes exist on at least three chromosomes, 2, 5, and 9 (64,65). Chromosomal mapping using blot hybridization has located the interferon β_2 gene on human chromosome 7 (66). Whether or not additional interferon β species can be matched to chromosomes 2 and 5 awaits further experimentation.

C. Interferon γ

Although interferon γ has been distinguished classically from type I interferons (IFNα and IFNβ) on the basis of its inactivation by pH 2 acid treatment or 56°C heat treatment (67), it differs in many of its physicochemical characteristics. It is typically produced during immune responses (67,68), hence its former name "immune interferon." T lymphocytes, including helper and cytotoxic/suppressor phenotypes, are the major producers of interferon γ (69-71), although production of IFNγ by natural killer (NK) cells has been reported (72-74). Interferon γ was originally estimated to have a molecular weight in the 36,000-58,000 range, however, it now is believed that these weights represent a complex of subunits, possibly dimers (9,75). SDS-PAGE of IFNγ reveals two proteins with molecular weights of 20,000 and 25,000 (76). Although two distinct bands are resolved in these gels, the observed difference in size between these forms represent differences in carbohydrate content only. Both forms are glycosylated (77) and analysis reveals identical amino acid sequences (78). Based on this sequence, the nonglycosylated form has a calculated molecular weight of 17,000 (8,79). Interferon γ shares little sequence homology with the type I interferons (8,9), although a common ancestral gene has been proposed, based on secondary structure and previously undetected sequence homologies (10). Unlike IFNα and IFNβ₁, the IFNγ peptide is only 143 amino acids in length (78). Screening of a human gene bank has revealed only one interferon γ gene (80), which contains three introns (79) and is located on human chromosome 12 (40,80), outside of the type I interferon gene cluster. Although interferon γ shares many biological activities with type I interferons, it has a distinct receptor (50).

III. INDUCTION OF INTERFERONS

Mechanisms of induction of interferons fall into two categories, classical and immune induction (81). Classical induction, or induction of interferon by viruses and polynucleotides, is exclusively a property of type I interferons. On the other hand, immune induction, or induction of interferon during the process of lymphoid cell activation, is the mechanism by which interferon γ is induced. Immune induction, however, is not restricted to IFNγ. Both IFNγ and IFNα are induced in sensitized lymphocytes exposed to viral antigens (82). Furthermore, phorbol ester (83) or lectin (84) stimulation of lymphoid cells results in simultaneous IFNα and IFNγ production.

A. Classical Induction

1. Double-Stranded RNA and the Induction Process

Although a variety of agents have been shown to induce type I interferons (Table 3) (reviewed in Refs. 2 and 85), most studies of classical induction of

Table 3 Inducers of Type I Interferon

Animal viruses, RNA and DNA
Plant viruses
Mycophages and bacteriophages
Double-stranded RNA, natural and synthetic
Unicellular agents
  Bacteria
  Rickettsia
  Mycoplasma
  Chlamydia
  Protozoa
  Tumor cells
Microbial and cellular products
  Bacterial polysaccharide, lipopolysaccharides, and proteins
  Fungal polysaccharide and peptidoglycan
  Viral glycoproteins
  Plant polysaccharide
  Nucleic acids, animal and viral
Synthetic chemical compounds
  High molecular weight polymers
    Polycarboxylates, polysulfates, and polyphosphates
    Synthetic single-stranded RNA
  Low molecular weight compounds
    Metabolic inhibitors
    Tricyclic compounds
    Pyrimidine derivatives

Source: Adapted from Ref. 85.

interferon have used viruses or double-stranded RNA as the inducer molecules. Induction of type I interferon by viruses and polynucleotides is thought to be mediated by basically identical mechanisms, although some differences exist in induction by these agents. Evidence suggests that viral induction of interferons requires the presence of double-stranded RNA. This has been demonstrated by comparing the interferon-inducing capacity of ultraviolet (UV) inactivated plus- and minus-stranded RNA viruses. Minus-stranded RNA viruses possess a transcriptase that generates complementary RNA in the absence of RNA replication; thereby enabling the production of double-stranded RNA and subsequent interferon induction despite UV inactivation (86). In plus-stranded RNA viruses, a polymerase required for the generation of double-stranded RNA replicative forms is coded for by the plus strand. UV-inactivation of plus-stranded RNA viruses destroys the viruses' ability to synthesize the polymerase, and thereby abolishes both the capacity to generate double-stranded RNA and to induce interferon (87). Induction of interferon by DNA viruses also has been explained by the presence of double-stranded RNA formed in virus-infected cells (88). However, the absolute requirement for double-stranded RNA in viral induction of interferon is not without controversy (for a more detailed review, see Ref. 2).

Although induction of type I interferons by polynucleotides and by viruses apparently share the requirement for double-stranded RNA, their respective induction processes are not identical. In studies comparing interferon induction by polynucleotides and viruses, different kinetics of induction have been observed. Poly(I):poly(C) induces rapid interferon production, with peak interferon concentrations seen 3-4 hours after induction. In contrast, virus-induced production of interferon has a longer lag period, peak interferon concentrations are achieved about 12 hours after induction, and interferon production is more protracted (89-91). Although exceptions to this kinetic relationship have been reported (91,92), it is tempting to speculate that the increased lag period observed in virus-induced interferon production is related to the time required to generate the double-stranded RNA molecule.

2. Induction Mechanism Hypotheses

Three alternative hypotheses have been proposed to explain the derepression of the interferon gene and the subsequent production of interferons. The double-stranded RNA hypothesis (93) defines double-stranded RNA as the proximal inducer of interferons. The double-stranded RNA binding enzymes, 2-5A synthetase and protein kinase, are proposed to function as receptors for double-stranded RNA, and the resulting complexes interact with a protein repressor bound to the interferon gene, resulting in derepression of the gene. This model is useful in explaining priming, the phenomenon in which pretreatment of cells with interferon results in enhanced interferon production in response to virus or polynucleotide exposure (91,94,95). It has been proposed that the addition of optimal amounts of interferons elevates intracellular levels of 2-5A synthetase and protein kinase, insuring maximal rates of double-stranded RNA-receptor interaction with the protein repressor and subsequent gene derepression. However, it is not known if protein kinase and 2-5A synthetase actually play a role in the induction process.

In the repressor-depletion hypothesis (96), it is proposed that reversible inhibitors of protein or RNA synthesis induce interferons, and inducers of interferons (virus, double-stranded RNA, and metabolic inhibitors) owe their ability to induce interferon to their capacity to inhibit synthesis of a rapidly turning-over protein which normally represses interferon genes. Evidence of a repressor molecule which binds to the regulatory region of the interferon β_1 enhancer element has been reported (97,98). Upon induction, the repressor is displaced by a transcription factor which activates the enhancer, allowing the synthesis of interferon mRNA. In addition to evidence on the molecular level in support of this hypothesis, it is attractive in that it better explains interferon induction by metabolic inhibitors than does the double-stranded RNA hypothesis.

The third hypothesis, the basal level interferon hypothesis, is related to the repressor-depletion hypothesis in that it is assumed that interferons themselves act as repressor molecules. Association of the inducing molecule with the interferon repressor molecule leads to derepression of the interferon genes and synthesis of new interferons (99). In support of this hypothesis, interferons bind selectively to columns of immobilized poly(A) or poly(U) (100), and thus interaction of an RNA inducer molecule with interferons as repressors could derepress interferon genes. However, interferon is not transcribed in uninduced cells (101). Thus, the repressor-depletion hypothesis remains the most attractive alternative.

3. Transcriptional Control of Interferon Production

Whatever mechanism is involved in interferon induction, the production of interferon requires de novo synthesis of interferon mRNA. Actinomycin D, a potent inhibitor of DNA-directed RNA synthesis, blocks the production of interferons if cells are treated before the first appearance of interferons (102). Interferon production is prevented by other inhibitors of RNA synthesis, including α-amanitin, an inhibitor of DNA polymerase II, an enzyme involved in mRNA synthesis (103). Interferon mRNA can be quantitated by in vitro translation in *Xenopus* oocytes. In fibroblasts induced with poly(I):poly(C), interferon mRNA correlates well with interferon synthesis (104-107). Interferon mRNA also has been measured using cloned interferon α and interferon β genes as hybridization probes (108-110). The kinetics of interferon mRNA production are similar to those of interferon production, and the kinetics of interferon α and interferon β mRNA production are similar. However, due to differences in gene copy number, IFNβ_1 may be more efficiently transcribed than IFNα (110).

That interferon α and interferon β share similar kinetics of induction is not surprising. Comparisons between the promotor, or 5'-distal DNA sequences of interferon β_1 and various interferon α species reveal extensive homology, from about 50% homogeneity in the 130 base pairs immediately preceding the transcription start, to about 60% homology between the -169 to -240 base pair region of IFNβ_1 and the corresponding region of IFNα (111). The similarity of these transcriptional control regions provides a basis for the observation that type I interferon often results in both interferon α and interferon β_1 production (112-114).

To dissect the importance of these promotor regions to normal interferon induction, genetic hybrids have been constructed. A DNA fragment from human interferon β_1, covering the region -280 to +20 base pairs, fused to a thymidine kinase gene confers virus-inducible promotor activity to the construct when it is integrated into a murine cell line (115). The rabbit β-globin gene, when joined to a 675 base pair fragment 5' to the IFNα_1 gene and integrated into a murine cell line, has been shown to be virally induced (116). Deletion analysis of the 5'-distal end of an IFNα_1 gene in a murine cell line has revealed that a fragment containing 117 base pairs 5' distal to the transcription start was functional and induced proper transcripts in response to viral challenge. However, another fragment containing only 74 base pairs 5' distal to the start was no longer inducible by virus (117).

The interferon β_1 gene has been shown to be regulated by an inducible enhancer element located in the region -36 to -77 base pairs upstream from the gene (97). Analysis of deletion mutants within this regulatory element has revealed that it consists of both a constitutive transcription element and a negative control sequence that prevents enhancer activity prior to induction. On the basis of DNAse I genomic footprint analysis of induced and uninduced cells, it has been shown that cellular factors bind to the negative regulatory element before induction, and to the constitutive transcription element after induction (98). Thus, in uninduced cells, a repressor molecule bound to the regulatory element apparently prevents binding of a constitutive transcription factor. Upon induction, the repressor dissociates and a transcription factor binds to and activates the enhancer.

4. Posttranscriptional Control of Interferon Production

Enhanced interferon production by cells induced with polynucleotides and subsequently treated with metabolic inhibitors, such as cycloheximide, has been termed superinduction (118). This observation has been interpreted as evidence of posttranscriptional control of interferon induction. In superinduced cells, increased levels of interferon mRNA have been demonstrated (105,119,120). Although the interferon mRNA in superinduced cells is produced with similar kinetics, it is degraded more slowly than interferon mRNA in normally induced cells. A 3-4-fold increase in the rate of transcription and a 14-fold increase in stability of interferon mRNA have been defined (121,122). While no interferon β mRNA is detectable in cellular cytoplasm after the cessation of interferon synthesis, mRNA synthesis continues in isolated nuclei for at least 24 hours (101). Furthermore, disappearance of cytoplasmic interferon mRNA is prevented by treatment with cycloheximide. This result suggests coinduction of a regulatory protein responsible for the specific degradation of interferon mRNA. The instability of interferon mRNA is apparently specified by its coding and/or 3' untranslated region (123). In transfected simian cells defective in interferon synthesis, induced human interferon β mRNA is quickly degraded whereas the half-life of a hybrid gene, in which the coding and $3'$ untranslated region of chloramphenicol acetyltransferase is regulated by the human interferon β promotor, is much longer.

Differential posttranscriptional regulation of interferon synthesis between classes also has been demonstrated. Although the kinetics of interferon α and interferon β_1 mRNA production in lymphoblastoid cells has been shown to be similar (109,110), IFNβ_1 has a higher turnover rate and a slower release than IFNα species, resulting in lower concentrations of IFNβ_1 in culture medium than predicted on the basis of the amount on IFNβ_1 mRNA synthesized (110).

B. Immune Induction

1. Induction of Interferon γ

In contrast to the induction of type I interferons, which requires the presence of double-stranded RNA, induction of interferon γ can result from mitogens, bacterial and viral antigens, allogeneic cells, interleukins, and T-cell surface component-specific antisera (Table 4). While diverse in nature, all interact with T-cell receptor molecules and can activate T cells in a polyclonal or clonally restricted manner. Thus, specific recognition by T cells may be part of the natural induction process.

Evidence that induction of interferon γ is a membrane surface-associated response has been demonstrated in experiments analyzing the effect of mitogens on the induction process. Since cleavage of galactose residues inhibits production of interferon γ in response to concanavalin A, phytohemagglutinin, staphylococcal enterotoxin, galactose oxidase, and sodium periodate, the signal for induction may be linked to terminal oligosaccharides of T-cell membranes. A functional calcium flux through the cell membrane is also required since depletion of calcium prevents interferon γ production (124). Furthermore, agents such as interleukin-2 (IL-2), calcium ionophores, and phorbol esters which affect calcium mobilization or activate protein kinase C, have been shown to induce interferon γ mRNA synthesis in an IL-2-independent murine T-cell line (125). Thus regulation of interferon γ gene expression may be correlated with activation of protein kinase C and calcium mobilization.

2. Genetically Restricted Interferon γ Induction

Specific induction of interferon γ is linked to the major histocompatibility locus. Murine cytotoxic T cells, derived from mice primed by infection with influenza virus and directed against virus-infected target cells, release interferon γ upon in vitro contact with target cells only if the target expresses the same major histocompatibility complex (MHC) antigens as the cytotoxic T cell (127). This MHC restriction is not limited to virally infected cells. Murine T-cell lines which mediate hapten-specific delayed-type hypersensitivity have been induced to

Table 4 Inducers of Interferon-γ

Inducer	Reference
Sensitized mononuclear cells plus antigen	
Virus-sensitized cells	126, 127
Tetanus toxoid	68
Diphtheria toxoid	68
Tubercular PPD	68
Alloantigen	128, 129
Phytohemagglutinin	67
Concanavalin A	130
Pokeweed mitogen	130
Streptolysin O	130
Bacterial lipopolysaccharides	131, 132
Antilymphocyte serum	133
Anti-T-cell monoclonal antibody	134, 135
Staphylococcal enterotoxin	136
Staphylococcal protein A	137
Phorbol esters	125, 138
Galactose oxidase	139
Calcium ionophores	125, 140
Interleukin-2	125, 141, 142

Source: Adapted from Ref. 85.

make interferon γ in the presence of antigen-presenting cells (143). Production of interferon γ by these cell lines was both antigen specific and restricted to antigen-presenting cells of the same H-2 type as the sensitized cells. Furthermore, clones of murine antigen-specific, H-2-restricted cytotoxic T cells release interferon-γ when cultivated with allogeneic cells presenting antigen (129). Induction of human interferon γ in mixed lymphocyte cultures has also shown to be under HLA control (144). Thus, MHC-restricted interferon γ production has been demonstrated with virus, hapten, and cellular antigen.

3. Role of Accessory Cells and Interleukins in Interferon Induction

a. Interferon-γ Activation of T cells by mitogens or specific antigens requires the presence of accessory cells such as macrophages (145,146). Macrophages serve two main functions in T-cell activation, the presentation of antigen (147) and the production of interleukin-1 (IL-1), a monokine that induces production of IL-2 (148). IL-2 is required for the proliferation of activated T cells (149,150). Since inducers of IFNγ also activate T cells, it follows that IFNγ induction also may require the presence of accessory cells or macrophages. Interferon γ production in human leukocyte cultures in response to the monoclonal antibody OKT3 depends on interaction between T cells and cells expressing OKM1 antigen (OKM1 is expressed predominantly on cells of monocyte/macrophage lineage) (134). However, production of IFNγ by murine T-cell lines or clones in response to mitogens or specific antigens in the absence of accessory cells also has been demonstrated (143,151). Freshly isolated leukocytes differ in their state of activation from T cells continuously grown in IL-2-conditioned medium. The former are in a resting state and require accessory cells or mitogen to enter into the cell cycle, whereas the latter are continuously cycling cells. Cells under these conditions also differ in their relative amounts of IFNγ production. T cells grown in IL-2-conditioned medium or prestimulated with mitogen produce high titers of interferon γ upon restimulation with mitogen, up to 30-fold more than cells receiving no prestimulation (152). The enhanced ability of spleen cells obtained from mice immunized with *Listeria monocytogenes* to produce interferon γ in response to mitogen may be due to analogous activation of resting T cells in vivo (153). Furthermore, the production of IFNγ by immune spleen cells stimulated with mitogen is rapid. Interferon γ can be detected as early as 4 hours after treatment, with peak levels occurring 18 hours posttreatment. The kinetics of IFNγ production by alloantigen-induced lymphocytes is similar (154).

Interleukin-2 can result in production of IFNγ in peripheral blood mononuclear cells, spleen cells, and T-cell clones, as well as in purified NK cells (73, 125,141,142). IL-2 can also replace macrophage accessory cells in restoring the IFNγ-producing ability of macrophage-depleted mixed lymphocyte cultures (155). In fact, it has been demonstrated that recombinant IL-2 induces the synthesis of IFNγ mRNA in an IL-2-independent murine cell line (125). However, there is evidence that IL-1 may still be required for maximal induction of interferon γ by IL-2 (142).

b. Other Interferons The induction of some type I interferons also may be regulated by cytokines. Interferon β_2 is induced in fibroblasts by TNF (60) and by a 22,000 dalton, leukocyte-derived protein (156,157). This purified protein has been identified biochemically and biologically as IL-1 (158,159).

Another cytokine, macrophage colony-stimulating factor (CSF-1), induces low levels of type I interferon in murine macrophages and bone marrow cells (160,161) and can augment type I interferon production by murine macrophages (162,163). Apparently, the induction of low levels of interferon is sufficient to prime the macrophages for enhanced interferon production upon further stimulation. CSF-1 may be responsible for the induction of IFNα in lymphoid cells exposed to antigens, phorbol esters, and lectins (82-84).

IV. CONCLUSION

In addition to viruses, a number of unicellular organisms can induce interferons. Among these organisms are included prokaryotes of the genera *Chlamydia* (164, 165), *Rickettsia* (166,167), *Listeria* (168), and family Enterobacteriaceae (169, 170), as well as eukaryotes of the genera *Toxoplasma* (171,172), *Plasmodium* (173), and *Trypanosoma* (174). Although the mechanism by which nonviral agents induce interferons remains unclear, a common characteristic shared by many of these organisms is that they are intracellular pathogens. Thus host immune responses may play a key role in the induction of interferon by these agents. The progress made in defining the molecular basis for viral induction of interferons may provide a framework upon which understanding of induction of interferons by pathogens, both bacterial and protozoan, can be built. It seems likely that a final common pathway will exist in regulation of expression of interferon genes for both viruses and other intracellular pathogens.

ACKNOWLEDGMENTS

We are indebted to Paula Pitha for her critical comments on the manuscript. Work in our laboratories is supported by Triton Biosciences and the University of Wisconsin Graduate School, Project Number 170492. ECB is an American Cancer Society Professor of Clinical Oncology.

REFERENCES

1. Isaacs, A., and J. Lindenmann. 1957. Virus interference. 1. The interferon. *Proc. R. Soc. London B. 147*:258–267.
2. Stewart II, W.E. 1979. *The Interferon System*. Springer-Verlag, Vienna and New York.
3. Finter, N.B. (Ed.). 1984. *Interferon*. Elsevier, Amsterdam.
4. Baron, S., and F. Dianzani. (Eds.). 1977. *Texas Reports on Biology and Medicine*, volume 35, *The Interferon System: A Current Review to 1978*. University of Texas Medical Branch, Galveston.
5. Baron, S., F. Dianzani, and G.J. Stanton (Eds.). 1981–1982. *Texas Reports on Biology and Medicine*, volume 41, *The Interferon System: A Review to 1982*—Part I and Part II. University of Texas Medical Branch, Galveston.
6. Borden, E.C., and L.A. Ball. 1981. Interferons: biochemical, cell growth inhibitory, and immunological effects. *Prog. Hemat. 12*:299–339.
7. Taniguchi, T., N. Mantei, M. Schwarzstein, S. Nagata, M. Muramatsu, and C. Weissmann. 1980. Human leukocyte and fibroblast interferons are structurally related. *Nature 285*:547–549.
8. Gray, P.W., D.W. Leung, D. Pennica, E. Yelverton, R. Najarian, C.C. Simonsen, R. Derynck, P.J. Sherwood, D.M. Wallace, S.L. Berger, A.D. Levinson, and D.V. Goeddel. 1982. Expression of human interferon cDNA in *E. coli* and monkey cells. *Nature 295*:503–508.

9. Devos, R., H. Cheroutre, Y. Taya, W. Degrave, H. Van Heuwerszywn, and W. Fiers. 1982. Molecular cloning of human interferon cDNA and its expression in eukaryotic cells. *Nucl. Acids Res. 10*:2487–2501.

10. DeGrado, W.F., Z.R. Wassermann, and V. Chowdhry. 1982. Sequence and structural homologies among type I and type II interferons. *Nature 300*: 379–381.

11. Gresser, I. 1961. Production of interferon by suspensions of human leukocytes. *Proc. Soc. Exp. Biol. Med. 108*:799–803.

12. Stewart II, W.E., and J. Desmyter. 1975. Molecular heterogeneity of human leukocyte interferon: two populations differing in molecular weights, requirements for renaturation, and cross species antiviral activity. *Virology 67*:68–73.

13. Törmä, E.T., and K. Paucker. 1976. Purification and characterization of human leukocyte interferon components. *J. Biol. Chem. 251*:4810–4816.

14. Chen, J.K., W.J. Jankowski, J.A. O'Malley, E. Sulkowski, and W.A. Carter. 1976. Nature of the molecular heterogeneity of human leukocyte interferon. *J. Virol. 19*:425–434.

15. Havell, E.A., Y.K. Yip, and J. Vilček. 1977. Correlation of physicochemical and antigenic properties of human leukocyte interferon subspecies. *Arch. Virol. 55*:121–129.

16. Stewart II, W.E., L.S. Lin, M. Wiranowska-Stewart, and K. Cantell. 1977. Elimination of size and charge heterogeneities of human leukocyte interferons by chemical cleavage. *Proc. Natl. Acad. Sci. (USA) 74*:4200–4204.

17. Rubinstein, M., S. Rubinstein, P.C. Familletti, R.S. Miller, A.A. Waldman, and S. Pestka. 1979. Human leukocyte interferon: production, purification to homogeneity, and initial characterization. *Proc. Natl. Acad. Sci. (USA) 76*:640–644.

18. Bridgen, P.J., C.B. Anfinsen, L. Corley, S. Bose, K.C. Zoon, U.T. Rüegg, and C.E. Buckler. 1977. Human lymphoblastoid interferon: large scale production and partial characterization. *J. Biol. Chem. 252*:6585–6587.

19. Berg, K., and I. Heron. 1980. The complete purification of human leukocyte interferon. *Scand. J. Immunol. 11*:489–502.

20. Zoon, K.C., M.E. Smith, P.J. Bridgen, D. ZurNeeden, and C.B. Anfinsen. 1979. Purification and characterization of human lymphoblastoid interferon. *Proc. Natl. Acad. Sci. (USA) 76*:5601–5605.

21. Rubinstein, M., W.P. Levy, J.A. Moschera, C.Y. Lai, R.D. Hershberg, R.T. Bartlett, and S. Pestka. 1981. Human leukocyte interferon: isolation and characterization of several molecular forms. *Arch. Biochem. Biophys. 210*: 307–318.

22. Allen, G., and K.H. Fantes. 1980. A family of structural genes for human lymphoblastoid (leukocyte-type) interferon. *Nature 287*:408–411.

23. Vilček, J. 1982. The importance of having gamma. In *Interferon 1982*, volume 4, I. Gresser (Ed.). Academic Press, London, pp. 129–154.

24. Preble, O.T., R.J. Black, R.M. Friedman, J.H. Klippel, and J. Vilček. 1982. Systemic lupus erythematosus: presence in human serum of an unusual acid-labile leukocyte interferon. *Science 216*:429–431.

25. DeStefano, E., R.M. Friedman, A.E. Friedman-Kien, J.J. Goedert, D. Henriksen, O.T. Preble, J.A. Sonnabend, and J. Vilček. 1982. Acid labile human leukocyte interferon in homosexual men with Kaposi's sarcoma and lymphadenopathy. *J. Infect. Dis. 146*:451–455.

26. Abb, J., M. Kochen, and F. Deinhardt. 1984. Interferon production in the male homosexual with the acquired immune deficiency syndrome (AIDS) or generalized lymphadenopathy. *Infection 12*:240–242.

27. Arvin, A.M., and J.J. Miller. 1984. Acid labile α-interferon in sera and synovial fluids from patients with juvenile arthritis. *Arthritis Rheum. 27*: 582–585.

28. Chadha, K.C. 1985. Acid labile human leukocyte interferon. In *The Interferon System*, F. Dianzani and G.B. Rossi (Eds.). Raven Press, New York, pp. 35–41.

29. Fischer, D.G., and M. Rubinstein. 1983. Spontaneous production of interferon-γ and acid-labile interferon-α by subpopulations of human mononuclear cells. *Cell Immunol. 81*:426–434.

30. Brack, C., S. Nagata, N. Mantei, and C. Weissmann. 1981. Molecular analysis of the human interferon-α gene family. *Gene 15*:379–394.

31. Nagata, S., C. Brack, K. Henco, A. Schamböck, and C. Weissmann. 1981. Partial mapping of ten genes of the human interferon-α family. *J. Interferon Res. 1*:333–336.

32. Lawn, R.M., J. Adelman, T.J. Dull, M. Gross, D.V. Goeddel, and A. Ullrich. 1981. DNA sequence of two closely linked human leukocyte interferon genes. *Science 212*:1159–1162.

33. Ullrich, A., A. Gray, D.V. Goeddel, and T.J. Dull. 1982. Nucleotide sequence of a portion of human chromosome 9 containing a leukocyte interferon gene cluster. *J. Mol. Biol. 156*:467–486.

34. Goeddel, D.V., D.W. Leung, T.J. Dull, M. Gross, R.M. Lawn, R. McCandliss, P.H. Seeburg, A. Ullrich, E. Yelverton, and P.W. Gray. 1981. The structure of eight distinct cloned human leukocyte interferon cDNAs. *Nature 290*: 20–26.

35. Lund, B., T. Edlund, W. Lindenmaier, T. Ny, J. Collins, E. Lundgren, and A. von Gabain. 1984. Novel cluster of α-interferon gene sequences in a placental cosmid DNA library. *Proc. Natl. Acad. Sci. (USA) 81*:2435–2439.

36. Weissmann, C., S. Nagata, W. Boll, M. Fountoulakis, A. Fujisawa, J.-I. Fujisawa, J. Haynes, K. Henco, N. Mantei, H. Ragg, C. Schein, J. Schmid, G. Shaw, M. Streuli, H. Tara, K. Todokoro, and U. Weidle. 1982. Structure and expression of human alpha-interferon genes. In *Chemistry and Biology of Interferons: Relationship to Therapeutics*, T. Merigan, R.M. Friedman, and C.F. Fox (Eds.). Academic Press, New York and London, pp. 295–326.

37. Mantei, N., M. Schwarzstein, M. Streuli, S. Panem, S. Nagata, and C. Weissmann. 1980. The nucleotide sequence of a cloned human leukocyte cDNA. *Gene 10*:1–10.

38. Streuli, M., S. Nagata, and C. Weissmann. 1980. At least three human type α interferons: structure of α$_2$. *Science 209*:1343–1347.

39. Owerbach, D., W.J. Rutter, T.B. Shows, P. Gray, D.V. Goeddel, and R.M.

Lawn. 1981. Leukocyte and fibroblast interferon genes are located in human chromosome 9. *Proc. Natl. Acad. Sci. (USA)* 78:3123–3127.

40. Trent, J.M., S. Olson, and R.M. Lawn. 1982. Chromosomal location of human leukocyte, fibroblast, and immune interferon genes by means of in situ hybridization. *Proc. Natl. Acad. Sci. (USA)* 79:7809–7813.

41. Havell, E.A., B. Berman, C.A. Ogburn, K. Berg, K. Paucker, and J. Vilček. 1975. Two antigenically distinct species of human interferon. *Proc. Natl. Acad. Sci. (USA)* 72:2185–2190.

42. Reynolds, F.H., and P.M. Pitha. 1975. Molecular weight study of human fibroblast interferon. *Biochem. Biophys. Res. Commun.* 65:107–114.

43. Knight, E., Jr. 1976. Interferon: purification and initial characterization from human diploid cells. *Proc. Natl. Acad. Sci. (USA)* 73:520–523.

44. Knight, E., Jr., and D. Fahey. 1981. Human fibroblast interferon: an improved purification. *J. Biol. Chem.* 256:3609–3611.

45. Taniguchi, T., S. Ohno, Y. Fuji-Kuryama, and M. Muramatsu. 1980. The nucleotide sequence of human fibroblast interferon cDNA. *Gene* 10:11–15.

46. Derynck, R., J. Content, E. DeClercq, G. Volckaert, J. Tavernier, R. Devos, and W. Fiers. 1980. Isolation and structure of a human fibroblast interferon gene. *Nature* 285:542–547.

47. Goeddel, D.V., H.M. Shepard, E. Yelverton, D. Leung, P. Crea, A. Sloma, and S. Pestka. 1980. Synthesis of human fibroblast interferon by *E. coli*. *Nucl. Acids Res.* 8:4057–4074.

48. Lawn, R.M., J. Adelman, A.E. Franke, C.M. Houck, M. Gross, R. Najarian, and D.V. Goeddel. 1981. Human fibroblast interferon gene lacks introns. *Nucl. Acids Res.* 9:1045–1053.

49. Tavernier, J., R. Derynck, and W. Fiers. 1981. Evidence for a unique human fibroblast interferon (IFN-β_1) chromosomal gene, devoid of intervening sequences. *Nucl. Acids Res.* 9:461–471.

50. Branca, A.A., and C. Baglioni. 1981. Evidence that types I and II interferons have different receptors. *Nature* 294:768–770.

51. Tan, Y.H., F. Barakat, W. Berthold, H. Smith-Johannsen, and C. Tan. 1979. The isolation and amino acid/sugar composition of human fibroblast interferon. *J. Biol. Chem.* 254:8067–8073.

52. Sehgal, P.B. 1982. How many interferons are there? In *Interferon 1982*, volume 4, I. Gresser (Ed.). Academic Press, London, pp. 1–22.

53. Sehgal, P.B., and A.D. Sagar. 1980. Heterogeneity of poly(I)·poly(C)-induced human fibroblast interferon mRNA species. *Nature* 288:95–97.

54. Weissenbach, J., Y. Chernajovsky, M. Zeevi, L. Shulman, H. Soreq, U. Nir, D. Wallach, M. Perricaudet, P. Tiollais, and M. Revel. 1980. Two interferon mRNAs in human fibroblasts: in vitro translation and *Escherichia coli* cloning studies. *Proc. Natl. Acad. Sci. (USA)* 77:7152–7156.

55. Content, J., L. DeWit, D. Pierard, R. Derynck, E. DeClercq, and W. Fiers. 1982. Secretory proteins induced in human fibroblasts under conditions used for the production of interferon-β. *Proc. Natl. Acad. Sci. (USA)* 79:2768–2772.

56. Haegeman, G., J. Content, G. Volckaert, R. Derynck, J. Tavernier, and W. Fiers. 1986. Structural analysis of the sequence coding for an inducible 26-kDa protein in human fibroblasts. *Eur. J. Biochem. 159*:625–632.

57. Zilberstein, A., R. Ruggieri, J.H. Korn, and M. Revel. 1986. Structure and expression of cDNA and genes for human interferon-beta-2, a distinct species inducible by growth stimulatory cytokines. *EMBO J. 5*:2529–2537.

58. Zilberstein, A., R. Ruggieri, and M. Revel. 1985. Human interferon-β_2: is it an interferon-inducer? In *The Interferon System*, F. Dianzani and G.B. Rossi (Eds.). Raven Press, New York, pp. 73–83.

59. Hirano, T., K. Yasikawa, H. Harada, T. Taga, Y. Watanabe, T. Matsuda, S. Kashiwamura, K. Nakajima, K. Koyama, A. Iwamatsu, S. Tsunasawa, F. Sakiyama, H. Matsui, Y. Takahara, T. Taniguchi, and T. Kishimoto. 1986. Complementary DNA for a novel interleukin (BSF-2) that induces B lymphocytes to produce immunoglobulin. *Nature 324*:73–76.

60. Kohase, M., D. Henriksen-DeStefano, L.T. May, J. Vilček, and P.B. Sehgal. 1986. Induction of β_2-interferon by tumor necrosis factor: a homeostatic mechanism in the control of cell proliferation. *Cell 45*:659–666.

61. Pitha, P.M., D.L. Slate, N.B.K. Raj, and F.H. Ruddle. 1982. Human β interferon gene localization and expression in somatic cell hybrids. *Mol. Cell. Biol. 2*:564–570.

62. Kelley, K.A., C.A. Kozak, and P.M. Pitha. 1985. Localization of the mouse interferon-β_1 gene to chromosome 4. *J. Interferon Res. 5*:409–413.

63. Kelley, K.A., C.A. Kozak, F. Dandoy, F. Sor, D. Skup, J.D. Windass, J. DeMaeyer-Guignard, P.M. Pitha, and E. DeMaeyer. 1983. Mapping of murine interferon-α genes to chromosome 4. *Gene 26*:181–188.

64. Tan, Y.H., R.P. Creagan, and F.H. Ruddle. 1974. The somatic cell genetics of human interferon: assignment of human interferon loci to chromosomes 2 and 5. *Proc. Natl. Acad. Sci. (USA) 71*:2251–2255.

65. Slate, D.L., and F.H. Ruddle. 1979. Fibroblast interferon in man is coded by two loci on separate chromosomes. *Cell 16*:171–180.

66. Sehgal, P.B., A. Zilberstein, R.-M. Ruggieri, L.T. May, A. Ferguson-Smith, D.L. Slate, M. Revel, and F.H. Ruddle. 1986. Human chromosome 7 contains the β_2 interferon gene. *Proc. Natl. Acad. Sci. (USA) 83*:5219–5222.

67. Wheelock, E.F. 1965. Interferon-like virus inhibitor induced in human leukocytes by phytohemagglutinin. *Science 149*:310–311.

68. Green, J.A., S.R. Cooperband, and S. Kibrick. 1969. Immune-specific induction of interferon production in cultures of human blood lymphocytes. *Science 164*:1415–1417.

69. Nathan, I., J.E. Groopman, S.G. Quan, N. Bersch, and D.W. Golde. 1981. Immune (γ) interferon produced by a human T-lymphoblast cell line. *Nature 292*:842–844.

70. Epstein, L.B., H.W. Kreth, and L.A. Herzenberg. 1974. Fluorescence-activated cell sorting of human T and B lymphocytes. II. Identification of the cell type responsible for interferon production and cell proliferation in response to mitogens. *Cell. Immunol. 12*:407–421.

71. Kasahara, T., J.J. Hooks, S.F. Dougherty, and J.J. Oppenheim. 1983. Inter-

leukin 2-mediated immune interferon (IFN-γ) production by human cells and T cell subsets. *J. Immunol. 130*:1784–1789.

72. Djeu, J.Y., N. Stocks, K. Zoon, G.J. Stanton, T. Timonen, and R.B. Herberman. 1982. Positive self regulation of cytotoxicity in human natural killer cells by production of interferon upon exposure to influenza and herpes viruses. *J. Exp. Med. 156*:1222–1234.

73. Handa, K., R. Suzuki, H. Matsui, Y. Shimizu, and K. Kumagai. 1983. Natural killer (NK) cells as a responder to interleukin 2 (IL-2). II. IL-2 induced interferon-γ production. *J. Immunol. 130*:988–992.

74. Kirchner, H., H.H. Peter, H.M. Hirt, R. Zawatzky, and P. Bradstreet. 1979. Studies on the producer cell of interferon in human lymphocyte cultures. *Immunobiology 156*:65–75.

75. Yip, Y.K., B.S. Barrowclough, C. Urban, and J. Vilček. 1982. Molecular weight of human gamma interferon is similar to that of other interferons. *Science 215*:411–413.

76. Yip, Y.K., B.S. Barrowclough, C. Urban, and J. Vilček. 1982. Purification of two subspecies of human γ (immune) interferon. *Proc. Natl. Acad. Sci. (USA) 79*:1820–1824.

77. Kelker, H.C., Y.K. Yip, P. Anderson, and J. Vilček. 1983. Effects of glycosidase treatment on the physicochemical properties and biological activity of human interferon-gamma. *J. Biol. Chem. 258*:8010–8013.

78. Rinderknecht, E., B.H. O'Conner, and H. Rodriguez. 1984. Natural human interferon-γ: complete amino acid sequence and determination of sites of glycosylation. *J. Biol. Chem. 259*:6790–6797.

79. Gray, P.W., and D.V. Goeddel. 1982. Structure of the human immune interferon gene. *Nature 298*:859–863.

80. Naylor, S.L., Sakaguchi, A.Y., T.B. Shows, M.L. Law, D.V. Goeddel, and P.W. Gray. 1983. Human immune interferon gene is located on chromosome 12. *J. Exp. Med. 157*:1020–1027.

81. Vilček, J. 1984. Interferon production and its regulation. In *Interferon*, volume 3, *Mechanisms of Production and Action*, N.B. Finter and R.M. Friedman (Eds.). Elsevier, Amsterdam, pp. 1–10.

82. Valle, M.J., A.M. Bobrove, S. Strober, and T.C. Merigan. 1975. Immune specific production of interferon by human T cells in combined macrophage-lymphocyte cultures in response to herpes simplex antigen. *J. Immunol. 114*:435–446.

83. Le, J., W. Prensky, D. Henriksen, and J. Vilček. 1982. Synthesis of alpha and gamma interferons by a human cutaneous lymphoma with helper T-cell phenotype. *Cell. Immunol. 72*:157–165.

84. Wiranowska-Stewart, M. 1981. Heterogeneity of human gamma interferon preparations: evidence for presence of alpha interferon. *J. Interferon Res. 1*:315–321.

85. Ho, M. 1984. Induction and inducers of interferon. In *Interferon*, volume 1, *General and Applied Aspects*, N.B. Finter and A. Billiau (Eds.). Elsevier, Amsterdam, pp. 79–124.

86. Clavell, L.A., and M.A. Bratt. 1971. Relationship between the ribonucleic

acid-synthesizing capacity of ultraviolet-irradiated virus and its ability to induce interferon. *J. Virol. 8*:500–508.

87. Marcus, P.I., and F.J. Fuller. 1979. Interferon induction by viruses. II. Sindbis virus: interferon induction requires one-quarter of the genome—genes G and A. *J. Gen. Virol. 44*:169–177.

88. Colby, C., and P.H. Duesberg. 1969. Double-stranded RNA in vaccinia virus infected cells. *Nature 222*:940–944.

89. Mozes, L.W., and J. Vilček. 1975. Distinguishing characteristics of interferon induction with poly I·poly C and Newcastle disease virus in human cells. *Virology 65*:100–110.

90. Tan, Y.H., J.A. Armstrong, and M. Ho. 1971. Intracellular interferon: kinetics of formation and release. *Virology 45*:837–840.

91. Stewart II, W.E., L.B. Gosser, and R.Z. Lockart, Jr. 1971. Distinguishing characteristics of the interferon responses of primary and continuous mouse cell cultures. *J. Gen. Virol. 13*:35–50.

92. Borden, E.C., B.W. Booth, and P.H. Leonhardt. 1978. Mechanistic studies of polyene enhancement of interferon production by poly(I)·poly(C). *Antimicrob. Agents Chemother. 13*:159–164.

93. Marcus, P.I. 1984. Interferon induction by viruses: double-stranded ribonucleic acid as the common proximal inducer molecule. In *Interferon*, volume 3, *Mechanisms of Production and Action*, N.B. Finter and R.M. Friedman (Eds.). Elsevier, Amsterdam, pp. 113–175.

94. Issacs, A., and D.C. Burke. 1958. Mode of action of interferon. *Nature 182*:1073–1074.

95. Rosztoczy, I., and I. Mecs. 1970. Enhancement of interferon synthesis by polyinosinic-polycytidylic acid in L-cells pretreated with interferon. *Acta Virol. 14*:398–400.

96. Tan, Y.H., and W. Berthold. 1977. A mechanism for the induction and regulation of human interferon genetic expression. *J. Gen. Virol. 34*:401–412.

97. Goodbourn, S., H. Burstein, and T. Maniatis. 1986. The human β-interferon gene enhancer is under negative control. *Cell 45*:601–610.

98. Zinn, K., and T. Maniatis. 1986. Detection of factors that interact with the human β-interferon regulatory region in vivo by DNAase I footprinting. *Cell 45*:611–618.

99. Kleinschmidt, W.J. 1972. Biochemistry of interferon and its inducers. *Ann. Rev. Biochem. 41*:517–542.

100. DeMaeyer-Guignard, J., M.N. Thang, and E. DeMaeyer. 1977. Binding of mouse interferon to polynucleotides. *Proc. Natl. Acad. Sci. (USA) 74*:3787–3790.

101. Raj, N.B.K., and P.M. Pitha. 1983. Two levels of regulation of β-interferon gene expression in human cells. *Proc. Natl. Acad. Sci. (USA) 80*:3923–3927.

102. Heller, E. 1963. Enhancement of Chikungunya virus replication and inhibition of interferon production by actinomycin D. *Virology 21*:652–656.

103. Atherton, K.T., and D.C. Burke. 1978. The effects of some different

metabolic inhibitors on interferon superinduction. *J. Gen. Virol. 41*: 229-237.

104. Cavalieri, R.L., E.A. Havell, J. Vilček, and S. Pestka. 1977. Induction and decay of human fibroblast interferon messenger RNA. *Proc. Natl. Acad. Sci. (USA) 74*:4415-4419.

105. Sehgal, P.B., B. Dobberstein, and I. Tamm. 1977. Interferon messenger RNA content of human fibroblasts during induction, shut-off, and super-induction of interferon production. *Proc. Natl. Acad. Sci. (USA) 74*: 3409-3412.

106. Raj, N.B.K., B.F. Fernie, and P.M. Pitha. 1979. Correlation between the induction of mouse interferon and the amount of its mRNA. *Eur. J. Biochem. 98*:215-221.

107. Pang, R.H.L., T.G. Hayes, and J. Vilček. 1980. Leukocyte interferon mRNA from human fibroblasts. *Proc. Natl. Acad. Sci. (USA) 77*:5341-5345.

108. Raj, N.B.K., and P.M. Pitha. 1981. Analysis of interferon mRNA in human fibroblast cells induced to produce interferon. *Proc. Natl. Acad. Sci. (USA) 78*:7426-7430.

109. Shuttleworth, J., J. Morser, and D.C. Burke. 1983. Coordinate expression of interferon-α and interferon-β mRNA in human fibroblastoid (Namalwa) cells. *Eur. J. Biochem. 133*:399-404.

110. Raj, N.B.K., M. Kellum, K.A. Kelley, S. Antrobus, and P.M. Pitha. 1985. Differential regulation of interferon synthesis in lymphoblastoid cells. *J. Interferon Res. 5*:493-510.

111. Collins, J. 1984. Interferon genes: gene structure and elements involved in gene regulation. In *Interferon*, volume 3, *Mechanisms of Production and Action*, N.B. Finter and R.M. Friedman (Eds.). Elsevier, Amsterdam, pp. 33-83.

112. Havell, E.A., Y.K. Yip, and J. Vilček. 1978. Characteristics of human lymphoblastoid (Namalva) interferon. *J. Gen. Virol. 38*:51-60.

113. Hayes, T.G., Y.K. Yip, and J. Vilček. 1979. Le interferon production by human fibroblasts. *Virology 98*:351-363.

114. Yamamoto, Y., and Y. Kawade. 1980. Antigenicity of mouse interferons: distinct antigenicity of the two L cell interferon species. *Virology 103*: 80-88.

115. Ohno, S., and T. Taniguchi. 1983. The 5'-flanking sequence of human interferon-β_1 gene is responsible for viral induction of transcription. *Nucl. Acids Res. 11*:5403-5412.

116. Weidle, U., and C. Weissmann. 1983. The 5'-flanking region of a human IFN-gene mediates the viral induction of transcription. *Nature 303*:442-446.

117. Ragg, H., and C. Weissmann. 1983. Not more than 117 base pairs of 5'-flanking sequences are required for inducible expression of a human IFN-α gene. *Nature 303*:439-442.

118. Myers, M.W., and R.M. Friedman. 1971. Potentiation of human interferon production by superinduction. *J. Natl. Cancer Inst. 47*:757-764.

119. Cavalieri, R.L., E.A. Havell, J. Vilček, and S. Pestka. 1977. Synthesis of human interferon by *Xenopus laevis* oocytes. 2. Structural genes for interferon in human cells. *Proc. Natl. Acad. Sci. (USA) 74*:3287-3291.

120. Raj, N.B.K., and P.M. Pitha. 1977. Relationship between interferon production and interferon messenger RNA synthesis in human fibroblasts. *Proc. Natl. Acad. Sci. (USA) 74*:1483-1487.

121. Sehgal, P.B., D.S. Lyles, and I. Tamm. 1978. Superinduction of human fibroblast interferon: further evidence for the increased stability of interferon mRNA. *Virology 89*:186-190.

122. Sehgal, P.B., and I. Tamm. 1979. Two mechanisms contribute to the superinduction of poly(I)·poly(C) induced human fibroblast interferon production. *Virology 92*:240-244.

123. Mosca, J.D., and P.M. Pitha. 1986. Transcriptional and posttranscriptional regulation of exogenous human beta interferon gene in simian cells defective in interferon synthesis. *Mol. Cell. Biol. 6*:2279-2283.

124. Dianzani, F., T.M. Monahan, and M. Santiano. 1982. Membrane alteration responsible for the induction of gamma interferon. *Infect. Immun. 36*: 915-917.

125. Farrar, W.L., M.C. Birchenall-Sparks, and H.B. Young. 1986. Interleukin 2 induction of interferon-γ mRNA synthesis. *J. Immunol. 137*:3836-3840.

126. Glasgow, L.A. 1966. Leukocytes and interferon in the host response to viral infections. II. Enhanced interferon response of leukocytes from immune animals. *J. Bacteriol. 91*:2185-2191.

127. Morris, A.G., Y.L. Lin, and B.A. Askonas. 1982. Immune interferon release when a cloned cytotoxic T-cell line meets its correct influenza-infected target cells. *Nature 295*:150-152.

128. Gifford, G.E., A. Tibor, and D.L. Peavy. 1971. Interferon production in mixed lymphocyte cell culture. *Infect. Immun. 3*:164-166.

129. Klein, J.R., D.H. Raulet, M.S. Pasternak, and M.J. Bevan. 1982. Cytotoxic T lymphocytes produce immune interferon in response to antigen or mitogen. *J. Exp. Med. 155*:1198-1203.

130. Friedman, R.M., and H.L. Cooper. 1967. Stimulation of interferon production in human lymphocytes by mitogens. *Proc. Soc. Exp. Biol. Med. 125*:901-905.

131. Youngner, J.S., and W.R. Stinebring. 1964. Interferon production in chickens infected with *Brucella abortus. Science 144*:1022-1023.

132. Ho, M. 1964. Interferon-like viral inhibitor in rabbits after intravenous administration of endotoxin. *Science 146*:1472-1474.

133. Falcoff, E., R. Falcoff, L. Catinot, A. Vomecourt, and J. Sanceau. 1972. Synthesis of interferon in human lymphocytes stimulated in vitro by anti-lymphocytic serum. *Rev. Eur. Etud. Clin. Biol. 17*:20-26.

134. Chang, T.W., D. Testa, P.C. Kung, L. Perry, H.J. Dreskin, and G. Goldstein. 1982. Cellular origin and interactions involved in γ-interferon production induced by OKT3 monoclonal antibody. *J. Immunol. 128*:585-589.

135. Pang, R.H.L., Y.K. Yip, and J. Vilček. 1981. Immune interferon induction

by a monoclonal antibody specific for human T cells. *Cell. Immunol. 64*: 304-311.

136. Osborne, L.C., J.A. Georgiades, and H.M. Johnson. 1979. Large scale production and partial purification of mouse immune interferon. *Infect. Immun. 23*:80-86.

137. Catalona, W.J., T.L. Ratliff, and R.E. McCool. 1981. γ interferon induced by *S. aureus* protein A augments natural killing and ADCC. *Nature 291*: 77-79.

138. Yip, Y.K., R.H.L. Pang, J.D. Oppenheim, M.S. Nachbar, D. Henriksen, I. Zerebeckyj-Eckhardt, and J. Vilček. 1981. Stimulation of human gamma interferon production by diterpene esters. *Infect. Immun. 34*: 131-139.

139. Dianzani, F., T.M. Monahan, A. Scupham, and M. Zucca. 1979. Enzymatic induction of interferon production by galactose oxidase treatment of human lymphoid cells. *Infect. Immun. 26*:879-882.

140. Dianzani, F., T.M. Monahan, J. Georgiades, and J.B. Alperin. 1980. Human immune interferon: induction in lymphoid cells by a calcium ionophore. *Infect. Immun. 29*:561-563.

141. Pearlstein, K.T., M.A. Palladino, K. Welte, and J. Vilček. 1983. Purified human interleukin-2 enhances induction of immune interferon. *Cell. Immunol. 80*:1-9.

142. Vilček, J., D. Henriksen-DeStefano, D. Siegel, and J. Le. 1985. IFN-gamma induction in peripheral blood leukocytes by interleukin 2: role of monocytes, interleukin 1 and IFN-gamma. In *The Interferon System*, F. Dianzani and G.B. Rossi (Eds.). Raven Press, New York, pp. 43-47.

143. McKimm-Breschkin, J.L., P.L. Mottram, W.R. Thomas, and J.F.A.P. Miller. 1982. Antigen-specific production of immune interferon by T cell lines. *J. Exp. Med.* 155:1204-1209.

144. Andreotti, P.E., and P. Cresswell. 1981. HLA control of interferon production in the human mixed lymphocyte culture. *Human Immunol. 3*: 109-120.

145. Oppenheim, J.J., B.G. Leventhal, and E.M. Hersh. 1968. The transformation of column-purified lymphocytes with nonspecific and specific antigenic stimuli. *J. Immunol. 101*:262-270.

146. Rosenstreich, D.L., J.J. Farrar, and S. Dougherty. 1976. Absolute macrophage dependency of T lymphocyte activation by mitogens. *J. Immunol. 116*:131-139.

147. Waldron, J.A., Jr., R.G. Horn, and A.S. Rosenthal. 1974. Antigen-induced proliferation of guinea pig lymphocytes in vitro: functional aspects of antigen handling by macrophages. *J. Immunol. 112*:746-755.

148. Smith, K.A., K.J. Gilbride, and M.F. Favata. 1980. Lymphocyte activating factor promotes T-cell growth factor production by cloned murine lymphoma cells. *Nature 287*:853-855.

149. Morgan, D.A., F.W. Ruscetti, and R. Gallo. 1976. Selective in vitro growth of T lymphocytes from normal human bone marrows. *Science 193*:1007-1008.

150. Gillis, S., and K.A. Smith. 1977. Long term culture of tumor-specific cytotoxic T cells. *Nature 268*:154-156.

151. Morris, A.G., J. Morser, and A. Meager. 1982. Spontaneous production of gamma interferon and induced production of beta interferon by human T-lymphoblastoid cell lines. *Infect. Immun. 35*:533-536.

152. Marcucci, F., B. Klein, H. Kirchner, and R. Zawatzky. 1982. Production of high titers of interferon-gamma by prestimulated murine spleen cells. *Eur. J. Immunol. 12*:787-790.

153. Havell, E.A., G.L. Spitalny, and P.J. Patel. 1982. Enhanced production of murine interferon γ by T cells generated in response to bacterial infection. *J. Exp. Med. 156*:112-127.

154. Perussia, B., L. Mangoni, H.D. Engers, and G. Trinchieri. 1980. Interferon production by human and murine lymphocytes in response to alloantigens. *J. Immunol. 125*:1589-1595.

155. Farrar, W.L., H.M. Johnson, and J.J. Farrar. 1981. Regulation of the production of immune interferon and cytotoxic T lymphocytes by interleukin 2. *J. Immunol. 126*:1120-1125.

156. Content, J., L. DeWit, P. Poupart, G. Opdenakker, J. VanDamme, and A. Billiau. 1985. Induction of a 26-kDa-protein mRNA in human cells treated with an interleukin-1-related, leukocyte-derived factor. *Eur. J. Biochem. 152*:253-257.

157. VanDamme, J., G. Opdenakker, A. Billiau, P. DeSomer, L. DeWit, P. Poupart, and J. Content. 1985. Stimulation of fibroblast interferon production by a 22K protein from human leukocytes. *J. Gen. Virol. 66*:693-700.

158. VanDamme, J., A. Billiau, M. DeLey, J. VanBeeumen, G. Opdenakker, and J. Content. 1985. Physicochemical and biological properties of a 22K interferon-β inducing lymphokine. In *The Interferon System*, F. Dianzani and G.B. Rossi (Eds.). Raven Press, New York, pp. 67-71.

159. VanDamme, J., M. DeLey, G. Opdenakker, A. Billiau, P. DeSomer, and J. VanBeeumen. 1985. Homogeneous interferon inducing 22K factor is related to endogenous pyrogen and interleukin-1. *Nature 314*:266-268.

160. Moore, R.N., H.S. Larsen, D.W. Horohov, and B.T. Rouse. 1984. Endogenous regulation of macrophage proliferative expansion by colony-stimulating factor-induced interferon. *Science 223*:178-181.

161. Moore, R.N., F.J. Pitruzzello, H.S. Larsen, and B.T. Rouse. 1984. Feedback regulation of colony-stimulating factor (CSF-1)-induced macrophage proliferation by endogenous E prostaglandins and interferon-α/β. *J. Immunol. 133*:541-543.

162. Moore, R.N., J.T. Hoffeld, J.J. Farrar, S.E. Mergenhagen, J.J. Oppenheim, and R.K. Shadduck. 1981. Role of colony-stimulating factors as primary regulators of macrophage function. In *Lymphokines*, volume 3, E. Pick (Ed.). Academic Press, New York, pp. 119-148.

163. Hilfiker, M.L., R.N. Moore, and J.J. Farrar. 1981. Biologic properties of chromatographically separated murine thymoma-derived interleukin 2 and colony-stimulating factor. *J. Immunol. 127*:1983-1987.

164. Merigan, T.C., and L. Hanna. 1966. Characteristic of interferon induced in vitro and in vivo by a TRIC agent. *Proc. Soc. Exp. Biol. Med. 122*:421–424.

165. Kozikowska, E.H., and H. Hahon. 1970. Interferon induction by psitticosis agent in guinea pig leukocyte cultures. *Infect. Immun. 2*:731–734.

166. Hopps, H.E., S. Kohno, M. Kohno, and J.E. Smadel. 1964. Production of interferon in tissue cultures infected with *Rickettsia tsutsugamishi. Bacteriol. Proc. 64*:115.

167. Kazar, J. 1966. Interferon-like inhibitor in mouse sera induced by rickettsia. *Acta Virol. 10*:277.

168. Lucas, B., and J. Hruskova. 1967. A virus inhibitor circulating in the blood of chickens induced by *Francisella tularensis* and *Listeria monocytogenes. Folio Microbiol. 12*:157–160.

169. Stinebring, W.R. and J.S. Youngner. 1964. Patterns of interferon appearance in mice injected with bacteria and bacterial endotoxin. *Nature 204*:712–715.

170. Kandefer-Szerszen, M. 1973. Typhoid-paratyphoid (TAB) vaccine as interferon inducer in mice. *Acta Microbiol. Pol. A. 5*:156–158.

171. Rytel, M.W., and T.C. Jones. 1966. Induction of interferon in mice infected with *Toxoplasma gondii. Proc. Soc. Exp. Biol. Med. 123*:859–862.

172. Freshman, M.M., T.C. Merigan, J.S. Remington, and I.E. Brownlee. 1966. In vitro and in vivo antiviral action of an interferon-like substance induced by *Toxoplasma gondii. Proc. Soc. Exp. Biol. Med. 123*:862–866.

173. Huang, K.Y., W.W. Schultz, and F.B. Gordon. 1968. Interferon induced by *Plasmodium berghei. Science 162*:123–125.

174. Rytel, M.W., and P.D. Marsden. 1970. Induction of an interferon-like inhibitor by *Trypanosoma cruzi* in mice. *Am. J. Trop. Med. Hyg. 19*:929–932.

2

Interferons and Their Mechanisms of Action

Donna M. Paulnock / University of Wisconsin Medical School, Madison, Wisconsin

I. INTRODUCTION

The interferons are a family of related glycoproteins synthesized in response to virus infection, which are capable of inhibiting viral replication in treated cells. Three major classes of interferons (type I: interferon α and interferon β; type II: interferon γ) were initially defined on the basis of physicochemical, antigenic, and biological differences. More recent molecular analyses have confirmed that these differences result from primary amino acid and nucleotide sequence differences among the three major classes (see Chap. 1 for review).

Although all members of the interferon family have at least some antiviral activity, the original biologic description of interferon (IFN) has now been considerably expanded, and it is clear that, in addition to inducing antiviral activity in interferon-treated cells, all three classes of interferons also can modulate the function of a number of cell types important in the immune response (1,2). Although a recapitulation of the results of numerous studies of the effects of interferons on the immune response is beyond the scope of this chapter, it can be said that the interferons generally have an enhancing effect on the activities of immune cells, particularly lymphocytes and macrophages. Such effects include potentiation of the functions of these cells and enhancing effects on their development, as well as modulation of expression of many cell surface molecules implicated in immune functions (1,2). Inhibitory effects, including in particular an inhibitory effect on cellular multiplication, also have been noted and are thought to play a role in some of the immunomodulatory activity of these molecules (3,4). These observations have provided the basis for the hypothesis that interferons represent important regulatory agents of the immune system.

It is the aim of this chapter to focus selectively on the nonantiviral properties induced in cells by IFN, and to discuss in particular those cellular alterations which may be of primary importance in the immune response to intracellular pathogens. These cellular responses can be grouped into two broad areas for the purposes of discussion: (a) alterations of the biological activities of both the lymphoid and macrophage cellular elements of the immune system which contribute to the development of protective immunity to intracellular pathogens, and (b) activation of particular biochemical events following IFN treatment which underlie the diverse biological changes induced. This review also will focus in particular on the mechanism of action of interferon gamma (IFNγ). Although experiments from numerous laboratories have demonstrated that all three classes of IFN have immunomodulating activity, the predominant production of IFNγ during the course of an antigen-specific immune response has suggested that this lymphokine represents the most promising candidate for an immunoregulatory molecule. As a result, much of the work to date concerning the immune mechanisms of IFN action has concentrated on studies of the mechanism of action of IFNγ, including numerous studies demonstrating the efficacy of IFNγ in restriction of the growth and replication of both obligate and facultative intracellular pathogens (for review see Ref. 5 and in this volume). These observations form the primary basis for this discussion.

II. CELLULAR ASPECTS OF MICROBIAL IMMUNITY

Before beginning a discussion of the mechanism of action of IFN in the development of immunity to intracellular pathogens, it may be useful to briefly review the nature of the immune responses which are thought to be effective against such organisms. Microbial and parasitic infections typically stimulate more than one immune defense mechanism, including both antibody-mediated and cell-mediated immunity, as well as activation of the cells of the mononuclear phagocyte system. As one would expect, the response that predominates depends on the nature of the infecting organism and the mosaic of antigens presented to the host immune system. However, in the case of most nonviral intracellular pathogens, cell-mediated immune responses are thought to play a central role in resistance to and recovery from infection (6,7). With regard to the role of other cellular elements of the immune response, it is clearly true that no cell-mediated response is likely to occur in the absence of antibody production, and that antibody can modify cellular responses in various ways. Some examples of the role of antibody in a host response to microorganisms include the ability of antibody to serve as an opsinin for organism binding and uptake by a variety of cell types, the formation of antigen-antibody complexes during a local inflammatory response, and even the potential ability of antibody to block organism determinants which might serve as recognition units for macrophages and T cells. However, in most responses to intracellular, as opposed to extracellular, pathogens, antibody appears to play a subordinate role in the development of a productive immune response (6,7). Similarly, although cytotoxic T cells might be expected to play a role in reactions against microbial intracellular infections, as they do in viral infections, to date this immune mechanism has been shown to occur only during a limited number of infections caused by intracellular pathogens (6,7). These observations thus suggest that T-cell cytotoxicity is not a predominant host defense mechanism against such organisms.

The T-cell responses thought to be of primary importance in the control of intracellular microbial infections involve the production of soluble factors, or lymphokines, by T cells bearing the L3T4 surface antigen. Effective stimulation of naive T cells requires antigen-specific triggering of the cells in an Ia-restricted manner, by antigen expressed in conjunction with I-region-encoded major histo-compatibility complex (MHC) molecules on the surface of an antigen-presenting cell, often a macrophage (8). In the presence of the appropriate stimulatory signals from the antigen-presenting cell (particularly interleukin-1), the activated T cell will proliferate in response to antigen and concomitantly produce a variety of lymphokine molecules, including interleukin-2 (IL-2) and IFNγ, which have effects on the recruitment, growth, and activation of diverse cell types (9). Recent work has subdivided the T-cell compartment into two phenotypically identical subsets, both of which produce factors which can modulate the activity of a variety of cells in the immune response, but only one of which produces IL-2 and IFNγ (10,11). The relative role of each of these subtypes in the response to specific antigens remains to be elucidated (11).

Thus, the stimulation of lymphokine production by T cells is antigen-specific in its induction but the produced molecules are nonantigen-specific in their action, in that they have the capacity to activate all cells which bear the appropriate factor receptors. In this way, the lymphokines produced have the ability to enhance the function of both the original stimulated T cell and additional cell types, and to amplify the initial immune response. The initial T-cell activation process leads to the production of regulatory molecules with a significant diversity of function which promote a continued response to antigen in the responsive host.

In the case of immunity to intracellular pathogens, the relatively low concentration of factors in biologically active preparations originally made it difficult to purify and identify those particular lymphokines which might play a role in the immune response to these organisms. Recent evidence from a variety of systems has suggested that a number of different lymphokines which can modulate the immune response to intracellular pathogens are present in crude factor preparations obtained following in vivo or in vitro T-cell stimulation by microbial antigens or mitogens. Some of these factors remain only partially characterized (12–15), although it is believed they may play a significant role in host defense against specific organisms. However, results of numerous studies have clearly demonstrated that IFNγ is the major lymphokine produced by T cells in response to nearly all intracellular pathogens, and that this molecule can significantly enhance the ability of immune cells to affect the growth and viability of these organisms (16-21). Thus, although other lymphokines can have activating properties for cells of the immune system, numerous studies with protozoa and intracellular bacteria have suggested that IFNγ plays a pivotal role in the acquired resistance to a variety of intracellular pathogens.

III. THE ROLE OF IFNγ IN THE DEVELOPMENT OF MICROBIAL IMMUNITY

At the current time, IFNγ is thought to have three principal effects on the cells of the immune system during the development of an immune response to intracellular organisms: a role in the maintenance of the lymphokine-producing T cells following antigen stimulation, a role in the recruitment of macrophages and other cell types in the formation of granulomas as part of the disease process, and a major role in macrophage activation. These are considered, in turn, below.

Interferon γ produced by antigen-activated T cells appears to play a critical role in the maintenance of responsiveness in the lymphokine-producing T-cell subset. As described above, the antigen-specific activation of L3T4-positive T cells stimulates both the proliferation of these cells and the release of lymphokines. Proliferation and expansion of an antigen-triggered T-cell clone is dependent on production by the cells of the mitogenic factor, interleukin-2 (22). A prerequisite to the proliferative effect of IL-2 is the expression of cell surface receptors for this molecule by the T cells (23). Initial receptor expression appears to be acquired following antigen activation of the cells; however, recent evidence suggests that receptor levels can be enhanced or maintained on the cells in response to IFNγ (24). Thus, IFNγ plays a key role in the activation of T cells, by stimulating increased expression of the IL-2 cellular receptor, which allows the continued proliferation of the T cells in response to IL-2. In turn, the T-cell response to IL-2 enhances the production of IFNγ (25-27), thereby creating an interdependent network of T-cell responses which allows the continued production of soluble factors. Recent work has indicated that some T cells proliferate in response to factors other than IL-2 (28,29). The relative role of IFNγ in promoting additional pathways to T-cell proliferation and expansion remains to be analyzed.

The formation of granulomas is characteristic of many infections caused by organisms which are obligate or facultative intracellular pathogens, including *Mycobacterium tuberculosis, M. leprae, Leishmania spp.*, and *Listeria monocytogenes* (6,7). Following the activation of T cells by specific microbial antigen(s) in restricted anatomical sites, the continuous release of lymphokines leads to the accumulation and proliferation of macrophages and other cell types at the focus of infection, with the eventual development of a nodular mass called a granuloma. This histological response represents a chronic cell-mediated immune response to local antigen. One purpose of this inflammatory lesion is thought to be the containment and concentration of microbes, preventing dissemination and allowing their elimination by activated macrophages. Although granulomas often resolve during the course of the disease, they represent both a site for the localization and destruction of the pathogen by activated macrophages (discussed below) and are a component of the immunopathology of these diseases. Studies of the role of T cells in granuloma production have demonstrated that the in vitro stimulation by specific antigen of cells from experimentally produced granulomas causes the release of a variety of mediators, including IFNγ, with macrophage-activating and chemotactic properties (30-32). More recently, using T-cell clones specific for *Listeria monocytogenes*, it has been shown that a single T-cell population which can initiate granuloma production and mediate antilisterial protection in vivo (33) also produces IFNγ after antigenic stimulation in vitro (34). These experiments suggest a key role for IFNγ in granuloma formation through the recruitment and activation of macrophages and other cell types.

IV. MODULATION OF MACROPHAGE FUNCTION BY IFNγ

Perhaps the key role for IFNγ in the resolution of infections by a variety of intracellular pathogens, however, lies in its potent ability to "activate" macrophages. Although a wide variety of cell types, including endothelial cells and fibroblasts, can support the replication of various intracellular pathogens, many of these organisms which cause infection in humans survive and proliferate within host macrophages. In addition, the cells of the mononuclear phagocyte system are exquisitely sensitive to the effects of many lymphokines, including IFNγ. Thus, the activation of macrophages by IFNγ is presently considered to be a major host defense mechanism against many such microorganisms. As a result, most experimental efforts in recent years have focused on characterization of the mechanisms of IFNγ-mediated macrophage activation. In some cases, these efforts have been extended to include investigations of the mechanisms of IFN-induced antimicrobial and antiparasitic activity in nonphagocytic cells that also serve as target cells for IFNγ binding.

The current experimental strategy for ascertaining the mechanistic bases of macrophage activity in resistance to intracellular pathogens involves the definition of changes in a variety of physiologic and metabolic characteristics of these cells, either following infection in vivo or following IFN activation of normal cells in vitro, and the subsequent association of such changes with enhancement of the ability of activated cells to restrict the viability or replication of an infecting organism.

Although the results of these studies (summarized below) have suggested that IFNγ-mediated macrophage activation may play a pivotal role in the acquisition of resistance to a variety of intracellular pathogens, it is of some interest that these studies also have suggested that not all cell types are equally susceptible to the effects of IFNγ. With respect to macrophages, several factors appear to regulate the response to this lymphokine, including the relative maturation stage and anatomical location of target cells, the potential requirement for multiple activation stages in the acquisition of the mature microbicidal phenotype, and genetic control of lymphokine responsiveness. Such factors are of interest in consideration of the biological mechanisms of action of IFN as they contribute to control of responsiveness to this lymphokine in the infected host.

Macrophages obtained from various tissue sites or representing various maturation stages exhibit a spectrum of functional capacities (35-37) and can manifest heterogeneous responses to IFNγ (38-40). One example of this heterogeneity is the observation that macrophages recruited by an inflammatory stimulus generally show an increased responsiveness to lymphokine stimulation when compared to the resident population (37,38). In this way, functions activated in response to an organismal stimulus would be influenced by the response capacity of infected or recruited cells. In a similar vein, growing evidence suggests that the induction of detectable macrophage microbicidal activity occurs in a series of stages or steps, each of which is principally induced by a distinct stimulus. Although this type of activation "pathway" was originally described for macrophage activation for tumor cell killing (41), much recent evidence suggests this may be the case for the development of other macrophage functions as well, including microbicidal activity (42-45). In a number of these pathways, IFNγ activation is an early step in the activation sequence (45,46), suggesting that not all possible functional responses will be activated by the action of IFNγ alone. Finally, some genetic control of IFNγ responsiveness has been observed. The qualitative and quantitative nature of the cellular response to IFNγ can be influenced by the animal species and strain in which the response occurs (47-49). For example, genetic control of lymphokine-induced macrophage antimicrobial and tumoricidal activity has been described for both P/J and A/J mouse strains (50,51). These strains appear to possess different defects in response capability to IFNγ (47,52); however, these distinct genetic defects result in an overall phenotypic reduction in macrophage activation in response to this lymphokine. Although the capability of IFN responsiveness is not eliminated by any of the defects or constraints described to date, these observations serve to point out the relatively complex regulation of macrophage responsiveness to IFNγ and suggest that it may be difficult to easily describe the mechanism of IFNγ action which leads to the development of functional capabilities, including the development of immunity to intracellular pathogens.

In spite of this apparent complexity of regulation, numerous attempts have been made to catalog the metabolic or biochemical alterations occurring in cells following IFNγ treatment which may have a direct relationship to the concomitant enhancement of functional activity induced by this lymphokine. It is clear that macrophages exhibit a number of changes in cell morphology, cell surface components, intracellular constituents, and secretory products following IFN stimulation (53). Efforts to correlate these cellular changes with the enhanced microbicidal and growth inhibitory activity induced by IFNγ-treatment have focused on changes which might be expected to affect the ability of activated cells to recognize and internalize an organism or which would result in the production of compounds within the cell which could have a direct lytic or a growth inhibitory effect on an organism. These changes fall into three categories: Lytic or growth inhibitory mechanisms dependent on the initiation of a respiratory burst in infected cells (that is, the reduction of the oxygen molecule to reactive intermediates which may have toxic properties), antimicrobial mechanisms shown to be independent of oxygen metabolism, including in particular the production and secretion of various enzymes with lytic potential, and the modulation of plasma membrane determinants which affect organism recognition and internalization by macrophages.

A. Regulation of the Macrophage Respiratory Burst by IFNγ

With respect to the killing mechanisms utilized most effectively by macrophages and other phagocytic cells, there has been much emphasis in the past on oxygen-dependent, or respiratory burst-dependent, microbicidal activity (54). Briefly, the synthesis of reactive oxygen intermediates (ROI) occurs as a result of the consumption and reduction of exogenous oxygen to superoxide anion (O_2^-) by an enzyme, possibly a cytochrome b-utilizing NADPH oxidase, present in the cellular and phagosomal membranes (55). This reaction is generally triggered by the phagocytic uptake of an organism and occurs spontaneously following internalization of the phagosome (56), although recent evidence suggests that ingestion is not necessarily required for effective activation of the pathway (57). The production of superoxide anion in turn gives rise to a number of chemically reactive, incompletely reduced metabolites of molecular oxygen, including hydroxyl radicals, singlet oxygen, and hydrogen peroxide, and all of these intermediates are potentially toxic for cells. If phagosome-lysosome fusion occurs following organism internalization, a further peroxidase-dependent reaction can generate additional toxic products (58). Thus, the net result of triggering of the respiratory burst is the production of a battery of reactive compounds that can be used to kill a variety of extra- and intracellular pathogens.

Several lines of evidence have suggested a key role for respiratory burst-dependent cytotoxicity in macrophage defense against intracellular pathogens.

Initial evidence was provided by the observations that cells from patients with chronic granulomatous disease lacked this pathway and that their cells were unable to kill certain microorganisms (59). Perhaps more convincing have been the repeated observations of an association between IFNγ-induced microbicidal activity of macrophages and the concomitant IFN-mediated increase in the respiratory activity of these cells, for which there is considerable evidence (60-66). This enhanced oxidative metabolism is reproducibly observed after incubation of macrophages with IFNγ, but not with a variety of other lymphokines which can affect macrophage function (67). Taken together, these results suggest that IFN-activated macrophages can use an oxygen-dependent mechanism to kill or inhibit the replication of diverse intracellular pathogens. More recent experiments have demonstrated that a monoclonal antibody capable of neutralizing the antiviral activity of IFNγ could inhibit the ability of recombinant IFNγ both to stimulate hydrogen peroxide release and to inhibit the antimicrobial activity of these cells against two intracellular pathogens (68) and have begun to link these events in a cause and effect relationship. The overall impact of this evidence has been the conclusion that the principal, if not only, lymphokine capable of the simultaneous activation of macrophage functional activity and oxygen metabolism is IFNγ.

B. Regulation of Oxygen-Independent Macrophage Activities by IFNγ

Although there is thus considerable evidence to support a role for oxygen-dependent mechanisms of microbial defense, there has been increasing interest in recent years in killing or growth inhibitory mechanisms which occur independently of oxygen utilization. This is due in large part to observations that lymphokines can activate macrophage and fibroblast antimicrobial activity for some organisms in the absence of detectable respiratory burst activity (69-73). A number of antimicrobial mechanisms not dependent on oxygen metabolism have been proposed to account for these observations.

Many changes in cell physiology have been documented in IFN-activated macrophages which might contribute to microbicidal or microbistatic effects. These include numerous changes in intracellular components, including an increased number of lysosomes, providing an increase in intracellular and secreted enzyme concentrations, an increased membrane turnover rate, enhancing phagocytic uptake, and an increased number of mitochondria (74-76). Similarly, analyses of changes in IFN-activated macrophages also have revealed that these cells show increased spreading in culture, accompanied by detectable changes in enzyme profile, including both secretory molecules such as plasminogen activator and lysozyme (77-81), and membrane-bound or cytoplasmic enzymes (82,

83). Although these changes have been associated with macrophage activation, they have not been correlated directly with increased macrophage antimicrobial activity. Considerable evidence suggests that extracellular components secreted by macrophages can have toxic effects on cells (84-86). However, at least some of the proteins secreted in abundance by activated cells are synthesized constitutively by all macrophages. One example of this is the lysozyme molecule, which is effective at digesting bacterial cell walls and which is produced continuously by all macrophage populations (83). However, the production of other secretory molecules, including neutral proteases, is tightly regulated and the release of these molecules is determined by the activation state of the cells (79). Thus, because few of the enzymes produced are unique to activated cells (81), it has not been clear how changes in the production of these molecules might correlate directly with antimicrobial activation.

Of particular interest in consideration of oxygen-independent pathways of macrophage activation are a number of recent studies which have demonstrated the ability of IFNγ to enhance indoleamine 2,3-dioxygenase (IDO) activity in a variety of cell types. IDO is the initial enzyme in the pathway of catabolism of the essential amino acid tryptophan, which is required for protein synthesis and synthesis of the nicotinamide coenzyme NAD. Although the induction of this enzyme in IFN-treated cells had been described previously (87), more recent reports have demonstrated that the depletion and degradation of this essential amino acid following IDO activation could have an effect on the replicative ability of the intracellular organism, *Toxoplasma gondii* (88,89). The induction of IDO activity and subsequent catabolism of tryptophan has also been shown to play a role in controlling the growth of the obligate intracellular pathogen *Chlamydia psittaci* (90), suggesting that the induction of this activity may be an important IFNγ-mediated event in controlling the intracellular replication of some organisms. Recent results suggesting that IDO activity can be enhanced in a number of cell types, including macrophages, following IFNγ treatment (67, 91), strengthens the hypothesis that IFNγ-mediated induction of IDO activity may be an effective antimicrobial mechanism.

The relationship, if any, between the respiratory burst-dependent and -independent microbicidal activities stimulated by IFNγ is unclear at this time. However, several investigators have speculated that the induction of oxygen metabolism by IFNγ treatment may play a role in both pathways (5,92), as IDO requires superoxide anion as a cofactor for initiation of tryptophan catabolism. Thus, reactive oxygen intermediates produced in cells by IFNγ treatment may serve both to directly kill a proportion of microorganisms following ingestion and to activate the initial catabolic enzyme in the tryptophan degradation pathway, leading to continued inhibition of microbe growth and/or additional microbicidal activity.

C. Alterations in Cell Surface Components
Following IFNγ Activation

The presence of reproducible and readily detectable changes in the plasma membrane components of macrophages following treatment of these cells with IFNγ has long been used to identify activated cells. Such modulation of the plasma membrane following activation by IFN has been felt to reflect alterations in the cells that are linked to enhanced functional activity; however, as in the case of other observations of morphologic changes induced by IFNγ, it has been difficult to directly correlate such changes with functional capacities. This may be due at least in part to the heterogeneity of lymphokine responsiveness present in normal macrophage populations. Although numerous studies of macrophage cell surface antigens have described alterations in surface antigenic determinants which are modulated during macrophage activation and differentiation (93-96), there has not been a single surface phenotype described to date which clearly distinguishes activated from untreated or inactive cells. A consistent phenotype associated with microbicidal activity has been described using cells harvested from animals infected with various organisms, including *Bacillus* Calmette-Guerin (BCG), *Corynebacterium parvum*, and *Trypanosoma brucei*, all of which induce sustained, enhanced oxidative antimicrobial or cytocidal activities in macrophages. These cells consistently exhibit enhanced levels of Ia antigens and increased levels of the Fc receptor (FcR) for IgG_1 and IgG_{2a} immunoglobulin isotypes, but reduced levels of the mannose-fucose receptor, the F4/80 antigen and the IgG_{2b} FcR. Receptors for the third component of complement (CR3) remain unchanged (97). This antigen profile has thus been associated with macrophages activated for enhanced microbial killing.

Recent studies of cells activated in vitro by IFNγ or partially purified lymphokine preparations have suggested that the use of changes in surface markers as a probe for changes in functional activity may be more complex than the in vivo situation might suggest. Further studies of macrophages activated in vivo by BCG confirmed that cells with this surface phenotype also were induced following adoptive transfer of sensitized T lymphocytes and specific antigen (98). However, subsequent in vitro studies in a number of laboratories of IFN-induced macrophage activation have demonstrated that, although the activation phenotype induced in vivo by BCG can be reproduced by treatment of some normal cells with highly purified or recombinant IFNγ preparations, different cell populations exhibit the induction of only some of these activation-associated markers following IFNγ treatment (39,99,100). One factor that may make a major contribution to the difficulties in the use of surface markers for analysis of activation is that the expression of many receptors is varied during both the development or maturation and the activation of these cells. Thus, expression of a particular surface molecule will be influenced by the source of the macrophage as well as the inducing agent. In addition, there is still uncertainty in the in vivo studies as to whether the observed results represent the modulation of surface phenotype on individual cells or a change in the cell population in response to activation signals.

In terms of a direct relationship between changes in specific macrophage membrane molecules and the development of immunity to intracellular pathogens, several observations can be made. Clearly, increased Ia antigen expression on macrophages and other antigen-presenting cells would allow the continued activation of specific T lymphocytes and the continued lymphokine production required for cell activation (101,102). Thus, Ia antigen expression, although not a direct reflection of microbicidal activity, would be critical for the successful initiation and continuing expansion of T-cell activation. In addition, a role for the Fc and complement receptors on activated macrophages in the uptake of opsonized organisms is well recognized. Ligand interactions with both these receptors have been shown to trigger the respiratory burst in macrophages, either alone (103) or in combination with other cellular receptors (104). This can result in lysis of organisms either extracellularly or following ingestion. Recent evidence also suggests that intracellular killing of ingested organisms by macrophages following FcR and CR3 ligand binding can result from mechanisms independent of respiratory burst triggering (105).

The relative role of IFN-induced changes in other plasma membrane ligands in microbial immunity is less well understood. Perhaps the best example of this lack of understanding is provided by observations of reductions in mannose-fucose receptor expression on cells activated in vivo. Although this change appears to be a sensitive and reproducible marker of in vivo macrophage activation (97), a reduction in this receptor would seem to be contrary to the functional goal of activation. It has been shown that this receptor contributes to the phagocytic recognition of certain microbial elements, and that it may act in concert with the CR3 receptor to trigger the respiratory burst in macrophages following ligand binding (98,100). However, expression of the mannose-fucose receptor has been demonstrated in some studies to vary inversely with respiratory burst activity in the same cells (98). Similarly, the biological nature of the F4/80 antigen remains unclear, although down-modulation of F4/80 expression generally accompanies both activation and differentiation of mononuclear phagocytes (107). These apparent contradictions suggest that the connections between IFN-induced changes in macrophage surface antigen profile and changes in the functional capacity of the cell remain poorly understood.

V. INDUCTION OF BIOCHEMICAL CHANGES IN IFNγ

The lack of a consistent relationship between the development of microbicidal activity following treatment with IFNγ and particular changes in macrophage cell biology, as described above, supports the general hypothesis that the pleio-tropic activites of IFNγ may reflect the induction of multiple response pathways in treated cells, an idea which is gaining increasing emphasis in the study of diverse IFN-mediated biological effects. With the recognition that there is a specific receptor for IFNγ that can regulate the activation process (108-110), recent studies have begun to dissect the molecular events that underlie the activation process in an effort to correlate the biochemical events with biological activation. However, much information remains to be elucidated concerning the mechanisms by which IFN mediates its biological effects. For example, little is known of the early cellular events which follow the binding of either class of IFN to plasma membrane receptors. A transient, calcium-dependent increase in guanylate cyclase has been observed to occur almost immediately following exposure of cells to IFN (111,112). Recent work also suggests that a rapid, transient increase in diacylglycerol is a key event in the transmembrane signaling process which follows ligand binding (113). It is not clear, however, whether IFNγ internalization following receptor interaction is required to stimulate the biological effects (114-117). The study of IFNγ-mediated activation of macrophages for tumoricidal activity has contributed additional information concerning the molecular events involved in this process, suggesting that protein kinase C activation and mobilization of intracellular calcium are essential steps in the IFN-dependent development of a functional response (118,119).

These relatively limited studies of molecular events occuring after ligand-receptor interaction have been complemented by a number of additional descriptions of alterations in cell structure and metabolism induced by IFNγ that are thought to be related to functional changes. These include, in particular, decreased membrane fluidity (120) and increased actin filament polymerization (121), which might influence biological activity dependent on membrane function or cell shape and motility (122); enhanced expression of adhesion molecules, including fibronectin (121) and glycoproteins in the LFA-1 family of molecules (123), which may influence both the uptake of microorganisms and the recruitment of monocyte/macrophages and granulocytes into inflammatory sites in vivo; and alterations in levels and methylation of RNA (124,125) and in protein phosphorylation (126,127), both of which have been documented in IFN-treated macrophages and proposed as contributors to the biochemical mechanisms of activation.

In addition to the descriptions of biochemical events occurring in response to IFN treatment, other experiments have focused on the description of changes in the protein mosaic of cells following IFNγ treatment. These studies have been prompted by early work which demonstrated a requirement for de novo protein synthesis and expression for development of the antiviral and antiproliferative functions of IFN (127-129). Four proteins with specific enzymatic activities have been identified in IFN-treated cells, including a protein kinase (130), 2'5'-oligoadenylate synthetase (131), a phosphodiesterase (132) and the indoleamine

2,3-dioxygenase described above (87). Both the $2'5'$-oligoadenylate synthetase and the protein kinase require double-stranded RNA for activation, and thus may be important factors in mediating the RNA changes described earlier. Changes in additional intracellular proteins with intriguing biological function have also been described following IFN treatment, including in particular the synthesis of a guanylate-binding protein that may function in the early IFN-mediated events and which is believed to play a regulatory role in cell metabolism (133). Finally, both one- and two-dimensional gel electrophoretic analysis of various cells following IFN treatment has shown that, depending on the cell type, between 8 and 12 new proteins are induced following IFN treatment (134–136), suggesting that some induced proteins remain to be identified.

Thus, the catalog of biochemical changes which are known to occur following IFN treatment is somewhat limited. Further information clearly is needed concerning the molecular events that effect the transduction of the IFN signal from the cell surface into the intracellular environment. In addition, the correlation between the protein changes observed in cell extracts and the proteins described on the basis of their activities remains to be accomplished. Recent studies using monoclonal antibodies directed against the IFNγ molecule have suggested that different epitopes on this glycoprotein may in fact stimulate separate cellular activities following ligand binding (137). Differences also have been observed in the cellular requirements for IFN-mediated triggering of distinct cellular responses. For example, the induction of antiviral activity and of tumoricidal capability in macrophages requires several hours of IFN exposure before responses can be detected (108,138). In contrast, induction of hydrogen peroxide production and modulation of FcR expression occur rapidly following IFN treatment (60,109). Independent regulation of macrophage Ia and LFA-1 surface proteins by IFNγ also has been documented (100). These observations suggest that the induction of individual functions in macrophages by IFNγ may require distinct intracellular signals. The recent observation that IFN-induced antiviral activity is not mediated by the same biochemical pathway as IFN-induced natural killer (NK) cell activation or IFN-induced antiproliferative effects (139) further supports this hypothesis and suggests that differential gene activation may be the basis of the multiple biologic responses observed in cells in response to IFN. Clear identification of the mechanism of action of IFN may finally depend on analysis of induction events at the molecular genetic level.

VI. CONCLUSIONS

Although interferons were originally described on the basis of antiviral activity, it is now clear that these proteins have a much broader role in regulation of the biological activity of a variety of cells. With regard to the role of IFN in the immunity of vertebrates to a wide variety of intracellular pathogens, most experimental work to date has concentrated on documentation of the consistent ability of IFNγ to induce a variety of antibacterial, antiprotozoal, and antifungal activities in a diversity of parasitized host cells. Particular emphasis has been given to description of the varied responses of macrophages to IFNγ, as many of these organisms can infect and reside in these cells. The production of reactive oxygen intermediates by macrophages following organism infection or uptake is clearly enhanced by treatment of the cells with IFNγ, and it is widely felt that these products of oxygen metabolism represent one important effector mechanism in the destruction and growth inhibition of many microbes. However, a number of organisms are resistant to the effects of these products and additional growth inhibitory and cytotoxic activities stimulated by IFNγ treatment may provide equally important contributions to host defense mechanisms. Many alterations in the protein mosaic of IFNγ-activated cells have been described to date. Although the role of a particular protein in the functional capacity of activated cells can be described in some cases, the majority of observed differences between activated and untreated cells remains obscure. It is expected that future studies will likely expand on the current preliminary evidence for multiple changes in macrophage membrane, secreted, and cytoplasmic proteins and attempt analysis at the molecular level of changes in gene expression which accompany the well-described changes in the functional state of many cells in response to IFNγ.

REFERENCES

1. Paulnock, D.M., and E.C. Borden. 1985. Modulation of immune functions by interferon. In *Interferon and Immunity to Cancer*, A.E. Reif and M.S. Mitchell (Eds.). Academic Press, Inc., New York, pp. 545–560.
2. Friedman, R.M., and S. Vogel. 1983. Interferons with particular emphasis on the immune system. *Adv. Immunol. 34*:97–140.
3. Sonnenfeld, G. 1981. Modulation of immunity by interferon. *Lymphokine Reports 1*:113–131.
4. Schultz, R.M. 1983. Modulation of immunity by interferons. *Lymphokine Reports 8*:303–322.
5. Nathan, C.E. 1983. Mechanisms of macrophage antimicrobial activity. *Trans. Roy. Soc. Trop. Med. Hyg. 77*:620–648.
6. Hahn, H., and S.H.E. Kaufmann. 1981. The role of cell-mediated immunity in bacterial infections. *Rev. Infect. Dis. 3*:1221–1250.
7. Mitchell, G.F. 1979. Effector cells, molecules and mechanisms in host-protective immunity to parasites. *Immunology 38*:209–223.
8. Unanue, E.R. 1983. The antigen-presenting cell function of the macrophage. *Ann. Rev. Immunol. 2*:101–195.
9. Waksman, B.H. 1981. Lymphokine networks in the immune response. In *The Lymphokines*, J.W. Hadden and W.E. Stewart II (Eds.), The Humana Press, pp. 1–10.
10. Mossmann, T.R., H. Cherwinski, M.W. Bond, M.A. Giedlin, and R.L. Coff-

man. 1986. Two types of murine helper T cell clone I. Definition according to profiles of lymphokine activities and secreted proteins. *J. Immunol. 136*: 2348-2357.

11. Cher, D.J., and T.R. Mosmann. 1987. Two types of helper T cell clone II. Delayed-type hypersensitivity is mediated by Th1 clones. *J. Immunol. 138*: 3688-3694.

12. Nacy, C.A., A.H. Fortier, M.S. Meltzer, N.A. Buchmeier, and R.D. Schreiber. 1985. Macrophage activation to kill Leishmania major: activation of macrophages for intracellular destruction of amastigotes can be induced by both recombinant interferon-γ and noninterferon lymphokines. *J. Immunol. 135*:3505-3511.

13. Kleinerman, E.S., R. Zicht, P.S. Sarin, R.C. Gallo, and I.J. Fidler. 1984. Constitutive production and release of a lymphokine with macrophage-activating factor activity distinct from gamma interferon by a human T-cell leukemia virus-positive cell line. *Cancer Res. 44*:4470.

14. Andrew, P.W., A.D.M. Rees, A. Scoging, N. Dobson, R. Matthews, J.T. Whittall, A.R.M. Coates, and B.B. Lowrie. 1984. Secretion of a macrophage-activating factor distinct from interferon-γ by a human T cell clone. *Eur. J. Immunol. 14*:962-975.

15. Byrne, G.I., and D.L. Kreuger. 1985. In vitro expression of factor-mediated cytotoxic activity generated during the immune response to Chlamydia in mice. *J. Immunol. 134*:4189-4193.

16. Horwitz, M.A., and S.C. Silverstein. 1981. Activated human monocytes inhibit the intracellular multiplication of Legionnaires disease bacilli. *J. Exp. Med. 154*:1618.

17. Rothermel, C.D., B.Y. Rubin, and H.W. Murray. 1983. Gamma-interferon is the factor in lymphokine that activates human macrophages to inhibit intracellular Chlamydia psittaci replication. *J. Immunol. 131*:2542-2544.

18. Murray, H.W., B.Y. Rubin, and C.D. Rothermel. 1983. Killing of intracellular Leishmania donovani by lymphokine-stimulated human mononuclear phagocytes. Evidence that interferon-γ is the activating lymphokine. *J. Clin. Invest. 72*:1506-1513.

19. Turco, J., and H.H. Winkler. 1984. Effect of mouse lymphokines and cloned mouse interferon-γ on the interaction of Rickettsia prowazekii with mouse macrophage-like RAW 264.7 cells. *Infect. Immun. 45*:303-308.

20. Kiderlen, A.F., S.H.E. Kaufmann, and M-L. Lohmann-Matthes. 1984. Protection of mice against the intracellular bacterium *Listeria monocytogenes* by recombinant immune interferon. *Eur. J. Immunol. 14*:964-967.

21. Havell, E.A., G.L. Spitalny, and P.J. Patel. 1982. Enhanced production of murine interferon by T cells generated in response to bacterial infection. *J. Exp. Med. 156*:112-123.

22. Smith, K.A. 1983. T cell growth factor. *Immunol. Rev. 51*:337-408.

23. Waldmann, T.A. 1986. The structure, function, and expression of interleukin-2 receptors on normal and malignant lymphocytes. *Science 232*: 727-732.

24. Johnson, H.M., and W.L. Farrar. 1983. The role of a gamma interferon-like

lymphokine in the activation of T cells for expression of interleukin 2 receptors. *Cell. Immunol. 75*:154–159.

25. Kasahara, T., J.J. Hooks, S.F. Dougherty, and J.J. Oppenheim. 1983. Interleukin 2-mediated immune interferon production by human T cells and T cell subsets. *J. Immunol. 130*:1784–1792.

26. Farrar, W.L., M.C. Birchenall-Sparks, and H.B. Young. 1986. Interleukin 2 induction of interferon-γ mRNA synthesis. *J. Immunol. 137*:3836–3840.

27. Wiskocil, R., A. Weiss, J. Imboden, R. Kamin-Lewis, and J. Stobo. 1985. Activation of a human T cell lines: a two-stimulus requirement in the pretranslational events involved in the coordinate expression of interleukin 2 and interferon-γ genes. *J. Immunol. 134*:1599–1611.

28. Grabstein, K.H., L.S. Park, P.J. Morrissey, H. Sassenfeld, V. Price, D.L. Urdahl, and M.B. Widmer. 1987. Regulation of murine T cell proliferation by B cell stimulatory factor-1. *J. Immunol. 139*:1148–1153.

29. Spits, H., H. Yssel, Y. Takebe, N. Arai, T. Yokota, F. Lee, K-I. Arai, J. Banchereau, and J.E. de Vries. 1987. Recombinant interleukin 4 promotes the growth of human T cells. *J. Immunol. 139*:1142–1147.

30. Larrick, L., and D.L. Boros. 1980. The artificial granuloma. 1. In vitro lymphokine production by pulmonary artificial hypersensitivity granulomas. *Clin. Immunol. Immunopathol. 17*:415–421.

31. Naher, H., U. Sperling, and H. Hahn. 1985. Role of T lymphocytes in granuloma formation. In *Mononuclear Phagocytes*, R. van Furth (Ed.). Martinus Nijhoff Publishers, The Hague, The Netherlands, pp. 561–568.

32. Kaufmann, S.H.E., and H. Hahn. 1982. Biological functions of T cell lines with specificity for the intracellular bacterium *Listeria monocytogenes* in vitro and in vivo. *J. Exp. Med. 155*:1754–1765.

33. Kaufmann, S.H.E. 1983. Effective antibacterial protection induced by a *Listeria monocytogenes*-specific T cell clone and its lymphokines. *Infect. Immun. 39*:1265–1270.

34. Kaufmann, S.H.E., H. Hahn, R. Berger, and H. Kirchner. 1983. Interferon-production by *Listeria monocytogenes*-specific T cells active in cellular antibacterial immunity. *Eur. J. Immunol. 13*:265–268.

35. Bordignon, D., R. Avallone, G. Peri, N. Polentarutti, C. Mangioni, and A. Mantovani. 1980. Cytotoxicity on tumor cells of human mononuclear phagocytes: defective tumoricidal capacity of alveolar macrophages. *Clin. Exp. Immunol. 41*:336–342.

36. Figdor, C.G., W.S. Bont, I. Touw, J. de Roos, E.E. Roosnek, and J.E. de Vries. 1982. Isolation of functionally different human monocytes by counterflow centrifugal elutriation. *Blood 60*:45–52.

37. Hoover, D.L., and C.A. Nacy. Macrophage activation to kill Leishmania tropica: defective intracellular killing of amastigotes by macrophages elicited with sterile inflammatory agents. *J. Immunol.*

38. Sorg, C. 1982. Heterogeneity of macrophages in response to lymphokines and other signals. *Mol. Immunol. 19*:1275–1278.

39. Vogel, S.N., K.E. English, D. Fertsch, and M.J. Fultz. 1983. Differential modulation of macrophage membrane markers by interferon: Analysis of

Fc and C3b receptors, Mac-1, and Ia antigen expression. *J. IFN. Res. 3*: 153–160.

40. Lee, S.H.S., and L.B. Epstein. 1980. Reversible inhibition by interferon of the maturation of human peripheral blood monocytes to macrophages. *Cell. Immunol. 50*:177–190.

41. Meltzer, M.S. 1978. Macrophage activation for tumor cytotoxicity: Development of macrophage cytotoxic activity requires completion of a sequence of short-lived intermediary reactions. *J. Immunol. 121*:2035–2042.

42. Buchmuller, Y., and J. Mauel. 1979. Studies on the mechanism of macrophage activation. II. Parasite destruction in macrophages activated by supernates from concanavalin A-stimulated lymphocytes. *J. Exp. Med. 150*:359–370.

43. Nathan, C.F., and R.K. Root. 1977. Hydrogen peroxide release from mouse peritoneal macrophages. Dependance on sequential activation and triggering. *J. Exp. Med. 146*:1649–1659.

44. Gordon, S., J.C. Unkeless, and Z.A. Cohn. 1974. Induction of macrophage plasminogen activator by endotoxin stimulation and phagocytosis. Evidence for a two-stage process. *J. Exp. Med. 140*:995–1011.

45. Krammer, P.H., C.F. Kubelka, W. Falk, and A. Ruppel. 1985. Priming and triggering of tumoricidal and schistosomulicidal macrophages by two sequential lymphokine signals: interferon-γ and macrophage cytotoxicity inducing factor 2. *J. Immunol. 135*:3258–3263.

46. Pace, J.L., S.W. Russell, B.A. Torres, H.M. Johnson, and P.W. Gray. 1983. Recombinant mouse γ-interferon induces the priming step in macrophage activation for tumor cell killing. *J. Immunol. 130*:2011–2014.

47. Meltzer, M.S., and C.A. Nacy. 1985. Macrophage cytotoxicity against tumor cell and microbial targets: genetic control of the activation network. In *Progr. Leuk. Biol.*, Vol. 3, *Genetic Control of Host Resistance to Infection and Malignancy*, E. Skamene (Ed.). Alan R. Liss, Inc., New York, pp. 595–604.

48. Skamene, E., S.L. James, M.S. Meltzer, and M.S. Nesbitt. 1984. Genetic control of macrophage activation for killing of extracellular and intracellular targets. *J. Leuk. Biol. 35*:65–70.

49. Kongshavn, P.A.L., C. Sandrarangani, and E. Skamene. 1980. Genetically determined differences in antibacterial activity of macrophages are expressed in the environment in which the macrophage precursors mature. *Cell. Immunol. 53*:341–349.

50. Nacy, C., M. Meltzer, and A. Fortier. 1984. Macrophage activation to kill *Leishmania tropica*: characterization of P/J mouse defects for lymphokine-induced antimicrobial activities against *Leishmania tropica amastigotes*. *J. Immunol. 133*:3344–3350.

51. Boraschi, D., and M. Meltzer. 1980. Defective tumoricidal capacity of macrophages from A/J mice. III. Genetic analysis of the macrophage defect. *J. Immunol. 124*:1050–1059.

52. Adams, D.O., P.A. Marino, and M.S. Meltzer. 1981. Characterization of genetic defects in macrophage tumoricidal capacity: identification of

murine strains with abnormalities in secretion of cytolytic protease and ability to bind neoplastic targets. *J. Immunol. 126*:1843-1847.

53. Adams, D.O., and T.A. Hamilton. 1984. The cell biology of macrophage activation. *Ann. Rev. Immunol. 2*:283-318.

54. Murray, H.W., and Z.A. Cohn. 1980. Macrophage oxygen-dependent antimicrobial activity. III. Enhanced oxidative metabolism as an expression of macrophage activation. *J. Exp. Med. 152*:1596-1600.

55. Babior, B.M., and W.A. Peters. 1981. The O_2-producing enzyme of human neutrophils. Further properties. *J. Biol. Chem. 256*:2321-2323.

56. Berton, G., and S. Gordon. 1985. Role of the plasma membrane in the regulation of superoxide anion release by macrophages. In *Mononuclear Phagocytes*, R. van Furth (Ed.). Martinus Nijhoff Publishers, The Hague, The Netherlands, pp. 435-443.

57. Leijh, P.C.J., C.F. Nathan, M.Th. van den Barselaar, and R. van Furth. 1985. Relationship between the extracellular stimulation of intracellular killing and oxygen-dependent microbicidal systems of monocytes. *Infect. Immun. 47*:502-507.

58. Crawford, D.R., and D.L. Schneider. 1983. Ubiquinone content and respiratory burst activity of latex-filled phagolysosomes isolated from human neutrophils and evidence for the probable involvement of a third granule. *J. Biol. Chem. 258*:5363-5367.

59. Wilson, C.B., V. Tsai, and J.S. Remington. 1980. Failure to trigger the oxidative burst by normal macrophages. Possible mechanism for survival of intracellular pathogens. *J. Exp. Med. 151*:328-336.

60. Nathan, C.F., H.W. Murray, M.E. Wiebe, and B.Y. Rubin. 1983. Identification of interferon-γ as the lymphokine that activates human macrophage oxidative metabolism and antimicrobial activity. *J. Exp. Med. 158*:670-689.

61. Nathan, C.F., N. Nogueira, C. Juangbhanich, J. Ellis, and Z.A. Cohn. 1979. Activation of macrophages in vivo and in vitro. Correlation between hydrogen peroxide release and killing of *Trypanosoma cruzi*. *J. Exp. Med. 149*: 1056-1065.

62. Sasada, M., and R.B. Johnston. 1980. Macrophage microbicidal activity. Correlation between phagocytosis-associated oxidative metabolism and the killing of *Candida* by macrophages. *J. Exp. Med. 152*:85-93.

63. Locksley, R.M., C.B. Wilson, and S.J. Klebanoff. 1982. Role for endogenous and acquired peroxidase in the toxoplasmacidal activity of murine and human mononuclear phagocytes. *J. Clin. Invest. 69*:1099-1112.

64. Walker, L., and D.B. Lowrie. 1981. Killing of *Mycobacterium microti* by immunologically activated macrophages. *Nature 293*:69-71.

65. Haidaris, C.G., and P.F. Bonventre. 1982. A role for oxygen-dependent mechanisms in killing of *Leishmania donovani* tissue forms by activated macrophages. *J. Immunol. 129*:850-861.

66. Locksley, R.M., C.B. Wilson, and S.J. Klebanoff. 1982. Role for endogenous and acquired peroxidase in the toxoplasmacidal activity of murine and human mononuclear phagocytes. *J. Clin. Invest. 69*:1099-1110.

67. Nathan, C.F. 1986. Peroxide and pteridine: A hypothesis on the regulation

of macrophage antimicrobial activity by interferon-γ. In *Interferon 7*, I. Gresser (Ed.). Academic Press Inc., London, pp. 125–143.

68. Murray, H.W., G.L. Spitalny, and C.F. Nathan. 1985. Activation of mouse peritoneal macrophages in vitro and in vivo by interferon-γ. *J. Immunol. 143*:1619–1622.

69. Byrne, G.I., and C.L. Faubion. 1983. Inhibition of *Chlamydia psittaci* in oxidatively active thioglycollate-elicited macrophages: distinction between lymphokine-mediated oxygen-dependent and oxygen-independent macrophage activation. *Infect. Immun. 40*:464–471.

70. Murray, H.W., G.I. Byrne, C.D. Rothermel, and D.M. Cartelli. 1983. Lymphokine enhances oxygen-independent activity against intracellular pathogens. *J. Exp. Med. 158*:234–239.

71. Sibley, L.D., J.L. Krahenbuhl, and E. Weidner. 1985. Lymphokine activation of J774G8 cells and mouse peritoneal macrophages challenged with *Toxoplasma gondii. Infect. Immun. 49*:760–764.

72. Murray, H.W., G.I. Byrne, C.D. Rothermel, and D.M. Cartelli. 1983. Lymphokine enhances oxygen-independent activity against intracellular pathogens. *J. Exp. Med. 158*:234–239.

73. Byrne, G.I., and C. Faubion. 1982. Lymphokine-mediated microbistatic mechanisms restrict Chlamydia psittaci growth in macrophages. *J. Immunol. 128*:469–473.

74. Steinman, R.M., I.S. Mellman, W.A. Muller, and Z.A. Cohn. 1983. Endocytosis and recycling of the macrophage plasma membrane. *J. Cell Biol. 96*:1–27.

75. Axline, S. 1970. Functional biochemistry of the macrophages. *Sem. Hematol. 7*:142–163.

76. Steinman, R.M., and Z.A. Cohn. 1974. The metabolism and physiology of the mononuclear phagocytes. In *The Inflammatory Process*, Vol. 1, B.W. Zweifach, L. Grant, and R.T. McCluskey (Eds.). Academic Press, New York, pp. 449–510.

77. Page, R.C., P. Davies, and A.C. Allison. 1978. The macrophage as a secretory cell. *Int. Rev. Cytol. 52*:119–132.

78. Davies, P., and R.J. Bonney. 1979. Secretory products of mononuclear phagocytes: a brief review. *J. Reticuloendothel. Soc. 26*:37–48.

79. Gordon, S. 1978. Regulation of enzyme secretion by mononuclear phagocytes: Studies with macrophage plasminogen activator and lysozyme. *Fed. Proc. 37*:2754–2758.

80. Gordon, S., J. Todd, and Z.A. Cohn. 1974. In vitro synthesis and secretion of lysozyme by mononuclear phagocytes. *J. Exp. Med. 139*:1228–1248.

81. Grand-Perret, T., J-F. Petit, and G. Lemaire. 1986. Modifications induced by activation to tumor cytotoxicity in the protein secretory activity of macrophages. *J. Leuk. Biol. 40*:1–19.

82. Morahan, P.S., P.J. Edelson, and K. Gass. 1980. Changes in macrophage ectoenzymes associated with anti-tumor activity. *J. Immunol. 125*:1312–1317.

83. Edelson, P.J. 1981. Macrophage plasma membrane enzymes as differentiation markers of macrophage activation. *Lymphokines 3*:57–71.

84. Adams, D.O., K.J. Kao, R. Farb, and S.V. Pizzo. 1980. Effector mechanisms of cytolytically activated macrophages. II. Secretion of a cytolytic factor by activated macrophages and its relationship to secreted neutral proteases. *J. Immunol. 124*:293–308.

85. Drysdale, B.E., C.M. Zacharchuk, and H.S. Shin. 1983. Mechanism of macrophage-mediated cytotoxicity: Production of a soluble cytotoxic factor. *J. Immunol. 131*:2362–2374.

86. Mannel, D.N., W. Falk, and M.S. Meltzer. 1981. Inhibition of non-specific tumoricidal activity by activated macrophages with antiserum against a soluble cytotoxic factor. *Infect. Immun. 33*:156–163.

87. Yoshida, R., J. Imanishi, T. Oku, T. Kishida, and O. Hayaishi. 1981. Induction of pulmonary indoleamine 2,3-dioxygenase by interferon. *Proc. Natl. Acad. Sci. (USA) 78*:129–132.

88. Pfefferkorn, E.R. 1984. Interferon γ blocks the growth of *Toxoplasma gondii* in human fibroblasts by inducing the host cells to degrade tryptophan. *Proc. Natl. Acad. Sci. (USA) 81*:908–912.

89. Pfefferkorn, E.R., and P.M. Guyre. 1984. Inhibition of growth of *Toxoplasma gondii* in cultured fibroblasts by human recombinant gamma interferon. *Infect. Immun. 44*:211–216.

90. Byrne, G.I., L.K. Lehmann, and G.J. Landry. 1986. Induction of tryptophan catabolism is the mechanism for gamma-interferon-mediated inhibition of intracellular *Chlamydia psittaci* replication in T24 cells. *Infect. Immun. 53*:347–351.

91. Carlin, J.M., E.C. Borden, P.M. Sondel, and G.I. Byrne. 1987. Interferon-mediated indoleamine 2,3-dioxygenase activity in human monocytes. *J. Immunol. 139*:2414–2418.

92. Byrne, G.I., L.K. Lehmann, J.G. Kirschbaum, E.C. Borden, C.M. Lee, and Raymond R. Brown. 1986. Induction of tryptophan degradation in vitro and in vivo: A γ-interferon-stimulated activity. *J. IFN. Res. 6*:389–396.

93. Ezekowitz, R.A.B., and S. Gordon. 1984. Alterations in surface properties by macrophage activation. Expression of receptors for Fc and mannose terminal glycoproteins and differentiation antigens. *Cont. Topics Immunobiol. 14*:33–56.

94. Yin, H.L., S. Aley, C. Bianco, and Z.A. Cohn. 1980. Plasma membrane polypeptides of resident and activated mouse peritoneal macrophages. *Proc. Natl. Acad. Sci. (USA) 77*:2188–2190.

95. Springer, T.A. 1981. Monoclonal antibody analysis of complex biological systems. Combination of cell hybridization and immunoadsorbents in a novel cascade procedure and its application to the macrophage cell surface. *J. Biol. Chem. 256*:3833–3839.

96. LeBlanc, P.A., S.W. Russell, and S. M-T. Chang. 1981. Mouse monoclonal phagocyte antigenic heterogeneity detected by monoclonal antibodies. *J. Reticuloendothel. Soc. 30*:439–445.

97. Ezekowitz, R.A.B., J.A. Austyn, P. Stahl, and S. Gordon. 1983. Surface

properties of Bacillus Calmette-Guerin-activated mouse macrophages. Reduced expression of mannose-specific endocytosis, Fc receptors and antigen F4/80 accompanies induction of Ia. *J. Exp. Med. 154*:60–76.

98. Ezekowitz, R.A.B., and S. Gordon. 1982. Down regulation of mannosyl receptor-mediated andocytosis and antigen F4/80 in BCG-activated mouse macrophages. Role of T lymphocytes and lymphokines. *J. Exp. Med. 155*: 1623–1637.

99. Lambert, L.E., J.S. Schrimpf, and D.M. Paulnock. 1986. Macrophage cell lines exhibit distinct patterns of responses to gamma interferon (abstr). *J. Leuk. Biol. 40*:256.

100. Strassmann, G., S.D. Somers, T.A. Springer, D.O. Adams, and T.A. Hamilton. 1986. Biochemical models of IFN-γ-mediated macrophage activation: independent regulation of lymphocyte function associated antigen (LFA)-1 and I-A antigen on murine peritoneal macrophages. *Cell. Immunol. 97*: 110–120.

101. Steinman, R.M., N. Nogueira, M.D. Witmer, J.D. Tydings, and I.S. Mellman. 1980. Lymphokine enhances the expression and synthesis of Ia antigens on cultured mouse peritoneal macrophages. *J. Exp. Med. 152*:1248–1259.

102. Scher, M.G., D.I. Beller, and E.R. Unanue. 1980. Demonstration of a soluble mediator that induces exudates rich in Ia-positive macrophages. *J. Exp. Med. 152*:1684–1691.

103. Leijh, P.C.J., MTh. van den Barsalaar, ThL. van Zwet, M.R. Daha, and R. van Furth. 1979. Requirement of extracellular complement and immunoglobulin for intracellular killing of microorganisms by human monocytes. *J. Clin. Invest. 63*:772–784.

104. Leijh, P.C.J., ThL. van Zwet, and R. van Furth. 1980. Effect of extracellular serum in the stimulation of intracellular killing of streptococci by human monocytes. *Infect. Immun. 30*:421–426.

105. Leijh, P.C.J., M.Th. van den Barselaar, Th.L. van Zwet, L.A. Ginzel, and R. van Furth. 1985. Membrane stimulation and intracellular killing of micro-organisms by human monocytes. In *Mononuclear Phagocytes*, R. van Furth (Ed.). Martinus Nijhoff Publishers, The Hague, The Netherlands, pp. 463–470.

106. Sung, S-S., T. Nelson, and R.S. Silverstein. 1983. Yeast mannans inhibit binding and phagocytosis of zymosan by mouse peritoneal macrophages. *J. Cell. Biol. 95*:160–166.

107. Hirsch, S., J.A. Austyn, and S.N. Gordon. 1981. Expression of macrophage-specific antigen F4/80 during differentiation of macrophage bone marrow cells in culture. *J. Exp. Med. 154*:713–725.

108. Celada, A., P.W. Gray, E. Rinderknecht, and R.D. Schreiber. 1984. Evidence for a gamma interferon receptor that regulates macrophage tumoricidal activity. *J. Exp. Med. 160*:55–63.

109. Celada, A., R. Allen, I. Esparza, P.W. Gray, and R.D. Schreiber. 1985. Demonstration and partial characterization of the interferon gamma receptor on human mononuclear phagocytes. *J. Clin. Invest. 2196*–3008.

110. Aiyer, R.A., L.E. Serrano, and P.P. Jones. 1986. Interferon-gamma binds to high and low affinity receptor components on murine macrophages. *J. Immunol. 136*:3329–3334.

111. Tovey, M.G., C. Rouchette-Egly, and M. Castagna. 1979. Effect of interferon on concentration of cyclic nucleotides in cultured cells. *Proc. Natl. Acad. Sci. (USA) 76*:3890–3893.

112. Vesely, D.L., and K. Cantell. 1980. Human interferon enhances guanylate cyclase activity. *Biochem. Biophys. Res. Commun. 96*:574–579.

113. Yap, W.H., T.S. Teo, and Y.H. Tan. 1986. An early event in the interferon-induced transmembrane signaling process. *Science 234*:355–358.

114. Anderson, P., Y.K. Yip, and J. Vilchek. 1983. Human interferon is internalized and degraded by human fibroblasts. *J. Biol. Chem. 758*:6497–6502.

115. Branca, A.A., C.R. Faltynek, S.B. D'Alessandro, and C. Baglioni. 1982. Interaction of interferon with cellular receptors. Internalization and degradation of cell-bound interferon. *J. Biol. Chem. 257*:13291–13296.

116. Ankel, H., C. Chany, B. Galio, M. Chevallier, and M. Robert. 1973. Antiviral effect of interferon covalently bound to sepharose. *Proc. Natl. Acad. Sci. (USA) 70*:2360–2363.

117. Auguet, M., and B. Blanchard. 1981. High-affinity binding of 125I-labeled interferon to a specific receptor. II. Analysis of binding properties. *Virology 115*:249–261.

118. Celada, A., and R.D. Schreiber. 1986. Role of protein kinase c and intracellular calcium mobilization in the induction of macrophage tumoricidal activity by interferon-γ. *J. Immunol. 137*:2373–2379.

119. Hamilton, T.A., D.L. Becton, S.D. Somers, P.W. Gray, and D.O. Adams. 1985. Interferon-γ modulates protein kinase C activity in murine pertoneal macrophages. *J. Biol. Chem. 260*:1378–1383.

120. Pfeffer, L.M., E. Wang, and I. Tamm. 1980. Interferon inhibits the redistribution of cell surface components. *J. Exp. Med. 152*:469–474.

121. Pfeffer, L.M., E. Wang, and I. Tamm. 1980. Interferon effects on microfilament organization, cellular fibronectin distribution, and cell motility in human fibroblasts. *J. Cell. Biol. 85*:9–17.

122. Mahoney, E.M., W.A. Scott, F.R. Landsberger, A.L. Hamil, and Z.A. Cohn. 1980. Influence of fatty acyl substitution on the composition and function of macrophage membranes. *J. Biol. Chem. 255*:4910–4918.

123. Anderson, D.C., and T.A. Springer. 1985. The importance of the MAC-1, LFA-1 glycoprotein family in adherence-dependent inflammatory functions: insights from an experiment of nature. In *Progr. Leuk. Biol.*, Vol. 3, *Genetic Control of Host Resistance to Infection and Malignancy*, E. Skamene (Ed.). Alan R. Liss, Inc., New York, pp. 611–621.

124. Varesio, L., M. Clayton, D. Radzioch, and E. Bonvini. 1987. Selective inhibition of 28S ribosomal RNA in macrophages activated by interferon-β or -γ. *J. Immunol. 138*:2332–2337.

125. Koblet, H., R. Wyler, and U. Kohler. 1979. Altered or increased transfer RNA methylation in the course of interferon action on cells in culture. *Experentia 35*:576–578.

126. Weiel, J., T.A. Hamilton, and D.O. Adams. 1986. LPS induces altered phosphate labeling of proteins in murine peritoneal macrophages. *J. Immunol. 136*:3012–3018.

127. LeBlue, B., G.C. Sen, S. Shaila, B. Cabrer, and P. Lengyel. 1976. Interferon, double-stranded RNA, and protein phosphorylation. *Proc. Natl. Acad. Sci. (USA) 73*:3107–3111.

128. Lockart, R.Z., Jr. 1964. The necessity for cellular RNA and Protein synthesis for viral inhibition resulting from interferon. *Biochem. Biophys Res. Commun. 15*:513–518.

129. Friedman, R.M., and J.A. Sonnabend. 1965. Inhibition of interferon action by puromycin. *J. Immunol. 95*:696–703.

130. Robert, W.K., A.G. Hovanessian, R.E. Brown, M.J. Clemens, and I.M. Kerr. 1976. Interferon-mediated protein kinase and low-molecular-weight inhibitor of protein synthesis. *Nature 264*:477–480.

131. Kerr, I.M., and R.E. Browm. 1978. pppA2′p5′A2′p5′A: an inhibitor of protein synthesis synthesized with an enzyme fraction from interferon-treated cells. *Proc. Natl. Acad. Sci. (USA) 75*:256–260.

132. Schmidt, A., A. Zilberstein, L. Shulman, P. Federman, H. Berissi, and M. Revel. 1978. Interferon action: isolation of nuclease F, a translation inhibitor activated by interferon-induced (2′,5′) oligo-isoadenylate. *FEBS Lett. 95*:257–264.

133. Mitchell, W.M., and R.L. Forti. 1983. Progress in the monitoring of human interferon in body fluids and the phenotypic expression of human interferon activity. *Prog. Clin. Pathol. 9*:101–119.

134. Rubin, B.Y., and S.L. Gupta. 1980. Differential efficacies of human Type I and Type II interferons as antiviral and antiproliferative agents. *Proc. Natl. Acad. Sci. (USA) 77*:5928–2932.

135. Weil, J., C.J. Epstein, L.B. Epstein, J.J. Sedmark, J.L. Sabran, and S.E. Grossberg. 1983. A unique set of polypeptides is induced by gamma interferon in addition to those induced in common with alpha and beta interferons. *Nature 301*:437–439.

136. MacKay, R.J., and S.W. Russell. 1986. Protein changes associated with stages of activation of mouse macrophages for tumor cell killing. *J. Immunol. 137*:1392–1398.

137. Schreiber, R.D., L.J. Hicks, A. Celada, N.A. Buchmeier, and P.W. Gray. 1985. Monoclonal antibodies to murine γ-interferon which differentially modulate macrophage activation and antiviral activity. *J. Immunol. 134*: 1609–1618.

138. Dianzani, F., L. Salter, W.R. Fleishman, and M. Zucca. 1978. Immune interferon activates cells more slowly than does virus-induced interferon. *Proc. Soc. Exp. Biol. Med. 159*:94–111.

139. Forti, R.L., W.M. Mitchell, W.C. Hubbard, R.J. Workman, and J.T. Forbes. 1984. Pleiotropic activities of human interferons are mediated by multiple response pathways. *Proc. Natl. Acad. Sci. (USA) 81*:170–174.

Part II

OBLIGATE INTRACELLULAR PATHOGENS

3

Interferons, Immunity, and Chlamydiae

Gerald I. Byrne / University of Wisconsin Medical School, Madison, Wisconsin

I. INTRODUCTION

A. Biology of the Chlamydiae

Chlamydiae are procaryotic organisms obligately restricted to an existence within susceptible eucaryotic cells due, at least in part, to their requirements for high energy intermediates of oxidative phosphorylation (adenosine triphosphate, ATP), which they must secure from preformed host cell stores (1). The cycle of intracellular chlamydial growth is unique in that the organisms undergo an orderly alternation between two distinct developmental forms, and this entire process of growth and differentiation occurs within the confines of a membrane-bound vesicle in the cytoplasm of susceptible host cells (2). There are two chlamydial species, *C. psittaci* and *C. trachomatis*. Both undergo similar morphologic alterations during the course of intracellular growth, but representatives of each species may be distinguished from each other and from members of the other species by distinct sero-specific epitopes present on the major outer membrane protein and on other minor surface-exposed polypeptides (3). The genus shares a lipopolysaccharide antigen, similar in nature to endotoxin lipopolysaccharide of gram-negative bacteria, but chlamydiae are set apart from gram-negative organisms by the lack of detectable peptidoglycan in their cell wall structure (4).

Morphologic differentiation by chlamydiae involves conversion of a rigid, metabolically inert infectious form of the organism (elementary body) to a metabolically active, noninfectious intracellular form of the organism (reticulate body). This differentiation takes place soon after uptake of the elementary body by susceptible host cells and is accompanied by changes in the protein and lipid

53

composition of the outer envelope, a loss of structural rigidity and an increase in porin activity (5--7). These changes result in a metabolically active reticulate body that replicates by binary fission within infected host cells. Chlamydial replication is always restricted to the confines of a membrane-bound vesicle that does not fuse with lysosomes (8), but may communicate with mitochondria or mitochondrial membranes (9). This aspect of the intracellular replication of chlamydiae exhibits similarities remarkably similar to *Toxoplasma* (see Chap. 8) despite the wide evolutionary distinctions between these two intracellular pathogens. Transmission of chlamydiae to a new host cell requires that a second round of differentiation takes place. Each reticulate body undergoes morphologic conversion to an elementary body at a point in the developmental cycle when the host cell cytoplasm is nearly filled with a bulging inclusion vesicle containing up to thousands of replicating chlamydiae. The events that trigger reticulate body to elementary body differentiation are not understood, but always occur after a considerable amount of chlamydial replication has taken place and never is 100% efficient. Once late differentiation becomes predominant, host cell lysis occurs, and released elementary bodies can initiate a new round of replication. There also is evidence that intact chlamydial inclusions can be released from host cells (10,11). This release may or may not be accompanied by host cell lysis. Presumably, the ultimate fate of a released intact inclusion is similar to what occurs when host cell and inclusion lysis occurs prior to inclusion release, in that elementary bodies eventually become free in the extracellular milieu.

It is important to recognize that events peculiar to the chlamydial developmental cycle may have profound ramifications on immunity to the chlamydiae and on the role of cytokines, including the interferons, in controlling chlamydial growth, development, and dissemination. These topics will be further developed as they arise in the course of the ensuing discussion.

B. Chlamydial Diseases

The chlamydiae are ubiquitously distributed in nature. Representatives of the genus are known to cause infections in molluscs (12), lower vertebrates (13), and avian species (14). Virtually every species of marsupial (15) and mammal (16) that have been investigated are susceptible to infection and disease by some strain of chlamydiae. Humans are subject to a wide variety of chlamydial infections (17,18). These range from acute systemic infections to chronic, protracted mucosal diseases. Humans can serve as hosts for *C. psittaci*, the most ubiquitously distributed of the chlamydial species, but *C. trachomatis* is virtually exclusively associated with humans and most chlamydial disease in humans is caused by one or another biovar of *C. trachomatis*. The classic human chlamydial disease that has been accurately described since the written word has been in use is trachoma, a chronic ocular chlamydial disease. Several serovars of *C. trachomatis* (A, B, Ba, and C) are associated principally with trachoma, and therefore, these particular serovars are restricted geographically to areas of the world where trachoma remains endemic. Trachoma is a disease associated with poverty, ill-housing, and poor sanitary and hygienic conditions. It remains the leading cause of preventable blindness in the world and is prevalent in rural areas of the Middle East, North Africa, the Near East, and some parts of the Far East. Trachoma was once endemic in native American populations of the Southwest, but improved hygiene and primary clinical care has eliminated trachoma from the United States. Trachoma is a disease of the ocular mucosal surface. Chlamydiae are isolated from conjunctival epithelial cells and the serious complications of trachoma are due to chronic inflammation, vascularization of the cornea, scarring of the conjunctiva, and distortion of the lid resulting in trichiasis and entropion. Although trachoma is not presently a problem in the United States, other chlamydial diseases of mucosal surfaces are currently among the most frequently encountered infectious diseases known in the developed world.

Chlamydial genital tract infections have been recognized for nearly 100 years, but the vast incidence, severity, and complications of these diseases have been appreciated only for the past 15 years. Cases of chlamydial urethritis now are recognized to exceed gonorrhea by at least twofold in the United States and Great Britain. A similar incidence exists for chlamydial cervicitis. Major complications of cervicitis include salpingitis and endometritis. These infections can result in fallopian tube damage and infertility. An infant born to a *C. trachomatis*-positive mother is at risk for conjunctival and respiratory disease. Infected neonates have been reported to harbor chlamydiae in the nasopharynx and gastro-intestinal tract for considerable periods of time and reactivation of these colonizations are possible.

In addition to the more commonly seen chlamydial infections, other chlamydial diseases are less frequently seen or at least less well appreciated. Lymphogranuloma venereum (LGV) is a sexually transmitted disease caused by a distinct biovar of *C. trachomatis* (serovars L_1, L_2, L_3). Although LGV is transmitted in a manner similar to other chlamydial genital tract infections, manifestations of the disease are quite distinct. A transient chancre appears at the initial site of infection and the organism quickly invades the lymphatic system, where it enters and multiplies within mononuclear phagocytes. This leads to dramatic lymphadenopathy of affected nodes. Typically, inguinal lymph nodes are involved, but systemic complications and replication of the organisms in extralymphatic tissue is not rare. LGV infections are common in some parts of Africa and Asia and has been principally associated with individuals of lower socioeconomic status.

A relatively more recently described chlamydial disease involves infection with yet another distinct chlamydial biovar that has not been adequately classified. The chlamydial strain has been referred to as TWAR (19) and causes human to human transmitted respiratory disease. Isolates of TWAR have been very

Table 1 Summary of Chlamydial Serovars and Human Diseases Caused by *Chlamydiae*

Species	Serovars	Disease	Complications
C. trachomatis	A, B, Ba, C	Trachoma	Blindness
C. trachomatis	D–K	Urethritis (males)	Epididymitis protitis
		Cervicitis (females)	Pelvic inflammatory disease neonatal infections
C. trachomatis	L_1, L_2, L_3	Lymphogranuloma venereum	Systemic spread
C. psittaci	Many unidentified serovars	Ornithosis (respiratory)	Systemic spread
Unclassified chlamydial strain	?	TWAR (respiratory)	None identified

difficult to cultivate in the laboratory, but indications are that TWAR organisms are more closely associated with *C. psittaci* than with *C. trachomatis* (20). Classic human *C. psittaci* infection (ornithosis) also is a respiratory disease, but virtually always has been associated with transmission from avian sources. The human to human spread described for TWAR establishes a mode that would lead to a much greater transmission potential than is possible for ornithosis. Indeed, if initial reports concerning TWAR incidence are correct (21), this disease may comprise as much as 20% of all atypical pneumonia cases reported. Table 1 provides a summary of chlamydial serovars and the human diseases with which they are principally associated.

It is clear that chlamydial infections represent a significant health concern both in the United States and worldwide. Not only do human chlamydial diseases compromise the health and well-being of hundreds of millions of individuals, but chlamydial infections of domestic and feral animals are of economic and humane importance. Features that are common to virtually all chlamydial infections is their tendency to persist if not treated, and involvement of immune responses in the pathogenesis of disease. The immunologic induction of cytokines, including interferons, appear to be contributing elements to protection, persistence, and disease pathology and a major purpose of this chapter is to document the mechanisms whereby these features of chlamydial infections can become manifest. The next section will present a general overview of what is known concerning immune responses to chlamydiae in both human disease and animal models of disease or protective immunity. This is followed by a more detailed account of the induction and mechanism of action of the interferon system in chlamydial pathogenesis and immunity. Special emphasis will be given to the role of gamma interferon in chlamydial infections, since this cytokine has been thus far more closely associated with both pathogenesis and recovery from acute stages of chlamydial disease. Finally, a prospectus will be given concerning how the interferon system functions within the context of chlamydial infections as well as infections by other intracellular pathogens and provide a working model of methods that can be applied to gain more insight into these important elements of pathogenesis and immunity.

II. IMMUNE RESPONSES TO CHLAMYDIAE

A. Immune Responses in Human Disease

Both antibodies and the induction of cell-mediated responses can be demonstrated after infection with chlamydiae (22), but the literature has not provided any consistent evidence that either of these responses is critical to protective immunity. There is some indication that a relative degree of resistence accompanies recovery from either mucosal or systemic chlamydial disease, but this immunity is apparently not long-lived and may be strain- or serovar-specific responses. Induction of cytokine responses has not been carefully studied during or following human chlamydial infections but indirect evidence is beginning to accumulate to support the hypothesis that gamma interferon is produced locally, at least during acute chlamydial ocular infections. Currently there is no reliable vaccine available for any human chlamydial disease, but research efforts with this goal in mind are being vigorously pursued. Twenty years ago, killed, whole cell suspensions of *C. trachomatis* were used in vaccine trials against trachoma (23). Results of this effort were disappointing in terms of trachoma control, but did provide important information concerning the potential for immune pathology contributing to the disease process and, at least for some of the trials, encouraging data vis à vis the possibility of some day achieving protective immunity against chlamydial disease.

B. Immune Responses in Animal Models

A variety of animal models are available to study immune responses to chlamydial infection or immunization protocols. Some of these models mimic human disease, others are designed to increase a basic understanding of immune responsiveness to chlamydiae in a more general way. Subhuman primates have been used to determine if protective immunity follows initial ocular infection with *C. trachomatis* (24,25). Relatively modest levels of protection have been observed in some studies, but protection from reinfection is frequently accompanied by increased hypersensitivity reactions that lead to exasercbated disease pathology. Apparently, antigens that mediate hypersensitivity are distinct from those that mediate protection since hypersensitivity responses are more long-lived and not as serovar specific as are protective responses (23).

Studies in guinea pigs and mice using either *C. psittaci* (26) or murine (27) and human (28) strains of *C. trachomatis* have provided evidence that protective immunity is possible to achieve, is dependent on induction of cell-mediated responses, the presence of secretory antibody and involves the production of cytokines (29).

III. INTERFERONS AND IMMUNE RESPONSES TO CHLAMYDIAE

A. Historical Review and Effects of Interferons α and β

The first report which suggested that chlamydiae were affected in cells treated with interferons (IFN) came in 1963 when Sueltenfuss and Pollard (30) reported that IFN induced in tissue culture by duck hepatitis virus curtailed chlamydial growth when added to cultured cells that were subsequently infected with *C. psittaci*. It was shown later that induction of IFN production in cell culture (31,

32) and in mice (32) could be mediated by chlamydiae. Thus it was established early on, at a time when the IFN system was ill-defined and thought only to inhibit viral replication, that chlamydiae were found to both induce and be inhibited by IFN.

In 1971, Kazar et al. (33) reported that treatment of mouse fibroblasts with Newcastle disease virus or poly I:C-induced IFN resulted in the inhibition of *C. trachomatis* growth when these cells were infected after exposure to IFN. Uptake of chlamydiae was not affected in IFN-treated cells, but since infectivity could not be recovered, the authors speculated that differentiation of elementary to reticulate bodies occurred and the effect of IFN was to restrict reticulate body replication. These authors also provided the first evidence that the IFN system might play a role in recovery from chlamydial disease when they showed that mice exposed to Newcastle disease virus aerosols were better able to survive pneumonia caused by exposing the animals to aerosolized preparations of the mouse pneumonitis agent, a murine strain of *C. trachomatis*.

An important factor that limited the early observations of the induction and effects of IFN on chlamydial development was that at the time that these initial studies were done it was not established that three main classes of IFN existed (IFNα, IFNβ, and IFNγ) and that these classes could be further categorized according to whether they were induced in response to viral infection (IFNα and β) or as part of the immune response (IFNγ). In addition, these early workers did not have highly purified preparations to work with. In these early studies it was not possible to evaluate if the reported observations truly reflected the effects of IFN or the effects of other cytokines produced concomitantly with and perhaps acting in concert with IFN.

As a consequence of these limitations, work in this field did not progress very rapidly for a number of years. It was not until 1983 that additional data on the effects of IFN on chlamydial growth was reported. Rothermel et al. (34) reaffirmed the work of Kazar et al. (33) using murine IFNα+β derived by poly I:C induction of murine fibroblasts. These authors went on to show that induction of the antichlamydial state in mouse fibroblasts was dependent on RNA and protein synthesis and, using electron microscopy, that differentiation of elementary to reticulate bodies occurred in IFN-treated cells, but the replication of reticulate bodies was curtailed in treated cells. Byrne and Rothermel (35) provided evidence that the effects of IFNα+β on chlamydial growth varied, depending on the particular strain employed. They showed that although the LGV biovar of *C. trachomatis* was greatly inhibited in IFNα+β-treated fibroblasts, replication of two strains of *C. psittaci* was unaffected after IFNα+β treatment of host cells.

At the same time that work was being done on the effects of virus or poly I:C-induced IFN on chlamydial growth in treated host cells, work was progressing along independent lines on the effects of so-called macrophage-

activating factors (MAF) on the replication of a variety of intracellular pathogens, including *Chlamydia*. It is the result of these latter studies, coupled with a
much better understanding of the molecular nature of MAF that has led to most
of the recent progress on the effects of IFN on intracellular chlamydial growth.

B. Macrophage-Activating Factors and IFNγ

In 1982, Byrne and Faubion (36) published a report showing that supernatant
fluids (lymphokine; LK) from mitogen- or antigen-stimulated murine spleen cells
taken from mice previously sensitized to chlamydiae activated peritoneal macrophages to restrict chlamydial replication. The inhibition was not a result of decreased uptake of the organisms by activated cells, and it resulted in inhibition
of growth in the absence of cidal activity. Subsequently (37), it was found that
although LK-treated murine macrophages exhibited enhanced oxidative activity,
inhibition of chlamydial growth occurred by mechanisms independent of the
respiratory burst. This finding was significant, because up until that time it was
thought that enhanced oxidative activity, with the accompanying elaboration of
toxic products of oxygen metabolism, was the principal antimicrobial mechanism in MAF-treated macrophages (38). Subsequently, Byrne and Krueger (39)
and Rothermel et al. (40) reported that macrophage-activating LK activity was
neutralized by anti-IFNγ immunoglobin and LK responses could be achieved by
substituting IFNγ for LK in the culture medium. In addition, it was found that
LK was capable of activating cells other than mononuclear phagocytes to restrict
chlamydial growth. It was at this point that work began in earnest to determine
the mechanisms involved in IFNγ-mediated inhibition of chlamydial growth in
both macrophages and epithelial cells.

C. Mechanism of IFNγ-Mediated Inhibition of Chlamydial Growth
in Epithelial Cells and Fibroblasts

Research directed at understanding the mechanism of IFNγ-mediated inhibition
of chlamydial growth remained unproductive until Pfefferkorn's study (41) concerning the effects of IFNγ-mediated inhibition on the intracellular growth of
Toxoplasma gondii, an obligate intracellular protozoan parasite. Pfefferkorn
showed that addition of exogenous tryptophan, one of the eight essential amino
acids, to human fibroblasts treated with human recombinant IFNγ resulted in a
reversal of the inhibition of *T. gondii* growth that was normally observed as a result incubating host cells with IFNγ prior to infection. He went on to show that
IFNγ-mediated induction of indoleamine 2,3-dioxygenase (IDO), an enzyme involved in the catabolism of tryptophan, was responsible for restricting the availability of tryptophan that was required for toxoplasma growth. The induction of
IDO activity in the presence of IFN was not a new observation. Previously,
Yoshida et al. (42) and Hayaishi (43) had described the induction of IDO ac-

tivity in mouse lung slices that had been incubated in the presence of virus-induced IFN, or bacterial LPS. IDO acts by cleaving the indole ring of tryptophan, resulting in the catabolic product, N-formylkynurenine. An additional constitutive enzyme then acts to cleave the formyl group leaving the diamine, kynurenine, as a product. This sequence of reactions is detailed in Figure 1. The critical connection that Pfefferkorn made in his work was relating this activity to an antimicrobial mechanism induced by IFN.

Soon after the publication of the Pfefferkorn study, research was initiated in my laboratory to determine if tryptophan catabolism was important in IFNγ-mediated inhibition of chlamydial growth (44). We examined the effects of human recombinant IFNγ on a human bladder carcinoma cell line (T24 cells), and found that incubation of these cells in the presence of various concentrations

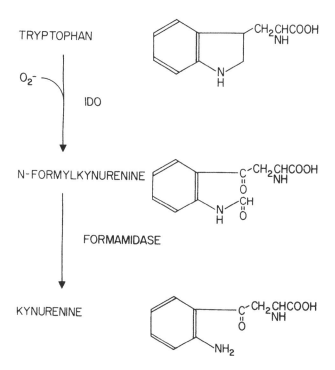

Figure 1 Initial events in the catabolism of tryptophan. The enzyme indole-amine 2,3-dioxygenase (IDO) causes cleavage of the indole ring of tryptophan in the presence of superoxide anion (O_2^-), resulting in the production of N-formyl-kynurenine. Formamidase then acts to cleave the formyl group resulting in the production of kynurenine. IDO is a nonconstitutive enzyme, induced by IFN. Formamidase is present constitutively.

of IFNγ for one or two days prior to infection with chlamydiae, resulted in an inhibition of chlamydial growth that was directly proportional to the amount of IFNγ present in the medium (Fig. 2). Furthermore, if exogenous tryptophan was added back to the treated, infected cells, then a reversal of the inhibitory process occurred which was directly proportional to the concentration of tryptophan added (Fig. 3). This finding was important because it not only established a means to determine the mechanism for IFNγ-mediated inhibition of chlamydial growth, but it also confirmed and explained how the microbistatic nature of this

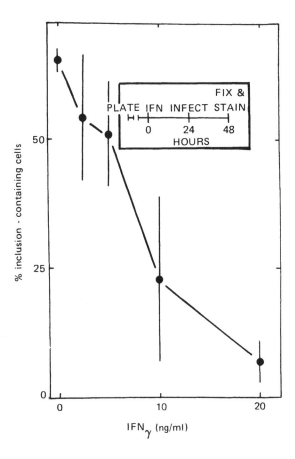

Figure 2 Effect of Hu-rIFNγ treatment on *C. psittaci* growth in T24 cells. Cells were plated, treated with the indicated amounts of Hu-rIFNγ, infected, stained, and examined for inclusion development as outlined in the inset. Each point represents the mean of triplicate determinations. Standard deviations are indicated by the bars. (Reprinted with permission from Ref. 44.)

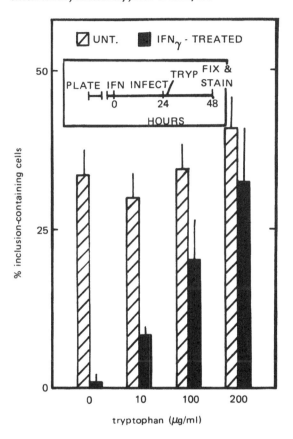

Figure 3 Effect of tryptophan on *C. psittaci* growth in Hu-rIFN-γ-treated T24 cells. Cells were plated, treated, infected, and processed as described in the legend to Figure 2 (inset). The indicated amounts of tryptophan were added 60 minutes after infection. Each bar represents the mean of triplicate determinations; vertical lines indicate standard deviations. Solid bars, cells that received 20 ng Hu-rIFNγ per ml; hatched bars, cells that were untreated but received an equivalent volume of growth medium at the indicated time. (Reprinted with permission from Ref. 44.)

inhibition operated. Addition of two other essential amino acids, isoleucine and lysine, to IFNγ-treated cells had no effect on the inhibition of chlamydial growth (Fig. 4), demonstrating that reversal of growth-restricting activity was specific for tryptophan.

When the transport of radiolabelled tryptophan was measured in T24 cells treated with IFNγ, it was found that intracellular levels of radioisotope reached

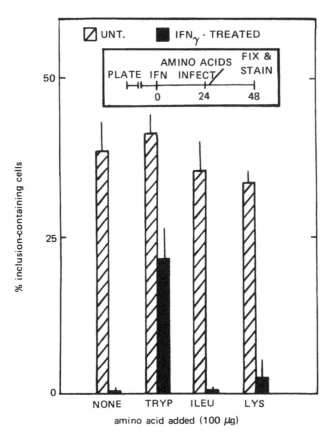

Figure 4 Comparison of the effects of tryptophan (TRYP), isoleucine (ILEU), and lysine (LYS) on *C. psittaci* growth in Hu-rIFNγ-treated T24 cells. Experimental conditions were similar to those described in the legend for Figure 3 and outlined in the inset.

steady-state levels 100 to 1000 times greater than was observed in untreated cells (Fig. 5). Transport of radiolabelled leucine was not altered in IFNγ-treated cells (Fig. 6), therefore once again demonstrating that this effect appeared to exhibit specificity with respect to tryptophan. The most straightforward explanation for elevated transport of tryptophan in IFNγ-treated cells was that as soon as tryptophan crossed the cytoplasmic membrane, it was converted to a product that was not metabolized and therefore the treated cells were continually taking in more tryptophan in an attempt to establish steady-state conditions. To determine if production of a tryptophan-degradative product was occurring in IFNγ-treated cells, overlay medium was collected several hours after treated cells

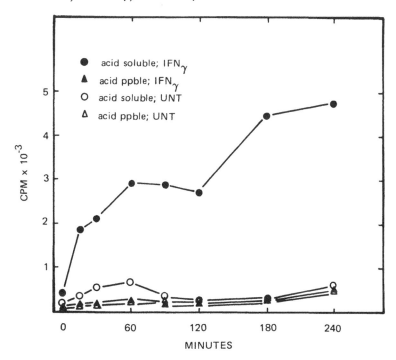

Figure 5 Tryptophan transport by Hu-rIFNγ-treated T24 cells. Cells were plated and treated with 20 ng of Hu-rIFNγ per ml; 44 hours later the medium was removed and replaced with L-[^{3}H] tryptophan. Uptake of labeled tryptophan into acid-soluble pools and into acid-precipitable material was measured at the indicated times. Circles represent acid-soluble counts, triangles acid-precipitable material. Filled symbols represent Hu-rIFNγ-treated cells. Open symbols represent untreated controls. (Reprinted with permission from Ref. 44.)

were incubated in Hank's balanced salt solution supplemented with radiolabeled tryptophan and a fixed amount of unlabeled carrier tryptophan. When this labeled medium was subjected to high performance liquid chromatography analysis, it was found that the amount of labeled material that cochromatographed with tryptophan decreased in proportion to the amount of IFNγ added to the cells, and the appearance of two peaks that cochromatographed with N-formylkynurenine and kynurenine increased in an IFNγ-dependent way (Fig. 7).

Thus, in a manner similar to that found by Pfefferkorn for IFNγ-mediated inhibition of *T. gondii* replication in human fibroblasts, the degradation of tryptophan via the induction of IDO was found to be responsible for the IFNγ-mediated inhibition of chlamydiae in human bladder carcinoma cells. An important aspect of these findings may relate directly to what is known concerning

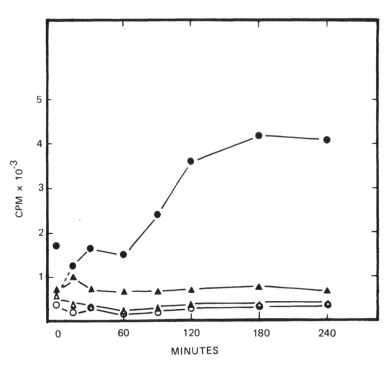

Figure 6 Comparison between tryptophan and leucine transport in Hu-rIFNγ-treated T24 cells. Cells were treated as described in the legend to Figure 5, then incubated for the indicated periods of time in the presence of either L-[^{3}H]-tryptophan or L-[^{3}H]leucine. Cell-associated counts from soluble pools were measured. Circles represent transported amino acids after Hu-rIFNγ treatment. Triangles represent transport by untreated controls. Filled symbols represent transport of tryptophan, open symbols represent transport of leucine.

the pathogenesis of chlamydial disease. Chlamydial infections, if left untreated, tend toward a persistent or chronic state. It is known that chlamydiae induce IFNγ in vivo, and the work concerning the effects of IFNγ on an epithelial cell line provided an explanation for how chlamydial persistence may be established. In this context, the presence of IFNγ at the site of infection during acute chlamydial disease would result in the induction of IDO in cells that also serve as host cells for chlamydiae. The subsequent reduction in tryptophan levels, at least locally, would result in an inhibition of chlamydial growth without eradicating the organisms within infected cells. This microbistatic action would continue for as long as infected cells were under the influence of IFNγ. It is known from other work (data not published) that IFNγ-secreting cells can remain at the site

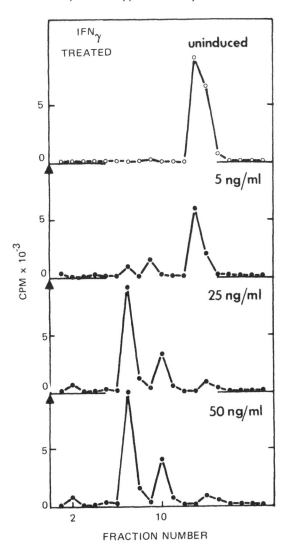

Figure 7   Reversed-phase high performance liquid chromatography of L-[³H]-tryptophan-containing overlay buffer after 4 hours of incubation with uninduced T24 cells or cells incubated with the indicated amounts of Hu-rIFNγ. The peak with the longest retention time (farthest to the right) cochromatographed with tryptophan. The major peak with the shortest retention time in treated samples (fraction 7) cochromatographed with kynurenine. The peak at fraction 10 cochromatographed with N-formylkynurenine. Each point represents the mean of triplicate determinations. (Reprinted with permission from Ref. 44.)

of chlamydial infection long after the acute stages of disease have been resolved. It also has been established that reactivation of chlamydial infections can occur in experimental animals that have recovered from chlamydial disease by impairing the immune system of these animals (45,46). It is therefore possible that the induction of the IFN system may, on the one hand, help resolve acute chlamydial disease, while at the same time promoting chlamydial persistence. More detailed in vivo work must be done to firmly establish this double-edged role for IFN in the pathogenesis of chlamydial disease.

IV. PROSPECTUS

In the past few years we have come a long way in understanding how the IFN system impacts on chlamydial disease and on the effects of IFN, especially IFNγ, on cells that also can serve as host cells for chlamydiae. Despite this progress, much more needs to be done. A notable deficiency is that virtually all the work done so far has been accomplished in cell culture systems. There is a real need to extend in vitro observations to experimental animal systems. Work in this area is currently underway, but we are far from showing whether predictions based on observations made in vitro will hold up when tested in models of chlamydial disease or systems designed to test protective immunity in animals.

We also need to determine if results obtained using one experimental system are compatible with other systems. For example, the work done to show that induction of IDO and the catabolism of tryptophan were responsible for IFNγ-mediated inhibition of chlamydial growth was done using human recombinant IFNγ and a human cell line. De la Maza et al. (47) have shown that the degradation of tryptophan was not involved in IFNγ-mediated inhibition of chlamydial growth in the murine system. We have subsequently confirmed these results (unpublished observations). These findings raise interesting possibilities concerning distinctions between IFN activities in different mammalian species, but they also force us to confront the fact that we do not yet know the mechanism for murine IFNγ-mediated inhibition of chlamydial growth in mice. Results obtained thus far support the hypothesis that IFNγ results in a microbistatic effect in both the murine and human systems, but the precise events that result in the inhibition of chlamydial growth in IFNγ-treated mouse cells still needs to be worked out.

An additional correlation that needs to be established is whether or not mechanisms of IFN action for one intracellular parasite are applicable to others. It is known now that some facultative and obligate intracellular pathogens are killed in activated macrophages by oxygen-dependent mechanisms (48,49), but oxygen-independent systems are important for inhibiting other intracellular organisms. *Toxoplasma* and *Chlamydia* both appear to be affected by oxygen-independent mechanisms, and for these organisms, at least in the human system, the induction of IDO is paramount to restricting pathogen replication in IFNγ-treated cells. Another important genus of intracellular pathogens that apparently are inhibited via oxygen-independent mechanisms in both macrophages and endothelial cells are the rickettsiae. Turco and Winkler (50) clearly have shown that the induction of IDO is unrelated to inhibition of *Rickettsia prowazekii* in both human and murine IFNγ-treated cells, but the precise mechanisms involved in the inhibition of rickettsiae have not as yet been established.

Finally, it will be important to establish if the induction of IFN or other cytokines is relevant to acquired resistance to the variety of infectious agents for which a role for IFN has been established. Does the IFN system contribute to protective immunity or to pathogenesis of disease?

These questions are currently being actively pursued and hopefully answers to at least some of them will be forthcoming within the next few years.

ACKNOWLEDGMENTS

I wish to thank Sherideen Stoll for her help in preparing this manuscript. Research in my laboratory has been supported by Public Health Services Grant number IA 19782 from the National Institute of Allergy and Infectious Diseases.

REFERENCES

1. Hatch, T.P. 1975. Utilization of of L-cell nucleoside triphosphates by *Chlamydia psittaci* for ribonucleic acid synthesis. *J. Bacteriol. 122*:393–400.

2. Moulder, J.W. 1969. A model for studying the biology of parasitism. *Chlamydia psittaci* and mouse fibroblasts (L cells). *BioScience 19*:875–881.

3. Batteiger, B.E., W.J. Newhall V, and R.B. Jones. 1985. Differences in outer membrane proteins of the lymphogranuloma and trachoma biovars of *Chlamydia trachomatis. Infect. Immun. 50*:488–494.

4. Barbour, A.G., K-I. Amano, T. Hackstadt, L. Perry, and H.D. Caldwell. 1982. *Chlamydia trachomatis* has penicillin-binding proteins but not detectable muramic acid. *J. Bacteriol. 151*:420–428.

5. Newhall, W.J., V. and R.B. Jones. 1983. Disulfide-linked oligomers of the major outer membrane protein of chlamydiae. *J. Bacteriol. 154*:998–1001.

6. Hatch, T.P., I. Allan, and J.H. Pearce. 1984. Structural and polypeptide differences between envelopes of infective and reproductive life cycle forms of *Chlamydia spp. J. Bacteriol. 157*:13–20.

7. Bavoil, P., A. Ohlin, and J. Schachter. 1984. Role of disulfide bonding in outer membrane structure and permiability in *Chlamydia trachomatis. Infect. Immun. 44*:479–485.

8. Friis, R.R. 1972. Interaction of L cells and *Chlamydia psittaci*: entry of the parasite and host response to its development. *J. Bacteriol. 110*:706–721.

9. Matsumoto, A. 1981. Isolation and electron microscopic observations of intracytoplasmic inclusions containing *Chlamydia psittaci. J. Bacteriol. 145*:605–612.

10. Doughri, A.M., J. Storz, and K.P. Altera. 1972. Mode of entry and release of chlamydiae in infections of intestinal epithelial cells. *J. Infect. Dis. 126:* 652–657.

11. Todd, W.J., and H.D. Caldwell. 1985. The interaction of *Chlamydia trachomatis* with host cells: ultrastructural studies of the mechanism of release of a biovar II strain from HeLa 229 cells. *J. Infect. Dis. 151*:1037–1044.

12. Harshbarger, J.C., S.C. Chang, and S.V. Otto. 1977. Chlamydiae (with phages), mycoplasmas, and rickettsiae in Chesapeake Bay bivalves. *Science 196*:666–668.

13. Eddie, B., F.J. Radovsky, D. Stiller, and N. Kumada. 1969. Psittacosis-lymphogranuloma venereum (PL) agents (*Bedsonia, Chlamydia*) in ticks, fleas and native mammals in California. *Am. J. Epidemiol. 90*:449–458.

14. Meyer, K.F., and B. Eddie. 1952. Reservoirs of the psittacosis agent. *Acta Trop. 9*:204–209.

15. Brown, A.S., and R.G. Grice. 1986. Experimental transmission of *Chlamydia psittaci* in the koala. In *Chlamydial Infections. Proceedings of the Sixth International Symposium on Human Chlamydial Infections*, D. Oriel, G. Ridgway, J. Schachter, D. Taylor-Robinson, and M. Ward (Eds.). Cambridge University Press, Cambridge, pp. 349–352.

16. Eugster, A.K. 1980. Chlamydiosis. In *CRC Handbook Series in Zoonoses*. J.H. Steele (Ed. in Chief). Section A: *Bacterial, Rickettsial, and Mycotic Diseases*, Volume II, H. Stoenner, W. Kaplan, and M. Torten (Eds.). CRC Press. Boca Raton, FL, pp. 357–417.

17. Schachter, J., and M. Grossman. 1981. Chlamydial Infections. *Ann. Rev. Med. 32*:45–61.

18. Ladany, S., and I. Sarov. 1985. Recent advances in *Chlamydia trachomatis. Eur. J. Epidemiol. 1*:235–256.

19. Saikku, P., S-P. Wang, M. Kleemola, E. Brander, E. Rusanen, and J.T. Grayston. 1985. An epidemic of mild pneumonia due to an unusual strain of *Chlamydia psittaci. J. Infect. Dis. 151*:832–839.

20. Wang, S-P., and J.T. Grayston. 1986. Microimmunofluorescence serological studies with the TWAR organism. In *Chlamydial Infections. Proceedings of the Sixth International Symposium on Human Chlamydial Infections*, D. Oriel, G. Ridgway, J. Schachter, D. Taylor-Robinson, and M. Ward (Eds.). Cambridge University Press, Cambridge, pp. 329–332.

21. Grayston, J.T., C-C. Kuo, S-P. Wang, and J. Altman. 1986. A new *Chlamydia psittaci* strain, TWAR, isolated in acute respiratory tract infections. *N. Engl. J. Med. 315*:161–168.

22. Kunimoto, D., and R.C. Brunham. 1985. Human immune response and *Chlamydia trachomatis* infection. *Rev. Infect. Dis. 7*:665–673.

23. Grayston, J.T., and S-P. Wang. 1978. The potential for vaccine against infection of the genital tract with *Chlamydia trachomatis. Sex. Trans. Dis. 5*:73–77.

24. Young, E., and H.R. Taylor. 1984. Immune mechanisms in chlamydial eye infection: cellular immune responses in chronic and acute disease. *J. Infect. Dis. 150*:745–751.

25. Taylor, H.R., A.B. MacDonald, J. Schachter, and R.A. Prendergast. 1987. Oral immunization against chlamydial eye infection. *Invest. Ophthal. Vis. Sci. 28*:249–258.

26. Howard, L.V., M.P. O'Leary, and R.L. Nichols. 1976. Animal model studies of genital chlamydial infections. Immunity to reinfection with guinea-pig inclusion conjunctivitis agent in the urethra and eye of male guinea pigs. *Br. J. Vener. Dis. 52*:261–265.

27. Williams, D.M., J. Schachter, J.J. Coalson, and B. Grubbs. 1984. Cellular immunity to the mouse pneumonitis agent. *J. Infect. Dis. 149*:630–639.

28. Chen, W.-J., and C-C. Kuo. 1980. A mouse model of pneumonitis induced by *Chlamydia trachomatis*. Morphologic, microbiologic, and immunologic studies. *Am. J. Pathol. 100*:365–382.

29. Byrne, G.I. 1986. Interferons and the immune response to chlamydial infections. In *Microbiology 1986*, P. Bonventre (Ed.). American Society for Microbiology Press, Washington, DC, pp. 99–102.

30. Sueltenfuss, E.A., and M. Pollard. 1963. Cytochemical assay of interferon produced by duck hepatitis virus. *Science 139*:595–596.

31. Jenkin, H.M., and Y.K. Lu. 1967. Induction of interferon by the Bour strain of trachoma in HeLa 229 cells. *Am. J. Ophthalmol. 63*:1110–1115.

32. Merigan, T.C., and L. Hanna. 1966. Characteristics of interferon induced in vitro and in vivo by a TRIC agent. *Proc. Soc. Exp. Biol. Med. 122*:421–424.

33. Kazar, J., J.D. Gillmore, and F.B. Gordon. 1971. Effect of interferon and interferon inducers on infections with a non-viral intracellular microorganism, *Chlamydia trachomatis. Infect. Immun. 3*:825–832.

34. Rothermel, C.D., G.I. Byrne, and E.A. Havell. 1983. Effect of interferon on the growth of *Chlamydia trachomatis* in mouse fibroblasts (L cells). *Infect. Immun. 39*:362–370.

35. Byrne, G.I., and C.D. Rothermel. 1983. Differential susceptibility of chlamydiae to exogenous fibroblast interferon. *Infect. Immun. 39*:1004–1005.

36. Byrne, G.I., and C.L. Faubion. 1982. Lymphokine-mediated microbistatic mechanisms restrict *Chlamydia psittaci* growth in macrophages. *J. Immunol. 128*:469–474.

37. Byrne, G.L., and C.L. Faubion. 1983. Inhibition of *Chlamydia psittaci* in oxidatively active thioglycolate-elicited macrophages: distinction between lymphokine-mediated oxygen-dependent and oxygen-independent macrophage activation. *Infect. Immun. 40*:464–471.

38. Murray, H.W., and Z.A. Cohn. 1980. Macrophage oxygen-dependent antimicrobial activity. III. Enhanced oxidative metabolism as an expression of macrophage activation. *J. Exp. Med. 152*:1596–1609.

39. Byrne, G.L., and D.A. Krueger. 1983. Lymphokine-mediated inhibition of *Chlamydia* replication in mouse fibroblasts is neutralized by anti-gamma interferon immunoglobulin. *Infect. Immun. 42*:1152–1158.

40. Rothermel, C.D., B.Y. Rubin, and H.W. Murray. γ-interferon is the factor in lymphokine that activates human macrophages to inhibit intracellular *Chlamydia psittaci* replication. *J. Immunol. 131*:2542–2544.

41. Pfefferkorn, E.R. 1984. Interferon blocks the growth of *Toxoplasma gondii* in human fibroblasts by inducing the host cells to degrade tryptophan. *Proc. Natl. Acad. Sci. (USA) 81*:908–912.

42. Yoshida, R., J. Imanishi, T. Oku, T. Kishida, and O. Hayaishi. 1981. Induction of pulmonary indoleamine 2,3-dioxygenase by interferon. *Proc. Natl. Acad. Sci. (USA) 78*:129–132.

43. Hayaishi, O. 1985. Indoleamine 2,3-dioxygenase—with special reference to the mechanism of interferon action. *Biken J. 28*:39–49.

44. Byrne, G.I., L.K. Lehmann, and G.J. Landry. 1986. Induction of tryptophan catabolism is the mechanism for gamma-interferon-mediated inhibition of intracellular *Chlamydia psittaci* replication in T24 cells. *Infect. Immun. 53*:347–351.

45. Stephens, R.S., W-J. Chen, and C.-C. Kuo. 1982. Effects of corticosteroids and cyclophosphamide on a mouse model of *Chlamydia trachomatis* pneumonitis. *Infect. Immun. 35*:680–684.

46. Yang, Y-S., C-C. Kuo, and W-J. Chen. 1983. Reactivation of *Chlamydia trachomatis* lung infection in mice by hydrocortisone. *Infect. Immun. 39*: 655–658.

47. de la Maza, L.M., E.M. Peterson, C.W. Fennie, and C.W. Czarniecki. 1985. The anti-chlamydial and anti-proliferative activities of recombinant murine interferon-γ are not dependent on tryptophan concentrations. *J. Immunol. 135*:4198–4200.

48. Murray, H.W., B.Y. Rubin, and C.D. Rothermel. 1983. Killing of intracellular *Leishmania donovani* by lymphokine-stimulated human mononuclear phagocytes. Evidence that interferon-γ is the activating lymphokine. *J. Clin. Invest. 72*:1506–1510.

49. Nathan, C.F., H.W. Murray, M.E. Wiebe, and B.Y. Rubin. 1983. Identification of interferon-γ as the lymphokine that activates human macrophage oxidative metabolism and antimicrobial activity. *J. Exp. Med. 158*:679–689.

50. Turco, J., and H.H. Winkler. 1986. Gamma-interferon-induced inhibition of the growth of *Rickettsia prowazekii* in fibroblasts cannot be explained by the degradation of tryptophan or other amino acids. *Infect. Immun. 53*:38–46.

4

Chlamydia trachomatis Growth Inhibition by IFNγ: Implications for Persistent Chlamydial Infections

Yonat Shemer and Israel Sarov / Ben-Gurion University of the Negev, Beer Sheva, Israel

I. INTRODUCTION

Chlamydiae are obligate intracellular bacteria which parasitize the host cell for nutrients and energy (1-5). In the course of infection, the endocytosed parasites convert from their infectious form, elementary bodies (EB), to their replicating form, reticulate bodies (RB). The RB multiply by means of binary fission within the inclusion present in the host cytoplasm until about 20 hours postinfection. Then some of the RB reorganize into EB, while others continue dividing. Two to three days after infection, the chlamydial inclusion may occupy a major part of the host cell cytoplasm, and contains mainly EB. At this stage, the host cell lyses, and the EB are released to start a new cycle (6,7).

Because of the unique developmental cycle of chlamydiae, a special order has been established consisting of one family, Chlamydiaceae, one genus, *Chlamydia*, and two species, *Chlamydia psittaci* (CP) and *Chlamydia trachomatis* (CT) (8,9).

Chlamydia psittaci, the etiologic agent of psittacosis/ornithosis, has a broad host range which includes avian and mammalian species as well as humans, in whom it causes pneumonia with systemic involvement (10,11). As for *C. trachomatis*, different serovars have been associated with clinically distinct infections ranging from hyperendemic trachoma to sexually transmitted infections such as lymphogranuloma venereum (LGV), nongonococcal urethritis, and postgonococcal urethritis, cervicitis, endometritis, neonatal conjunctivitis, pneumonitis, and epididymitis (12). Recent studies have also implicated *C. trachomatis* as one of the major causes of pelvic inflammatory disease (PID) which may lead to infertility.

73

Chlamydial infection can be persistent and recurrent infection is common. Hanna et al. (13) have described cases of persistent ocular infection. Individuals who have left trachoma-endemic areas and who have not had active disease since childhood may develop acute trachoma in their sixth or seventh decade (14). Henry-Suchet et al. (15) have recovered *C. trachomatis* from the fallopian tubes and/or peritoneal cavity in 15% to 35% of infertile women with tubal obstruction. Moreover, infertile women with tubal obstruction and patients with salpingitis or epididymitis characteristically have elevated IgG and IgA antibodies to chlamydia when compared to matched controls (16-19). In LGV, high IgG and IgA titers to chlamydia can be found 40 years after acute infection. It has been suggested that continuous antigenic stimulation results in high antibody titers which might reflect chronicity, persistence, or even latency of LGV infections (20). This assumption is further supported by Dan et al. (21), who isolated LGV agents from a patient 20 years after onset of disease. *C. trachomatis* latent genital infection has been described, and it has been suggested that such latent infections can be reactivated in some women by *Neisseria gonorrhoeae* (22).

The immune response to chlamydial infections and the role it plays in resistance to reinfection, resolution of ongoing infection, and development of persistent chlamydial infection is poorly understood. However, evidence showing that reactivation of apparently resolved infection occurs under conditions of immunosuppression (23) implicates the immune system in the maintenance of persistent infection.

II. CHLAMYDIAL INFECTION AND INTERFERONS

It has been suggested that in vivo interferon (IFN) protects many cells from the cytopathic effects of virus infection, but it does not eliminate viral infections completely (24). As a result, interferon may be a factor in the establishment and maintenance of persistent viral infections.

It was first found by Sueltenfuss and Pollard (25) that *C. psittaci* resembles a virus not only because it is an intracellular parasite, but also in its susceptibility to crude IFN. Later, Merigan and Hanna demonstrated that *C. trachomatis* is not only susceptible to the action of interferon but it is capable of stimulating the production of IFN (26).

Rothermel et al. have shown that murine fibroblast IFN caused, in cell cultures, a reduction in the rate of *C. trachomatis* RB replication and in the number of *C. trachomatis* inclusion-bearing cells (27). Comparison between *C. trachomatis* and *C. psittaci* showed that murine fibroblast IFN (alpha and beta) reduced the intracellular development of *C. trachomatis* by 90%, whereas *C. psittaci* growth was not affected (28).

IFNγ activates professional phagocytes to eliminate intracellular parasites such as rickettsiae (29), toxoplasmae (30), and chlamydiae (31), but it is also

Table 1 Dose Response for α, β, and γ Human IFN Inhibition of *C. trachomatis*

IFN concentrations IU/ml	% of inhibition[a]		
	IFNα	IFNβ	IFNγ
0	0	0	0
7.2	0	13	59
36	0	82	90
180	10	80	78
900	59	82	98

[a]HEp-2 cells were treated with threefold dilutions of each of the three interferons 24 hr before infection, then infected with serial dilutions of *C. trachomatis*. The chlamydial titer for each IFN-treatment was determined, and the titer reduction was calculated as percent of the control.

capable of inducing resistance to various intracellular parasites in nonprofessional phagocytes (32–34). It has been reported that lymphokines produced by mouse spleen lymphocytes can suppress the growth of *C. psittaci* in mouse fibroblast cell lines. These lymphokines were induced by stimulating spleen cell suspensions from CP-immune animals with concanavalin-A antigen. Antimurine IFNγ immunoglobulin neutralized the lymphokine-mediated inhibition of *C. psittaci* (35).

As for the immune system, the "immune" (Gamma) IFN is more efficient (relative to its antiviral activity) than alpha and beta IFNs in inducing some changes on the cell surface which makes cells less susceptible to killing by the natural killer (NK) cells (36). Other studies have suggested that the immune IFN is also more effective than the IFNα or β in suppressing antibody formation (37, 38). The foregoing implies that under certain in vivo conditions IFNγ might be involved in the establishment of persistent chlamydial infection (27).

When the effect of purified human α, β, and γ interferon on the infectivity of *C. trachomatis* (L$_2$/434 serovar) in HEp-2 cell culture was investigated in our laboratory, the *C. trachomatis* was found to be more sensitive to inhibition by recombinant IFNγ than to α and β IFN (Table 1).

III. ANTICHLAMYDIAL ACTIVITY OF HUMAN GAMMA INTERFERON

The protection of the host cell by IFN is a function of the interferon concentration and the multiplicity of infection (MOI). The effect of IFN concentration on *C. trachomatis* infectious particle production was measured by one-step growth experiments. The yield of *C. trachomatis* from treated cells was titrated on HEp-2 cells to determine the inclusion-forming units (IFU) by immunoperoxidase

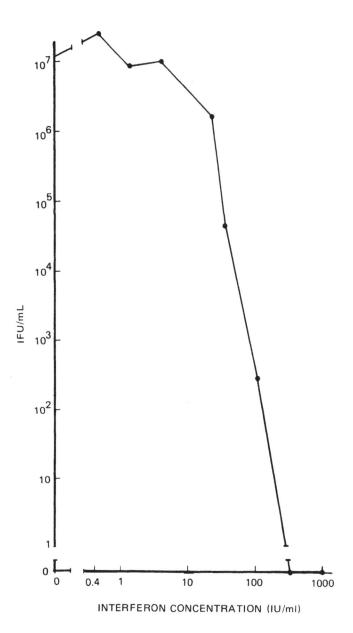

Figure 1 *C. trachomatis* inhibition by gamma IFN as measured by the assay of infectious particle yield (one step growth). After 24 hr of HEp-2 cell exposure to growth medium containing IFNγ at various concentrations, the cells were infected at an MOI of 1; at 48 hr postinfection samples of the infected cells were collected, sonicated, and titrated to determine the IFU/ml for each of the IFN concentrations. (From Ref. 39.)

76

assay (IPA) (39). The curve in Figure 1 demonstrates a yield reduction in *C. trachomatis* infectious particles with increasing IFNγ concentration. At a concentration of 12.3 IU/ml of IFNγ, there is already a one log reduction in *C. trachomatis* infectious particle yield, while 37 IU/ml reduced the yield by 3 logs. The antichlamydial effects of IFNγ were similar in some respects to the antiviral activity induced by IFN and other inhibitors of viral replication. The sigmoidal dose-response curve, with its linear portion over a range corresponding to 25–75% inhibition in viral systems (40), was also observed for *C. trachomatis* inhibition in IFNγ-treated cells (Fig. 2). We have found that 100 IU/ml of human IFNγ caused almost complete inhibition (97%) and 400 IU/ml inhibited chlamydiae completely (Fig. 2).

The effect of IFNγ on *C. trachomatis* replication was influenced by the MOI used, except for MOI less than 1 where there was no effect on the ED_{50} (the dose of IFN required to reduce the yield by 50%) value. Due to the toxic effect

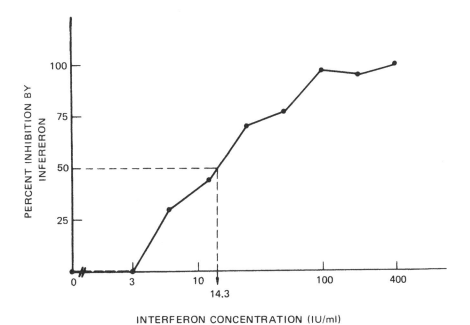

Figure 2 Dose-response curve for IFN. After 24 hr of HEp-2 cell exposure to growth medium or medium containing IFNγ at serial dilutions, the cells were infected with 10-fold dilutions of purified CT. The CT titer for each IFN concentration was determined and the CT titer reduction was calculated as percent of the control.

of *C. trachomatis* (41) the study was restricted to MOIs equal to and lower than 10.

IV. NEUTRALIZATION OF IFNγ ACTIVITY BY SPECIFIC ANTIBODIES

To determine whether the antichlamydial effect of IFNγ could be neutralized, polyclonal antibodies (Abs) to human IFNγ were incubated with IFN as follows: 1000 neutralizing units of Abs and 1000 IU of IFNγ per ml (a ratio of 1) and 100 neutralizing units of Abs with 1000 IU of IFNγ per ml (0.1 ratio). After 1 hr of incubation at room temperature, the mixtures of IFNγ-Abs were titrated in threefold dilutions on HEp-2 cells. Samples were run in triplicate. After 24 hr, the cells were infected at chlamydial dilution that resulted in 50% infection in untreated cells. After an additional 48 hr, the microplates were fixed and subjected to the IPA test. At a ratio of 1 neutralizing unit of Abs per 10 IU of IFN, the inhibitory effect of IFN on chlamydial growth was effective only above 666 IU/ml; the effective dose of IFN which reduced *C. trachomatis* replication by 50% of IFNγ of the control without Abs was 14.5 IU/ml.

V. KINETICS OF *C. TRACHOMATIS* PRODUCTION AFTER GAMMA IFN REMOVAL

IFN-induced antiviral activity is known to decline after IFN is removed from the cells. To determine whether the same phenomena exist for *C. trachomatis*, HEp-2 cells were treated with various gamma IFN concentrations in twofold dilutions (from 400 IU/ml to 25 IU/ml) for 24 hr. For preinfection treatment (24 hr of exposure before infection), the IFNγ was removed from the cells just before infection; for IFN continuous exposure, the media added after the infection contained IFN at the appropriate concentration. The degree of chlamydial infection (expressed as a percentage of control IFN-untreated cells) was determined by IPA for each time point. The cells were initially infected at a very low multiplicity of infection of *C. trachomatis* (0.001). The experiment was designed so that more than two cycles of intracellular development occurred; the control was almost completely infected by 5 days after infection. Continuous exposure to IFNγ (at concentrations greater than 25 IU/ml) resulted in almost complete inhibition of chlamydial infection. In contrast, preinfection treatment of IFNγ, resulted in the progression of chlamydial infection (Fig. 3). The lower the preinfection IFNγ concentration, the less time it took for the chlamydial infection to overcome the inhibitory effect of the IFN. At an IFN preinfection treatment of 25 IU/ml, the inhibitory effect of IFNγ was completely overcome within 48 hr of infection. After preinfection treatment with 100 or 200 IU/ml, inhibition was overcome to the extent of 40% chlamydial production only after 96–120 hr. At

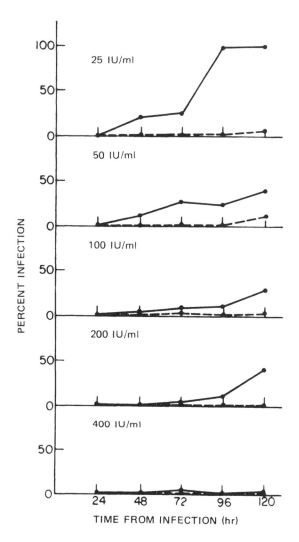

Figure 3 Kinetics of *C. trachomatis* production after IFNγ removal. HEp-2 cells were either exposed to IFN for 24 hr prior to infection (———), or continuously, starting from 24 hr prior to infection (— — —). IFN-untreated cells served as controls. For each IFN concentration the percent infection (of control) was determined at: 24 hr, 48 hr, 72 hr, 96 hr, and 120 hr postinfection. (From Ref. 39.)

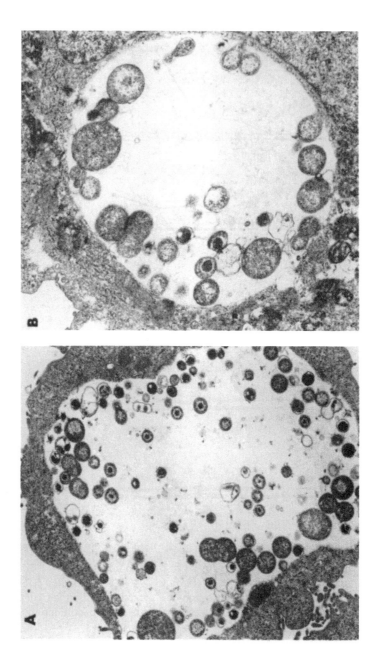

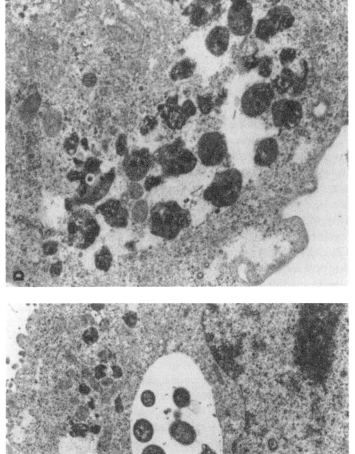

Figure 4 Electron micrograph of HEp-2 cells infected with *C. trachomatis*, *A* to *D* 48 hr after infection *A*. Control, no interferon treatment; x11,000; *B*. IFN treatment of 14 IU/ml; x15,000; *C*. IFN treatment of 70 IU/ml, x11,000; *D*. IFN treatment of 350 IU/ml; x15,000. (From Ref. 39.)

a 400 IU/ml concentration of IFNγ preinfection treatment, there was no sign of chlamydial production through 120 hr.

VI. MODULATION OF *C. TRACHOMATIS* INFECTION BY IFNγ

To determine whether IFNγ concentration had influenced transformation of EB to RB and, vice versa, along the developmental cycle of chlamydia, HEp-2 cells were treated with IFNγ 24 hr before infection and examined by electron microscopy (EM).

We observed that the stage of the chlamydial developmental cycle reached was dependent upon the IFN concentration present in the culture. Under conditions of complete inhibition of *C. trachomatis* infectivity by IFN (350 IU/ml) at 24 and 48 hr, there were only damaged forms of *C. trachomatis* (Fig. 4D). At an IFN dose effecting almost complete *C. trachomatis* inhibition (70 IU/ml), the developmental cycle of the *C. trachomatis* at 48 hr had only reached the stage of RB formation (Fig. 4C); in IFN 50% effective dose treatment (14.5 IU/ml), the developmental cycle reached the formation of EBs as well (Fig. 4B). In contrast to the control (Fig. 4A), however, in which the ratio of EB to RB was 3:1, in the IFN 50% effective dose treatment RBs outnumbered EBs threefold. Thus, the IFN concentration seems able to modulate *C. trachomatis* replication, which may indicate a possible role for IFN in the maintenance of chronic *C. trachomatis* infection in vivo (35,39).

Inhibition of chlamydial growth as a result of deprivation of necessary metabolites, such as certain amino acids, has been described previously and shown to be serovar specific (42,43).

The effect of 70 IU/ml IFN on *C. trachomatis* development as observed in the EM studies (Fig. 4C) resembles the effect observed in studies with cysteine omission from *C. trachomatis* (E/DK-20) growth medium. Deprivation of cysteine severely retarded differentiation of reproductive RBs to EBs (44). Pfefferkorn has reported that human IFNγ blocks the growth of *Toxoplasma gondii* in human fibroblasts by inducing the host cells to degrade tryptophan (32), probably through induction of the enzyme indolamine 2-3-dioxygenase by interferon (45).

We have studied the influence of tryptophan (try) on the inhibition of *C. trachomatis* (L_2/434 serovar) growth induced by recombinant human gamma IFN (46).

VII. THE EFFECT OF TRYPTOPHAN ON *C. TRACHOMATIS* INHIBITION BY HUMAN IFNγ

In the presence of increasing concentrations of tryptophan, the inhibitory effect of 100 IU/ml of IFNγ on chlamydial growth was reversed (Fig. 5). A minimal

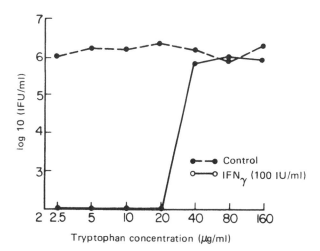

Figure 5 The effect of tryptophan concentration on the inhibition of *C. trachomatis* growth in cells treated with IFNγ. The one-step growth experiment was done on HEp-2 cultures pretreated with either control growth medium or medium containing 100 IU/ml recombinant human IFNγ for 24 hr. The cultures were then infected at MOI of 1. After 48 hr, cultures were harvested, frozen, and titrated on HEp-2 cells to determine the yield in IFU/ml. (From Ref. 46.)

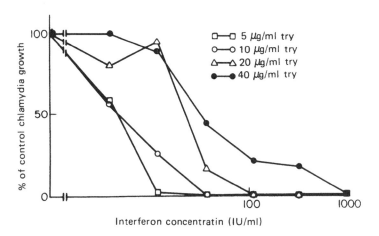

Figure 6 Dose-response curve for IFNγ in media with four tryptophan concentrations. The one-step growth experiment was carried out on HEp-2 cell cultures treated with 3-fold dilutions of human IFNγ in four media that differed only in tryptophan content. The results are expressed as percent of IFU/ml observed in control cultures not treated with interferon. (From Ref. 46.)

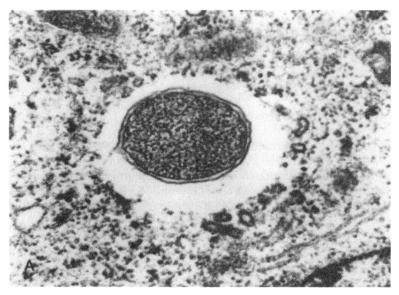

Figure 7   Electron micrograph of HEp-2 cells infected with *C. trachomatis* 24 hr after infection. *A*. Treatment with tryptophan (10 μg/ml) and human IFNγ (100 IU/ml); x45,000. *B*. Treatment with tryptophan (60 μg/ml) and human IFNγ (100 IU/ml); x7,000. *C*. Control, no interferon treatment, tryptophan 10 μg/ml x7000. (From Ref. 46.)

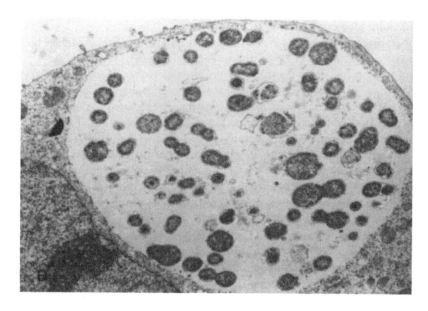

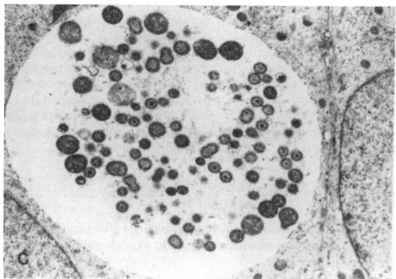

tryptophan concentration of 40 μg/ml was required to achieve almost complete *C. trachomatis* recovery from IFN inhibition.

Progressively more gamma IFN was required to inhibit the chlamydia as the tryptophan concentration in the medium was raised (Fig. 6), specifically, at 5 μg/ml of tryptophan complete inhibition of *C. trachomatis* by interferon occurred at 36 IU/ml, while at 40 μg/ml the same effect was obtained with 100 IU/ml. The ED_{50} levels for the recombinant human IFNγ were elevated as the tryptophan present in the reaction medium was raised. In the presence of 20 and 40 μg/ml tryptophan, the ED_{50} were elevated to 23 and 33 IU/ml, respectively.

An experiment was done in which HEp-2 cells pretreated for 24 hr with 100 IU/ml of recombinant IFNγ in the presence of either 10 or 60 μg/ml tryptophan were infected with *C. trachomatis* at MOI of 1. The infected cells were examined by EM at 24 and 48 hr postinfection. The controls were cultures that received 10 or 60 μg/ml tryptophan without gamma IFN. At 24 hr postinfection, 70% of the control cells had inclusion bodies which contained appropriate RBs and EBs at a ratio of 8:1 (Fig. 7C). In most of the cells treated with 100 IU/ml IFNγ, and 10 μg/ml tryptophan, infection could not be detected by EM, although occasionally a reticulate body (Fig. 7A) or more rarely, an elementary body could be found. In contrast, when the cells were pretreated with 100 IU/ml IFNγ and 60 μg/ml of tryptophan (Fig. 7B), 50% contained inclusion bodies with an RB/EB ratio of 21:1. At 48 hr postinfection, the control cells exhibited inclusion bodies that occupied most of the cell and had RB/EB ratios of 1:3. In IFN-treated cells that received 10 μg/ml tryptophan, no infection could be demonstrated and there were only a few cells with inclusions that contained a reticulate body. In contrast, cells treated with 100 IU/ml of interferon and 60 μg/ml tryptophan showed large inclusion bodies with RB/EB ratios of 2:3. It thus appears that the elevation of tryptophan concentration in the growth media of IFN-treated cells resulted in inclusion bodies formation and development of elementary bodies (only 15% less than the control).

VIII. TIME COURSE OF TRYPTOPHAN REVERSION OF
C. TRACHOMATIS INHIBITION BY IFNγ

We have found that *C. trachomatis* inhibition by 100 IU/ml IFNγ could be fully reversed even when the tryptophan was added up to 24 hr after infection (Fig. 8, A–C).

When tryptophan was added 48 hr after infection, *C. trachomatis* could be recovered to the extent of 30% of optimal yield (Fig. 8D). When tryptophan was added 72 hr postinfection, the recovery of *C. trachomatis* was less than 1% of optimal (Fig. 8E).

The reversal effect of tryptophan on *C. trachomatis* replication in cells pretreated with IFNγ has a number of interesting features. When tryptophan was

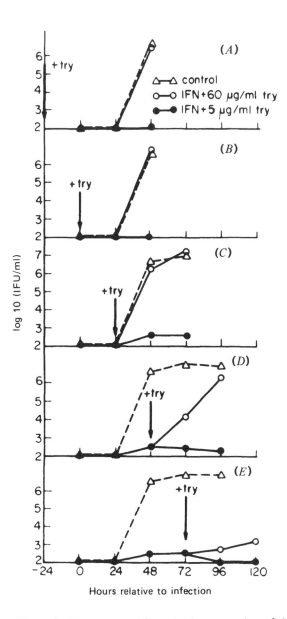

Figure 8 Time course of tryptophan reversion of *C. trachomatis* inhibition by IFNγ in HEp-2 cells expressed as titer of chlamydia produced in the culture. Cultures were pretreated with 100 IU/ml IFNγ in the presence of 5 μg/ml tryptophan for 24 hr and infected at an MOI of 1. Cells not treated with interferon served as controls. At various times, cultures (control and IFNγ pretreated) received an additional 5 or 60 μg/ml tryptophan (final concentration) as follows: *A*. 24 hr before infection; *B*. On the same day as infection; *C*. 24 hr after infection; *D*. 48 hr after infection; *E*. 72 hr after infection. (From Ref. 46.)

87

added 24 hr postinfection, the recovery of C. trachomatis infectivity was faster, as compared with the addition of tryptophan at the day of infection or at the day of IFN pretreatment (Fig. 8). This might imply that the inhibition by IFNγ is not limited to the EB that had entered the cells, but to a later stage in the C. trachomatis developmental cycle.

This is further supported by our EM data. We have detected reticulate bodies more frequently than elementary bodies with both recombinant and lymphocyte human IFNγ-treated cells 24 and 48 hr postinfection (39,46).

When the cells were pretreated with a relatively high concentration of IFNγ (100–1000 IU/ml) its inhibitory effect could not be completely reversed with tryptophan (Fig. 6). The latter results may be explained by the possibility that at high concentrations of IFNγ, the phagosome-enclosed chlamydia may be harmed, thereby permitting phagolysosome fusion and the destruction of the parasite. A second possibility is that in the cells pre-treated with high concentrations of interferon γ, an irreversible inhibitory effect on cell metabolism results in inhibition of C. trachomatis replication.

The possibility that tryptophan degradation may be related to the mechanism of IFNγ inhibition of chlamydia is further supported by the study of Byrne et al. (47), who demonstrated that human IFNγ-mediated inhibition of C. psittaci growth in T24 cells is reversed by tryptophan.

These results differ from those of de la Maza et al. (48) who have shown that the antiproliferative activity and antichlamydial activity of recombinant murine interferon γ was independent of tryptophan concentration. However, their work was carried out under different conditions which included the C. trachomatis L$_1$ serovar and murine IFNγ in McCoy cells.

Turco et al. (49) examined the role of amino acid deprivation in IFNγ-induced suppression of the growth of Rickettsia prowazekii in mouse L929 and human fibroblasts by measuring the amino acid pools of untreated and IFNγ-treated cells. They found that reconstitution of the tryptophan pool did not relieve IFNγ-induced inhibition of rickettsial growth. There was a difference between the intracellular and extracellular tryptophan pools of human and mouse cells. While in human fibroblasts treated with human IFNγ there was tryptophan degradation, in mouse L-929, 3T3-A31 cells and mouse embryo fibroblasts there was no degradation of tryptophan with IFN treatment.

Other studies by Yoshida et al. have demonstrated the induction of indoleamine 2,3-dioxygenase by mouse IFN in mouse lung slices (45), but in this work the tryptophan level in the treated cells was not examined.

It thus appears that the mechanisms involved in IFNγ-induced inhibition of intracellular microorganism multiplication, like the mechanisms involved in other IFN-induced effects, may vary with the host cell and with the parasite.

IX. CURRENT PROBLEMS

Persistent and recurrent chlamydial infections are common. They were recognized in various animal infections with *C. psittaci* (11) and were later observed in trachoma, LGV, in genital, and recently in respiratory tract infections (12). The frequency of occurrence of persistent infection in man is difficult to assess, partly because of the problem of distinguishing between persistent, chronic inapparent, and recurrent infection. Likewise, the nature of the chlamydia organism in the dormant stage is unknown. Is it an RB, an EB, or a specific cryptic form, as suggested by Moulder et al. (50)? What are the factors that reactivate the dormant stage so that the number of infectious particles increases and clinical signs of disease appear?

That the immune system plays a role in the process of reactivation is demonstrated by the appearance of active infection under conditions of immunosuppression (23). The immune system may control latent or chronic inapparent infection by modulation of the chlamydial developmental cycle in the cell harboring the parasite and/or by inhibition of chlamydial replication and spread.

The immune/IFNγ, which is produced by antigen-activated T lymphocytes, has been shown to activate professional phagocytes against intracellular pathogens (29–31).

Most intriguing are the observations that gamma interferon is also capable of acting upon nonprofessional phagocytes to inhibit the growth of intracellular parasites such as: *Toxoplasma gondii* (32), *Rickettsia prowazekii* (34), *C. psittaci* (47), and *C. trachomatis* (39,46).

Growth inhibition by gamma IFN could be reversed by tryptophan (32,46, 47) or removal of IFNγ after infection (39). Furthermore, since the concentration of IFN seems able to modulate intracellular parasite replication in HEp-2 cells, IFN may have a role in the maintenance of persistent infection in vivo.

Such a model would require, on the one hand, continuous stimulation of IFN secretion and, on the other, transmission of the persistent or latent parasite to both daughter cells at cell division. Early in chlamydial replication, genus-specific chlamydial antigen can be demonstrated by immunoelectron microscopy at the surface of infected host cells and in their immediate vicinity (51).

The secretion of chlamydial immunogenic glycolipid to the media of infected cell cultures was recently reported (52). It may be that these antigens are involved in ongoing stimulation of the immune system in vivo. Another possible trigger of a chronic inflammatory sequence might be poorly degraded nonviable microbial constituents which may persist for long periods both extracellularly and within phagocytic cells (53).

Ongoing stimulation of IFNγ is not enough to establish persistent infection. Preservation of the chlamydiae in the host demands either chlamydial existence in nondividing cells or multiplication of the chlamydial inclusion with the host. It has recently been suggested that *C. trachomatis* cytoplasmic inclusions undergo division in replicating endometrial epithelial cells (54). It still has to be determined whether inclusion division and transmission within dividing eukaryotic cell populations occur in the presence of IFN.

Recently, Oriel and Ridgway (22) reported the possibility of reactivation of latent chlamydial infection by secondary *Neisseria gonorrheae* infection. What is the possible mechanism by which superinfection with other microorganisms may reactivate or aggravate the course of chronic trachoma and delay healing? It may be that the bacterial superinfection is associated with elevated levels of prostaglandins (PG) that block the inhibitory effect of IFN (55) and suppress natural killer cell function (56), enabling *C. trachomatis* to convert from its dormant to an activated state. It has been shown indirectly that endogenous PG synthesis may support *C. trachomatis* infection (57). Perhaps it is possible that an autoregulatory mechanism involving both PG and IFN controls the body's initial defense systems.

As for the different results reached in different cell systems with various intracellular parasites concerning the tryptophan reversal of chlamydial IFNγ inhibition, it would seem that tryptophan degradation is not the only mechanism of IFN action. Further studies are required to explore the significance of different factors, such as chlamydial strain, cell employed, and interferon origin on the tryptophan reversal of gamma IFN inhibition.

Our studies have shown (39) that high IFN concentrations inhibit inclusion formation and result in complete and irreversible restriction of chlamydial growth with damage to the inclusion.

Still to be clarified is the question of whether the metabolic events induced by high IFN concentrations affect the nutrient sources of the chlamydia, resulting in irreversible parasite inhibition and whether these events harm the phagosome-enclosed chlamydia. In that case, what are the consequences? Does it permit phagolysosome fusion and the destruction of the parasite?

Knowledge of the mechanism by which irreversible inhibition of chlamydia can be achieved by IFNγ may be useful in the development of an efficient chlamydiacidic drug.

Since most drugs in use today fail to prevent persistent chlamydial infections and many leave the patient symptomless but still infected (58), the search for a new drug is imperative.

An additional possibility that needs to be investigated is whether IFN and various chlamydiastatic drugs may, under certain conditions, act synergistically for complete eradication of the infection.

In vivo studies in animal models are required to evaluate the potential application of IFNγ as an antichlamydial drug.

REFERENCES

1. Moulder, J.W. 1964. *The Psittacosis Group as Bacteria*. John Wiley & Sons Inc., New York.
2. Sarov, I., and Y. Becker. 1968. RNA in the elementary bodies of trachoma agent. *Nature 217*:849–852.
3. Sarov, I., and Y. Becker. 1969. Trachoma agent DNA. *J. Mol. Biol. 42*:581–589.
4. Sarov, I., and Y. Becker. 1971. Deoxyribonucleic acid-dependent ribonucleic acid polymerase activity in purified trachoma elementary bodies: effect of sodium chloride on ribonucleic acid transcription. *J. Bacteriol. 107*:593–598.
5. Tamura, A. 1967. Isolation of ribosome particles from meningopneumonitis organisms. *J. Bacteriol. 93*:2009–2016.
6. Schachter, J., and H.D. Caldwell. 1980. Chlamydiae. *Ann. Rev. Microbiol. 34*:285–309.
7. Ward, M.E. 1983. Chlamydial classification, development and structure. *Br. Med. Bull. 39*:109–115.
8. Page, L.A. 1968. Proposal for recognition of two species in the genus *Chlamydia*. Jones, Racke and Stearns, 1945. *Int. J. Syst. Bacteriol. 18*:51–66.
9. Storz, J., and L.A. Page. 1971. Taxonomy of the chlamydiae: reasons for classifying organisms of the genus *Chlamydia*, family Chlamydiaceae, in a separate order, *Chlamydiales. Int. J. Syst. Bacteriol. 21*:332–334.
10. Kuo, C.C., H.H. Chen, S.P. Wang, and J.T. Grayston. 1986. Characterization of TWAR strains, a new group of chlamydia psittaci. In *Chlamydial Infections*, D. Oriel, G. Ridgway, J. Schachter, D. Taylor-Robinson, and M. Ward (Eds.). Cambridge University Press, Cambridge, pp. 321–324.
11. Schachter, J., and M. Grossman. 1981. Chlamydial infections. *Ann. Rev. Med. 32*:45–61.
12. Ladany, S., and I. Sarov. 1985. Recent advances in *Chlamydia trachomatis*. *Eur. J. Epidemiol. 1(4)*:235–256.
13. Hanna, L., C.R. Dawson, O. Briones, P. Thygeson, and E. Jawetz. 1968. Latency in human infections with tric agents. *J. Immunol. 101*:43–50.
14. Schachter, J. 1978. Chlamydial infections. *New Engl. J. Med. 298*:428–435, 490–495, 540–549.
15. Henry-Suchet, J., F. Catalan, V. Loffredo, M.J. Sanson, C. Debache, F. Pigeau, and R. Coppin. 1981. *Chlamydia trachomatis* associated with chronic inflammation in abdominal specimens from women selected for tuboplasty. *Fertil. Steril. 36*:599–605.
16. Mardh, P-A., I. Lind, L. Svensson, L. Westrom, and B.R. Moller. 1981. Antibodies to *Chlamydia trachomatis, Mycoplasma hominis* and *Neisseria gonorrhoeae* in sera from patients with acute salpingitis. *Br. J. Vener. Dis. 57*:125–129.
17. Moore, D.E., H.M. Foy, J.R. Daling, J.T. Grayston, L.R. Spadony, S.-P. Wang, C.-C. Kuo, and D.A. Eschenbach. 1982. Increased frequency of

serum antibodies to *Chlamydia trachomatis* in infertility due to distal tubal disease. *Lancet 2*:574–577.

18. Piura, B., I. Sarov, B. Sarov, D. Kleinman, W. Chaim, and V. Insler. 1985. Serum IgG and IgA antibodies specific for *Chlamydia trachomatis* in salpingitis patients as determined by the immunoperoxidase assay. *Eur. J. Epidemiol. 1*:110–116.

19. Sarov, I., D. Kleinman, G. Holcberg, G. Potashnik, V. Insler, R. Cevenini, and B. Sarov. 1986. Specific IgG and IgA antibodies to *Chlamydia trachomatis* in infertile women. *Int. J. Fertil. 31 (3)*:193–197.

20. Meurman, O., P. Terho, and C.E. Sonck. 1982. Type-specific IgG and IgA antibodies in old lymphogranuloma venereum determined by solid phase radioimmunoassay. *Med. Microbiol. Immunol. 170*:279–286.

21. Dan, M.M., H.H. Rothmensch, E. Eylan, A. Rubinstein, R. Ginsberg, and M. Liron. 1980. A case of lymphogranuloma venereum of 20 years duration. Isolation of *Chlamydia trachomatis* from perianal lesions. *Br. J. Ven. Dis. 56*:344–346.

22. Oriel, J.D., and G.L. Ridgway. 1982. Epidemiology of chlamydial infection of the human genital tract: Evidence for the existence of latent infections. *Eur. J. Clin. Microbiol. 1*:69–75.

23. Yang, Y.S., C.-C. Kuo, and W.J. Chen. 1983. Reactivation of *Chlamydia trachomatis* lung infection in mice by cortisone. *Infect. Immun. 39*:655–658.

24. Joklik, W.K. Interferons. In *Virology*, B.N. Fields, D.M. Knipe, J.L. Melnick, R.M. Chanock, B. Roizman, and R.E. Shope (Eds.). Raven Press, New York, 1985, pp. 281–307.

25. Sueltenfuss, E.A., and M. Pollard. 1963. Cytochemical assay of interferon produced by duck hepatitis virus. *Science 139*:595–596.

26. Merigan, T.C., and L. Hanna. 1966. Characteristics of interferon induced in vitro and in vivo by a TRIC agent. *Proc. Soc. Exp. Biol. Med. 122*:421–424.

27. Rothermel, C.D., G.I. Byrne, and E.A. Havell. 1983. Effect of interferon on the growth of *Chlamydia trachomatis* in mouse fibroblasts (L cells). *Infect. Immun. 39*:362–370.

28. Byrne, G.I., and C.D. Rothermel. 1983. Differential susceptibility of chlamydiae to exogenous fibroblast interferon. *Infect. Immun. 39*:1004–1005.

29. Nacy, C.A., and J.V. Osterman. 1979. Host defenses in experimental *Scrub typhus*: Role of normal and activated macrophages. *Infect. Immun. 26*:744–750.

30. Nathan, C.F., H.W. Murray, M.E. Wiebe, and B.Y. Rubin. 1983. Identification of interferon gamma as the lymphokine that activates human macrophages oxidative metabolism and antimicrobial activity. *J. Exp. Med. 158*:670–689.

31. Rothermel, C.D., B.Y. Rubin, E.A. Jaffe, and H.W. Murray. 1986. Oxygen-independent inhibition of intracellular *Chlamydia psittaci* growth by human monocytes and interferon-gamma-activated macrophages. *J. Immunol. 137*:689–692.

32. Pfefferkorn, E.R. 1984. Interferon gamma blocks the growth of *Toxoplasma gondii* in human fibroblasts by inducing the host cells to degrade tryptophan. *Proc. Natl. Acad. Sci. (USA) 81*:908–912.

33. Turco, J., and H.H. Winkler. 1983A. Comparison of the properties of antirickettsial activity and interferon in mouse lymphokines. *Infect. Immun. 42*:27–32.

34. Turco, J., and H.H. Winkler. 1983B. Inhibition of the growth of *Rickettsia prowazekii* in cultured fibroblasts by lymphokines. *J. Exp. Med. 157*:974–986.

35. Byrne, G.I., and D.A. Krueger. 1983. Lymphokine-mediated inhibition of *Chlamydia* replication in mouse fibroblasts is neutralized by anti-gamma interferon immunoglobulin. *Infect. Immun. 42*:1152–1158.

36. Wallach, D. 1982. Regulation of susceptibility to natural killer cells' cytotoxicity and regulation of HLA synthesis: Differing efficacies of alpha, beta and gamma interferons. *J. Interferon Res. 2*:329–338.

37. Sonnenfeld, G., A.D. Mandel, and T.C. Merigan. 1977. The immunosuppressive effect of type II mouse interferon preparations on antibody production. *Cell Immun. 34*:193–206.

38. Virelizier, J.L., E.L. Chan, and A.C. Allison. 1977. Immunosuppressive effects of lymphocyte (type II) and leucocyte (type I) interferon on primary antibody responses in vivo and in vitro. *Clin. Exp. Immunol. 30*:299–304.

39. Shemer, Y., and I. Sarov. 1985. Inhibition of growth of *Chlamydia trachomatis* by human gamma interferon. *Infect. Immun. 48*:592–596.

40. Taylor, J.L., W.J. O'Brien, and A.I. Goldman. 1984. The determination of effective antiviral doses using a computer program for sigmoid-response curves. *J. Virol. Meth. 8*:225–232.

41. Kellog, K.R., K.D. Horoschak, and J.W. Moulder. 1977. Toxicity of low and moderate multiplicities of *Chlamydia psittaci* for mouse fibroblasts (L cells). *Infect. Immun. 18*:531–541.

42. Allan, I., and J.H. Pearce. 1983. Differential amino acid utilization by *Chlamydiae psittaci* (strain guinea pig inclusion conjunctivitis) and its regulatory effect on chlamydial growth. *J. Gen. Microbiol. 129*:1991–2000.

43. Allan, I., T.P. Hath, and J.H. Pearce. 1985. Influence of cysteine deprivation on chlamydial differentiation from reproductive to infective life-cycle forms. *J. Gen. Microbiol. 131*:3171–3177.

44. Stirling, P., I. Allan, and J.H. Pearce. 1983. Interference with transformation of chlamydiae from reproductive to infective body forms by deprivation of cysteine. *FEMS Microbiol. Lett. 19*:133–136.

45. Yoshida, R., J. Imanishi, T. Oku, T. Kishida, and O. Hayaishi. 1981. Induction of pulmonary indoleamine 2,3-dioxygenase by interferon. *Proc. Natl. Acad. Sci. (USA) 78*:129–132.

46. Shemer, Y., R. Kol, and I. Sarov. 1987. Tryptophan reversal of recombinant human gamma interferon inhibition of *C. trachomatis* growth. *Curr. Microbiol. 16*:9–13.

47. Byrne, G.I., L.K. Lehmann, and G.I. Landry. 1986. Induction of tryptophan catabolism is the mechanism for gamma-interferon-mediated inhibition of intracellular *Chlamydia psittaci* replication in T24 cells. *Infect. Immun. 53*:347–351.

48. de la Maza, L.M., E.M. Peterson, C.W. Fennie, and C.W. Czarniecki. 1985. The anti-chlamydial and anti-proliferative activities of recombinant murine interferon-gamma are not dependent on tryptophan concentrations. *J. Immunol. 135*:4198–4200.

49. Turco, J., and H.H. Winkler. 1986. Gamma-interferon-induced inhibition of the growth of *Rickettsia prowazekii* in fibroblasts cannot be explained by the degradation of tryptophan or other amino acids. *Infect. Immun. 53*:38–46.

50. Moulder, J.W., N.J. Levy, and L.P. Schulman. 1980. Persistent infection of mouse fibroblasts (L cells) with *Chlamydia psittaci*: Evidence for a cryptic chlamydial form. *Infect. Immun. 30*:874–883.

51. Richmond, S.J., and P. Stirling. 1981. Localization of chlamydial group antigen in McCoy cell monolayers infected with *Chlamydia trachomatis* or *Chlamydia psittaci. Infect. Immun. 34*:530–561.

52. Stuart, E.S., and A.B. Macdonald. 1986. Genus glycolipid exoantigen from *Chlamydia trachomatis*: component preparation, isolation and analysis. In *Chlamydial Infections*, D. Oriel, G. Ridgway, J. Schachter, D. Taylor-Robinson, and M. Ward (Eds.). Cambridge University Press, Cambridge, pp.122–125.

53. Ginsburg, I., and M. Lahav. 1983. Lysis and biodegradation of microorganisms in infectious sites may involve cooperation between leukocyte, serum factors and bacterial wall autolysis: A working hypothesis. *Eur. J. Clin. Microbiol. 2*:186–191.

54. Richmond, S.J., E. Crosdale, M. Lusher, and P.J. Haynes. 1986. Inclusion division and pleomorphism in cultured human endometrial cells. In *Chlamydial Infections*, D. Oriel et al. (Eds.). Cambridge University Press, Cambridge, pp. 63–66.

55. Trofatter, K.F., and C.A. Daniels. 1980. Effect of prostaglandins and cyclic adenosine $3', 5'$-monophosphate modulators on herpes simplex virus growth and interferon response in human cells. *Infect. Immun. 27*:158–167.

56. Bankhurst, A.D. 1982. The modulation of human natural killer cell activity of prostaglandins. *J. Clin. Lab. Immunol. 7*:85–91.

57. Ward, M.E., and H.S. Salari. 1982. Control mechanisms governing the infectivity of *Chlamydia trachomatis* for HeLa cells: Modulation by cyclic nucleotides, prostaglandins and calcium. *J. Gen. Microbiol. 128*:639–650.

58. Oriel, J.D. 1986. Chemotherapy. In *Chlamydial Infections*, D. Oriel (Ed.). Cambridge University Press, Cambridge, pp. 513–523.

5

Interactions Between *Rickettsia prowazekii* and Cultured Host Cells: Alterations Induced by Gamma Interferon

Jenifer Turco and Herbert H. Winkler / University of South Alabama College of Medicine, Mobile, Alabama

I. INTRODUCTION

A. Rickettsiae

Bacteria of the genus *Rickettsia* are obligate intracellular microorganisms which grow directly in the cytoplasm of their host cells, unbounded by a phagosomal or phagolysosomal membrane (1,2). There are three groups within the genus: the typhus group (*R. prowazekii, R. typhi,* and *R. canada*), the spotted fever group (*R. rickettsii, R. conorii, R. akari,* and other species), and the scrub typhus group (*R. tsutsugamushi*) (2). The rickettsiae cause several diseases in humans, and arthropods play an important role in the transmission of these diseases. For example, *R. prowazekii* is the etiological agent of epidemic typhus, and is transmitted by human lice. In the human, *R. prowazekii* grows mainly within the endothelial cells that line the small blood vessels (3); it can also grow within macrophages (4).

B. Host Defense Against *R. prowazekii* and Other Rickettsial Species

Historically, epidemic typhus has been responsible for much human suffering; but even without antibiotic treatment, this disease is not uniformly fatal. Therefore, the immune response of the host must limit the growth of the rickettsia and lead to its elimination or to its persistence in an inactive form. Evidence for the persistence of *R. prowazekii* in the host in the absence of disease derives from the occurrence of Brill-Zinsser disease, a recrudescence of a latent infection with *R. prowazekii* (2).

Both humoral immunity and cell-mediated immunity play roles in defending the host against rickettsial infections. When *R. prowazekii* is treated with immune serum, it fails to grow within human monocyte-derived macrophages (5). In addition, administration of immune serum has had a protective effect in human cases of epidemic typhus (6). These data suggest that specific antibody can provide some protection; however, other studies underscore the importance of cell-mediated immunity (7-13). For example, adoptive transfer of spleen cells (but not serum) collected from *R. typhi*-immune guinea pigs to nonimmune guinea pigs can protect the nonimmune animals against intradermal challenge with *R. typhi* (7,8). Similarly, an established splenic *R. typhi* infection in mice is controlled by administration of T lymphocytes collected from *R. typhi*-immune mice, but not by administration of serum from *R. typhi*-immune mice (9). Other studies have provided evidence that T lymphocytes also play a role in host defense against murine infections with *R. tsutsugamushi, R. akari*, and *R. conorii* (10-13).

Exactly how T lymphocytes function in host defense against rickettsial infections is a matter of interest, and both cytotoxic T lymphocytes and lymphokines produced by T lymphocytes may be important. Cells that are infected with *R. typhi* bear rickettsial antigens on their surfaces (14), and the spleens of rickettsia-infected mice contain T lymphocytes which are capable of lysing rickettsia-infected cells in an H-2-restricted fashion (15). The fate of the rickettsiae in an infected cell that is lysed by a cytotoxic T lymphocyte is not known. However, since a rickettsia does not have a virus-like eclipse phase, it is probable that most of these rickettsiae survive and reinfect other cells. Thus, the cytotoxic T lymphocyte may simply be accelerating the normal bursting of the infected cell. It may be that this reaction is more important in immunopathology than in host defense.

Studies describing the ability of lymphokines to activate mouse macrophages for destruction of *R. tsutsugamushi* first indicated the importance of lymphokines in host defense against rickettsial infections (16-18). When mouse macrophage cultures are treated with lymphokines after infection with *R. tsutsugamushi*, the percentage of macrophages free of rickettsiae in the treated cultures at 24 hr after infection is dramatically increased in comparison with that in untreated control cultures (17). In addition, when macrophages are treated with lymphokines before infection with *R. tsutsugamushi* and examined for rickettsiae immediately after the infection period, the percentage of cells that are infected with rickettsiae is suppressed in the lymphokine-treated cultures in comparison with the untreated cultures (17).

These studies with *R. tsutsugamushi* and macrophages caused us to question whether lymphokines might alter *R. prowazekii* infections in cells other than macrophages, particularly cells that are not specialized for the phagocytosis and destruction of bacteria. This question seemed important since a major site of growth of *R. prowazekii* in the host is the endothelial cell, and because a plausible explanation for the ability of the host to recover from infection with *R. prowazekii* would appear to require an effector mechanism which acts against rickettsiae sequestered within their host cells (19).

C. Early Studies with Interferon (IFN) and Rickettsiae

The relationship between rickettsial infections and acid-stable, type I interferons (IFNβ and IFNα) (20) was first examined about 20 years ago (21,22). Hopps et al. (21) reported that chick embryo cells infected with *R. tsutsugamushi* produce an acid-stable IFN. In addition, Kazàr (22) detected an acid-stable IFN in the sera of mice that had been injected intravenously with *R. prowazekii*. The highest titers of IFN were found at 3 and 5 hours after injection (22). A later study provided evidence that the growth of *R. akari* in mouse L929 cells is inhibited by virus-induced (type I) IFN (23). These studies with IFN and rickettsiae, as well as early studies with IFN and other nonviral infectious agents, have been reviewed (24,25).

D. Scope

The purpose of this chapter is to discuss (1) how lymphokines and IFNγ alter the interactions between *R. prowazekii* and cultured cells (both fibroblasts and macrophages) and (2) what is known about the mechanisms by which these alterations are induced. Another chapter in this book focuses on IFNγ and experimental rickettsial infections in animals.

II. STUDIES WITH FIBROBLASTS

A. Lymphokine-Mediated Inhibition of Rickettsial Growth

Crude lymphokine preparations inhibit the growth of *R. prowazekii* in fibroblasts, and these lymphokine preparations manifest species preference in their action on the host cells (26) (Fig. 1). When mouse lymphokines (supernatant fluids collected from mouse spleen cells cultured with concanavalin A or antigen) are added to X-irradiated cultures of mouse L929 cells and human fibroblasts 24 hours before infection and are continuously present after infection, growth of *R. prowazekii* is inhibited in the mouse cells, but is unaffected in the human cells. Similarly, human lymphokines (supernatant fluids collected from human peripheral blood mononuclear cells cultured with concanavalin A) inhibit the growth of *R. prowazekii* in human fibroblasts but not in mouse L929 cells (Fig. 1). In the cultures treated with homologous lymphokine preparations, the average number of rickettsiae per infected cell increased less than fourfold as compared with the 16- to 23-fold increases observed in the untreated control cultures during the 48 hr period following infection (26). During the same time

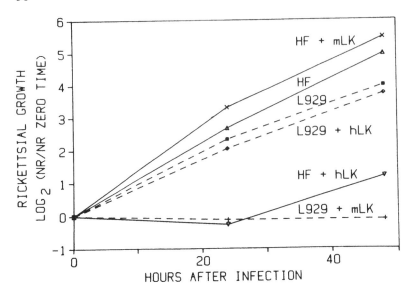

Figure 1 Ability of crude mouse and human lymphokines to inhibit rickettsial growth in cells of the homologous species and failure of the lymphokines to inhibit rickettsial growth in cells of the heterologous species. Growth of *R. prowazekii* was monitored in X-irradiated cultures of mouse cells (L929) and human foreskin fibroblasts (HF) that were treated with the following materials for 24 hr before infection and the entire period after infection: serum-supplemented medium (MS), MS plus 20% crude mouse lymphokines (mLK), and MS plus 20% crude human lymphokines (hLK). The average numbers of rickettsiae per cell (NR) at 24 hr and 48 hr were divided by the values at 0 hr and the logarithms of these numbers to the base 2 are plotted. Each point represents the mean for two or more experiments. The standard errors of NR in these experiments averaged 16% of the mean values. At 0 hr, the percentage of cells infected ranged from 35 to 90; the average number of rickettsiae per infected cell ranged from 2.1 to 4.6; and NR ranged from 0.8 to 4.1. (From Ref. 26.)

period, there was also a decrease in the percentage of cells infected in the cultures treated with homologous lymphokine preparations, as compared with no change or a slight increase in the percentage of cells infected in the untreated control cultures. These results indicate not only that the rate of rickettsial growth was much lower in the lymphokine-treated cultures, but also that the rickettsiae actually disappeared from some of the lymphokine-treated host cells (26).

The ability of mouse lymphokines to inhibit rickettsial growth is concentration dependent, and the avirulent E strain and the virulent Breinl strain of *R. prowazekii* have similar susceptibilities to the lymphokines (26). The growth of organisms of another species and another genus of rickettsiae, *R. conorii* (26) and *Coxiella burnetti* (27), in L929 cells is also suppressed by mouse lymphokines.

The time of addition of lymphokines to L929 cell cultures affects the degree of inhibition of rickettsial growth observed (26). Inhibition of rickettsial growth is most pronounced when the lymphokines are added 24 hours before infection and are continuously present after infection as described above. When the cells are treated with lymphokines before infection only, inhibition of rickettsial growth is much less pronounced, and the antirickettsial effect of the lymphokines is greater during the interval from 0 to 24 hr after infection than during the interval from 24 to 48 hr after infection. In contrast, when lymphokines are present only after infection, the antirickettsial effect of the lymphokines is greater during the interval from 24 to 48 hr after infection. These data suggest that substantial time is required for the lymphokines to induce antirickettsial activity in L929 cells and that removal of the lymphokines results in a decay of the antirickettsial effect (26).

When an inhibitor of eucaryotic protein synthesis (cycloheximide or emetine) is added to the mouse fibroblasts at the same time as the lymphokines, the ability of the lymphokines to inhibit the growth of *R. prowazekii* is suppressed (26). These data suggest that host cell protein synthesis is required for lymphokine-mediated induction of antirickettsial activity in fibroblasts.

Because interferons also require host cell protein synthesis for activity and manifest species preference in exerting their effects on cells, these features of lymphokine-mediated induction of antirickettsial activity suggest the participation of IFN. The antiviral and antirickettsial activities present in the lymphokine preparations are both destroyed by trypsin, by exposure to pH 2, or by heating at 80°C (28). Loss of activity at pH 2 is a characteristic of IFNγ but not of IFNβ and IFNα (20). Both the antiviral activity and the antirickettsial activity in the lymphokine preparations are neutralized by rabbit antisera prepared against partially purified murine IFNγ (28), and by a monoclonal antibody against murine IFNγ (29,30). These characteristics of the antirickettsial activity are all consistent with the idea that IFNγ is responsible for the ability of mouse lymphokines to inhibit rickettsial growth in fibroblasts.

Wisseman and Waddell (31) demonstrated that supernatant fluids collected from cultures of antigen- or phytohemagglutinin-stimulated human leukocytes inhibit the growth of *R. prowazekii* in cultured human endothelial cells and human monocyte-derived macrophages, as well as cultured human fibroblasts. The active material in these supernatant fluids is labile at pH 2, and inhibition of rickettsial growth by these fluids requires synthesis of host cell mRNA and protein (31). Furthermore, these human supernatant fluids are not active in chicken embryo fibroblasts (31). These characteristics of the active material in the human leukocyte supernatant fluids are consistent with those of IFNγ (31).

B. IFNγ-Mediated Inhibition of Rickettsial Growth

The time and effort required for demonstrating that IFNγ is the principal antirickettsial effector in the lymphokines was remarkably shortened by the availability of recombinant IFNγ (rIFNγ) (Genentech, Inc., South San Francisco,
CA). Experiments with murine rIFNγ demonstrated that IFNγ, in the absence of
other lymphokines, inhibits the growth of R. prowazekii in mouse L929 cells
(32). When L929 cells are infected with R. prowazekii and treated with rIFNγ
after infection, concentrations as low as 23 U/ml inhibit rickettsial growth: in
three experiments the average number of rickettsiae per infected cell increased
only 6.2 ± 2.4-fold in the treated cultures, but it increased 12.6 ± 2.8-fold in the
control cultures (mean ± standard error) (32). Like the crude mouse lymphokines, the murine rIFNγ also manifests species preference and its antirickettsial
action is inhibited by cycloheximide (19,32). Experiments with human rIFNγ
likewise demonstrated that it inhibits the growth of R. prowazekii in human
fibroblasts (33).

In contrast to the marked inhibitory effect of murine rIFNγ, there is little
or no inhibition of the growth of R. prowazekii in L929 cells treated with crude
mouse type I IFN (IFNβ and IFNα) (28). However, Kazàr et al. (23) found that
growth of R. akari in L929 cells is inhibited by type I IFN. This difference may
reflect a difference in the two rickettsial species and/or in the experimental
methods used in the two studies.

Assessment of the inhibition of rickettsial growth that occurred in infected
L929 cells that were treated with various concentrations of IFN as either crude
lymphokines or murine rIFNγ indicated that similar degrees of inhibition of
rickettsial growth are induced by equivalent (in terms of antiviral activity)
amounts (units) of IFN as either crude lymphokines or rIFNγ (32). Native IFNγ
is a glycosylated protein, but rIFNγ made in Escherichia coli is not. However,
both the nonglycoslyated rIFNγ and glycoslyated rIFNγ made in animal cells are
equally effective (32). These data indicate that most of the antirickettsial activity of crude mouse lymphokine preparations can be explained by the IFNγ
that is present in these preparations.

C. Effect of Lymphokines and rIFNγ on the Ability of L929
Cells to Be Initially Infected with R. prowazekii

In spite of the exceptional correspondence described above in the inhibition of
rickettsial growth, crude lymphokines and rIFNγ differ in their abilities to alter
the interaction of R. prowazekii and L929 cells in at least one way. High concentrations of crude lymphokines (IFN, ≥ 83 U/ml) suppress the initial infection of
L929 cells by R. prowazekii; however, rIFNγ does not have this effect (34)
(Table 1, left). This lymphokine activity is not found in supernatant fluids collected from unstimulated cultures of normal mouse spleen cells, similar super-

Table 1   Effect of Lymphokines and rIFNγ on the Ability of L929 Cells and RAW264.7 Cells to Be Initially Infected with *R. prowazekii*[a]

Cell treatment	IFN concentration (U/ml)	Initial rickettsial infection in the following cell lines (% of control value)			
		L929		RAW264.7	
		%R	RI	%R	RI
Control	0	100	100	100	100
		(82–99)	(2.6–5.8)	(60–99)	(2.0–6.8)
LK	2.5	94[b]	97	100 ± 2	88 ± 8
	10	95[b]	100	84 ± 8	87 ± 5
	40	102 ± 6	96 ± 16	75 ± 9	57 ± 4
	83	64 ± 17	69 ± 3	56 ± 3	56 ± 3
	167	66 ± 11	61 ± 5	49 ± 5	42 ± 4
	423	54 ± 7	54 ± 5	79 ± 1	41 ± 1
	700	57 ± 5	47 ± 6	50 ± 4	36 ± 4
rIFNγ	10	98[b]	111	95 ± 1	91 ± 8
	40	101[b]	86	90 ± 11	67 ± 10
	167	99[b]	103	68 ± 5	67 ± 2
	313	98 ± 1	97 ± 0	ND[c]	ND[c]
	700	95 ± 0	106 ± 2	50 ± 6	58 ± 7
	2130	101 ± 2	110 ± 3	77 ± 4	38 ± 5

[a]X-irradiated cells were treated for 24 hr with control medium, lymphokines (LK) diluted in control medium, or rIFNγ diluted in control medium. The cells were then washed and infected with *R. prowazekii* strain E or strain Breinl for 1 hr. After washing, the cells were dried, fixed, and stained. The percentage of cells infected (%R) and the average number of rickettsiae per infected cell (RI) in the treated cultures are expressed as percentages of the values in the control cultures. Each value represents the mean ± standard error for two or more experiments, except as noted. The ranges for the actual values of %R and RI in the control cultures are given in parentheses.

[b]Each value represents the mean of duplicate determinations in one experiment.

[c]ND, not determined.

Sources: From Refs. 34 and 35.

natant fluids to which concanavalin A has been added, or mixtures of these supernatant fluids and rIFNγ (34) (data not shown). These results with L929 cells contrast with the results of similar experiments with macrophage-like RAW264.7 cells: rIFNγ as well as crude lymphokines can inhibit the initial rickettsial infection in RAW264.7 cells (35) (Table 1, right, and Section III.D.).

Interestingly, the ability of the crude lymphokines to inhibit the initial rickettsial infection in L929 cells is neutralized by a monoclonal antibody against

Table 2 Neutralization of Lymphokine-Mediated Inhibition of the Initial
Rickettsial Infection in L929 Cells by a Monoclonal Antibody Against Murine
IFNγ[a]

Experi-ment	Addition of lymphokines	Concentration of antibody (μg/ml)	IFN (antiviral activity) remaining (U/ml)	Initial rickettsial infection[b]	
				%R	RI
1	−	0	0	89 ± 6	3.7 ± 0.3
	+	0	294	40 ± 8	1.8 ± 0.1
	+	0.2	13	83 ± 3	3.2 ± 0.2
	−	0.2	0	76 ± 2	2.4 ± 0.2
2	−	0	0	92 ± 5	4.1 ± 0.5
	+	0	679	48 ± 10	1.9 ± 0.1
	+	2.2	9	94 ± 2	4.5 ± 0.4
	−	2.2	0	88 ± 11	4.1 ± 1.0

[a]Mixtures of lymphokine and monoclonal antibody were incubated for 1.5 hr at 34°C.
After incubation, the mixtures were assayed for antiviral activity and added to X-irradiated
L929 cell cultures 24 hr before infection of the cells with R. prowazekii. After infection
and washing, slides were stained and examined for rickettsiae.
[b]%R, percentage of cells infected; RI, average number of rickettsiae per infected cell. Each
value represents the mean ± standard deviation of triplicate determinations.

murine IFNγ (34) (Table 2). The monoclonal antibody was the gift of Dr. George
Spitalny (30). In addition, lymphokine-induced suppression of the initial rickettsial infection in L929 cells is partially inhibited by adding cycloheximide to the
cells at the same time that the lymphokines are added, and suppression of the
initial rickettsial infection does not occur in mouse lymphokine-treated human
fibroblasts (34). There are several possible explanations for these results (34).
One possibility is that IFNγ and another factor in the crude lymphokines are
both required for inhibition of the initial rickettsial infection in L929 cells. This
factor might be produced by macrophage-like cells in response to rIFNγ.
Another possibility is that a non-IFNγ factor in the lymphokines that shares an
epitope with murine IFNγ is responsible for the ability of the lymphokines to
inhibit the initial rickettsial infection in L929 cells. Experiments are in progress
to explore these possibilities.

D. Evaluation of Amino Acid Deprivation as a Possible Mechanism for IFNγ-Induced Inhibition of Rickettsial Growth

Amino acid deprivation was of particular interest because Pfefferkorn had shown
that degradation of tryptophan by the host cell is responsible for IFNγ-induced

inhibition of the growth of *Toxoplasma gondii* in human fibroblasts (36). In IFNγ-treated human fibroblasts, tryptophan is degraded to *N*-formylkynurenine and kynurenine (36). In order to evaluate amino acid deprivation as a possible mechanism for IFNγ-induced inhibition of the growth of *R. prowazekii* in fibroblasts, the amino acid pools were measured in untreated cells and in L929 cells and human fibroblasts treated with homologous rIFNγ (33). The data indicated that tryptophan is markedly reduced in IFNγ-treated human fibroblasts in comparison with untreated human fibroblasts: it is not detected in either the intracellular pool or the extracellular medium of IFNγ-treated human fibroblast cultures (33). In contrast, the pools of tryptophan in untreated and IFNγ-treated mouse L929 cells are similar (33). None of the other amino acids measured in the study was severely reduced in human fibroblasts or L929 cells that had been treated with IFNγ. Examination of cell extracts for indoleamine 2,3-dioxygenase activity (ability to convert tryptophan into compounds that co-chromatographed with *N*-formylkynurenine and kynurenine) revealed activity in extracts prepared from IFNγ-treated human fibroblasts, but no activity in extracts prepared from untreated human fibroblasts, untreated L929 cells, and IFNγ-treated L929 cells (33).

To determine whether the difference in the responses of human fibroblasts (which were derived from primary cultures) and mouse L929 cells (a continuous cell line) to homologous IFNγ was related to the passage history of the cells or to the human versus mouse derivation, human HeLa cells (a continuous cell line), mouse embryo cells (derived from primary cultures), and mouse 3T3-A31 cells (a continuous cell line), were examined. HeLa cells, like human fibroblasts, degrade tryptophan after IFNγ treatment; mouse embryo cells and mouse 3T3-A31 cells, like mouse L929 cells, do not degrade tryptophan (33).

The tryptophan pool was reconstituted in cultures of IFNγ-treated human fibroblasts by adding extra tryptophan to the culture medium, and rickettsial growth was monitored in these cells (33) (Fig. 2). Although the tryptophan pool in these IFNγ-treated cultures was demonstrably reconstituted, the inhibition of rickettsial growth by IFNγ was not reversed. Growth of *R. prowazekii* in the untreated control cultures to which extra tryptophan had been added was similar to that observed in the untreated control cultures without extra tryptophan (33).

The failure of tryptophan pool reconstitution to reverse IFNγ-induced inhibition of the growth of *R. prowazekii* in human fibroblasts suggests that either tryptophan depletion is not involved or that tryptophan depletion and some other mechanism(s) in addition to tryptophan depletion are responsible for the inhibition of rickettsial growth in IFNγ-treated human fibroblasts (33). The data for the L929 cell cultures provides no evidence to support the hypothesis that amino acid deprivation is responsible for IFNγ-induced inhibition of rickettsial growth in these mouse cells. Furthermore, tryptophan degradation cannot be the mechanism for inhibition of any intracellular microorganism by IFNγ in the

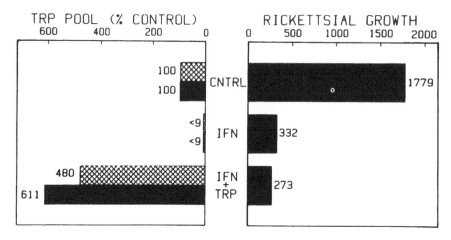

Figure 2 Failure of reconstitution of the tryptophan pool in IFNγ-treated human fibroblasts to reverse the inhibition of rickettsial growth. X-irradiated human fibroblasts were treated with human rIFNγ (188 U/ml) for 24 hr before infection or mock infection. Supplemental tryptophan (100 μg/ml) was added to some of the IFNγ-treated cultures both before and after infection (or mock infection). The intracellular tryptophan pools in the treated cultures at 48 hr after infection (or mock infection) are expressed as percentages of the tryptophan pool in the untreated control cultures, which averaged 1.24 nmol/mg of protein (left). For the mock-infected cultures (crosshatched bars), n = 3 and each bar represents the mean; for the infected cultures (solid bars), n = 1. Rickettsial growth was also monitored, and the average number of rickettsiae per cell at 48 hr after infection is expressed as a percentage of the number present at 0 hr (4.2 ± 0.3) (right). Each bar represents the mean of two experiments. (From Ref. 33.)

three mouse cell lines that were studied because these cells do not degrade tryptophan after treatment with IFNγ (33).

E. Toxicity of *R. prowazekii* and IFNγ for Fibroblasts

The combination of *R. prowazekii* and lymphokines (or rIFNγ) has a relatively rapid toxic effect on macrophage-like cells (35) (Section III.B.), but *R. prowazekii*-infected, lymphokine-treated L929 cells remain viable at 24 hr after infection (35). In contrast, Wisseman and Waddell (31) noted that supernatant fluids collected from stimulated human leukocytes have a cytolytic effect on *R. prowazekii*-infected human fibroblasts, and suggested that this effect is due to IFNγ.

Further examination of the viability of L929 cells that were infected with *R. prowazekii* and treated with IFNγ after infection indicated that the cells re-

Table 3 Toxicity of IFNγ Treatment and *R. prowazekii* Infection for Mouse L929 Cells and Human Fibroblasts (HF)[a]

Cell line and rickettsial burden[b]	rIFNγ (U/ml)	% Dead cells 24 hr	% Dead cells 48 hr
L929, none	0	0 ± 0	0 ± 0
(%R = 0)	15	0 ± 0	0 ± 0
	100	0 ± 0	0 ± 0
L929, low	0	0 ± 0	0 ± 0
(%R = 48 ± 2; RI = 2.2 ± 0.1)	15	0 ± 0	2 ± 0
	100	0 ± 0	4 ± 1
L929, medium	0	0 ± 0	0 ± 0
(%R = 74 ± 2; RI = 5.8 ± 1.2)	15	0 ± 0	19 ± 1
	100	0 ± 0	23 ± 3
L929, high	0	0 ± 0	0 ± 0
(%R = 98 ± 1; RI = 12.4 ± 0.2)	15	1 ± 0	76 ± 3
	100	0 ± 0	81 ± 6
HF, none	0	0 ± 0	
(%R = 0)	2	1 ± 0	
	15	0 ± 0	
	100	1 ± 0	
HF, low	0	2 ± 1	
(%R = 46 ± 5; RI = 3.4 ± 1.3)	2	11 ± 4	
	15	12 ± 2	
	100	33 ± 1	
HF, medium	0	3 ± 2	
(%R = 84 ± 3; RI = 7.1 ± 2.0)	2	56 ± 2	
	15	56 ± 5	
	100	58 ± 2	
HF, high	0	2 ± 0	
(%R = 86 ± 9; RI = 12.4 ± 1.2)	2	ND[c]	
	15	82 ± 6	
	100	76 ± 2	

[a]Suspensions of mouse L929 cells and human foreskin fibroblasts (HF) were mock-infected or infected with various concentrations of *R. prowazekii*. The cells were then washed and planted in serum-supplemented medium with or without homologous rIFNγ. After incubation for 1 or 2 days, the attached and detached cells were harvested and stained with trypan blue. Each value represents the mean ± the standard deviation for 1 to 3 experiments. The number of replicates per experiment ranged from 2 to 4.

[b]Immediately after infection and washing, samples of the cells were cytocentrifuged onto slides and stained for determination of the rickettsial burden. %R, percentage of cells infected; RI, average number of rickettsiae per infected cell.

[c]ND, not determined.

mained viable at 24 hr after infection; however, some of the cells were dead by 48 hr after infection (Table 3). The percentage of cells killed was directly related to the number of rickettsiae that initially infected the cells. Similar experiments with human fibroblasts suggested that these cells were more sensitive to the toxic effect of *R. prowazekii* and IFNγ: the human fibroblasts were killed within 24 hr and were susceptible to fewer rickettsiae than the L929 cells (Table 3).

III.   STUDIES WITH MACROPHAGE-LIKE CELLS

A.   Use of RAW264.7 Cells for Studies of *R. prowazekii*-Macrophage Interactions

Mouse macrophage-like RAW264.7 cells have been used as host cells in studies designed to determine how lymphokines and rIFNγ alter the interaction of *R. prowazekii* and macrophages (35). The RAW264.7 cells constitute a good model because they are very similar to human monocyte-derived macrophages in their responses to *R. prowazekii* (37). Specifically, the avirulent E strain of *R. prowazekii* fails to grow in most of the RAW264.7 cells, but the virulent Breinl strain proliferates very well (37). In addition, clearance of virulent rickettsiae from the RAW264.7 cells occurs if the rickettsiae have been treated with specific antiserum (37).

B.   Toxicity of IFNγ and *R. prowazekii* for RAW264.7 Cells

Treatment of macrophage-like RAW264.7 cells with lymphokines or rIFNγ for 24 hr, followed by infection with *R. prowazekii* (E or Breinl strains), leads to killing of a substantial proportion of the RAW264.7 cells within 4 to 6 hr after infection (35) (Table 4). Many of the RAW264.7 cells are also killed if untreated cultures are first infected with *R. prowazekii* and then treated with lymphokines for 4 to 6 hours (35). Killing of macrophage-like RAW264.7 cells by IFNγ and *R. prowazekii* is much more rapid than the killing of fibroblastic L929 cells by IFNγ and *R. prowazekii*, which was described in Section II.E. The percentage of RAW264.7 cells killed is directly proportional to the number of rickettsiae that initially infected the cells; and heat-killed rickettsiae are not toxic for the treated cells (35). Neither treatment with IFNγ alone nor *R. prowazekii* alone results in this toxicity: both agents are required for killing of the RAW264.7 cells.

Whether true macrophages, like macrophage-like cells, are susceptible to killing by *R. prowazekii* and IFNγ (or crude lymphokines) has not been reported. Wisseman and Waddell (31), who documented the cytolytic action of supernatant fluids collected from antigen or phytohemagglutin-stimulated human leukocyte cultures on *R. prowazekii*-infected human fibroblasts, presented no data concerning the possible cytolytic effect of these supernatant fluids on the human monocyte-derived macrophages used in their study. In

Table 4 Killing of Macrophage-Like RAW264.7 Cells after Treatment with Crude Lymphokines or rIFNγ and Infection with *R. prowazekii*[a]

Cell treatment	Rickettsiae	Rickettsial infection at 0 hr[b]		% Dead at 4–6 hr after infection
		%R	RI	
Control	–	0		5 ± 1
	+	77 ± 6	3.4 ± 0.3	9 ± 1
LK (IFN, 10 U/ml)	–	0		8 ± 2
	+	78 ± 7	3.6 ± 0.4	80 ± 3
rIFNγ (10 U/ml)	–	0		8 ± 3
	+	70 ± 5	2.6 ± 0.2	74 ± 5

[a]X-irradiated cells were treated with lymphokines (LK) or rIFNγ for 24 hr. The cells were washed and then mock-infected or infected with *R. prowazekii* Breinl strain for 1 hr. After washing, slides were dried and stained for determination of the initial rickettsial infection. The remaining cultures were incubated at 34°C for 4 to 6 hr and stained with trypan blue. Each value represents the mean ± the standard error for three experiments.

[b]%R, percentage of cells infected; RI, average number of rickettsiae per infected cell.

Source: From Ref. 35.

addition, studies with other rickettsial species and mouse macrophages treated with crude lymphokines or rIFNγ have not reported such a toxic effect (17,29).

C. Inhibition of Rickettsial Growth

The growth of *R. prowazekii* (Breinl strain) is inhibited in the RAW264.7 cells that survive the toxic effect of the lymphokines (or rIFNγ) and *R. prowazekii* (35). In contrast, untreated RAW264.7 cells support good growth of the Breinl strain of *R. prowazekii* (35,37).

D. Inhibition of the Initial Rickettsial Infection in Lymphokine- or rIFNγ-Treated RAW264.7 Cells

High concentrations of IFNγ as either crude lymphokines or rIFNγ inhibit the ability of RAW264.7 cells to be initially infected with *R. prowazekii* (35) (Table 1, right). As noted previously, Nacy and Meltzer (17) had earlier described a lymphokine-induced suppression of the ability of mouse peritoneal macrophages to be infected with *R. tsutsugamushi*.

The results with RAW264.7 cells differ from the results with L929 cells (Section II.C), since crude lymphokine, but not rIFNγ, suppresses the ability of fibroblastic L929 cells to be initially infected with *R. prowazekii* (34) (Table 1). One possible explanation for the data is that there is another factor in the crude lymphokines which, together with the IFNγ, is necessary for inhibition of the initial rickettsial infection. Production of this factor by the RAW264.7 cells in response to IFNγ would explain the fact that rIFNγ inhibits the initial rickettsial infection in the RAW264.7 cells. Experiments are in progress to explore these possibilities.

E. Toxicity of IFNγ and *R. prowazekii* for Macrophage-Like Cells with a Defect in Oxygen Metabolism

Because IFNγ enhances macrophage oxidative metabolism in response to stimulants such as phorbol myristate acetate (38), it seemed reasonable to investigate the possible involvement of the macrophage respiratory burst in the rapid killing

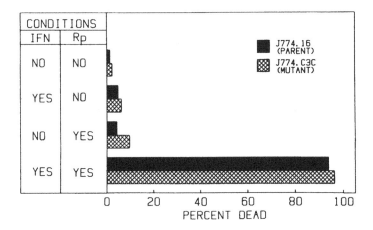

Figure 3 Toxicity of IFNγ and *R. prowazekii* for macrophage-like J774.16 cells and J774.C3C cells. X-irradiated cells were planted in 24-well tissue culture plates. Before addition of the cells, coverslips were placed in those wells that were to be stained for rickettsiae. After incubation overnight, murine rIFNγ (15 U/ml) was added to some of the wells, and the cells were incubated for an additional 24 hr. After washing, the cells were mock-infected or infected with *R. prowazekii* for 1 hr at 34°C. After additional washing, the coverslips were removed and stained for determination of the initial rickettsial infection. The remaining wells were incubated for 4 hr and were then stained with trypan blue. Each bar represents the mean of two experiments. The standard errors averaged 12% of the mean values. In the untreated and IFNγ-treated J774.16 cell cultures, respectively, 97 ± 1% and 98 ± 1% of the cells were infected and there were 8.0 ± 1.0 and 8.2 ± 0.4 rickettsiae per infected cell. In the untreated and IFNγ-treated J774.C3C cell cultures, respectively, 94 ± 3% and 98 ± 1% of the cells were infected and there were 6.5 ± 0.5 and 9.0 ± 0.7 rickettsiae per infected cell.

of macrophage-like cells by IFNγ and *R. prowazekii*. This problem was studied by using the mouse macrophage-like cell line J774.16 and a variant macrophage-like cell line derived from it, J774.C3C (39,40). These cell lines were kindly given to us by Dr. Barry R. Bloom. After stimulation, J774.16 cells produce superoxide and hydrogen peroxide and metabolize increased quantities of glucose by way of the hexose monophosphate shunt; however, J774.C3C cells do not do so (39,40). The defect in the cell line J774.C3C is thought to be associated with a b-type cytochrome that plays a role in the generation of superoxide (40). As expected, both untreated and IFNγ-treated J774.16 cells produced superoxide after exposure to phorbol myristate acetate or opsonized zymosan, but neither untreated nor IFNγ-treated J774.C3C cells produced superoxide after exposure to these stimulants (data not shown). Thus, J774.C3C cells, unlike the parent cell line, do not undergo a respiratory burst after exposure to stimulants.

Like RAW264.7 cells, both J774.16 cells and J774.C3C cells were killed after treatment with IFNγ and infection with *R. prowazekii* (Fig. 3). Therefore the function(s) that is defective in J774.C3C cells is not required for the killing of macrophage-like cells by IFNγ and *R. prowazekii*.

IV. SUMMARY, CONCLUSIONS, AND PERSPECTIVE

The interaction of *R. prowazekii* and host cells is altered dramatically by crude lymphokines or rIFNγ in at least three ways. First, lymphokine or rIFNγ-treated cells (both fibroblasts and macrophages) become poor host cells for supporting growth of *R. prowazekii* (26,31,32,35). Second, some of the IFNγ-treated, *R. prowazekii*-infected host cells are killed (31,35). This killing requires both IFNγ and rickettsiae and occurs very rapidly with mouse macrophage-like cells and more slowly with mouse fibroblasts (35) (Section II.E.). Finally, in some instances, treatment with lymphokines or rIFNγ inhibits the ability of cells to be initially infected with *R. prowazekii* (34,35). Treatment with *either* crude lymphokines or rIFNγ at high concentrations inhibits the ability of mouse macrophage-like cells to be initially infected with *R. prowazekii* (35); however, only crude lymphokines (and not rIFNγ) are effective at inhibiting the ability of mouse fibroblasts to be initially infected with *R. prowazekii* (34).

Little is known about the mechanisms by which IFNγ alters the interaction of *R. prowazekii* and host cells. The antirickettsial effect of IFNγ manifests species preference and requires protein synthesis by the host cell (19,32); therefore it is unlikely that the IFNγ itself has a direct antirickettsial action. It is reasonable to hypothesize that proteins produced by the host cell after IFNγ treatment play some role in the antirickettsial effect of IFNγ. For example, an IFNγ-induced protein might be toxic for the rickettsiae, or might cause the host cell to produce a material that is toxic for the rickettsiae. Alternatively, an

IFNγ-induced protein might alter the host cell or induce the production of a material that alters the host cell in such a way that the growth requirements of the rickettsiae are not satisfied. Since murine IFNγ has a marked inhibitory effect on the growth of *R. prowazekii* (32) whereas murine type I IFN has little effect (28), proteins that are induced only by IFNγ or proteins that are induced more strongly by IFNγ than by type I IFNs are likely to be important in the inhibition of rickettsial growth.

The possibility that the depletion of a particular amino acid(s) in the IFNγ-treated host cells might explain the inhibition of rickettsial growth has been examined (33). The recent finding that the serine and glycine pools in Vero cells are too low to support the growth of *R. prowazekii*, although the host cell grows fine, provides an example of how important such changes might be (41). In the case of the intracellular protozoan, *Toxoplasma gondii*, and the intracellular bacterium, *Chlamydia psittaci*, lack of the amino acid tryptophan is the mechanism for inhibition of parasite growth in IFNγ-treated human cells (36,42). In contrast, in IFNγ-treated mouse L929 cells, none of the amino acids measured (including tryptophan) was severely depleted, and there was no evidence to support amino acid deprivation as the mechanism for inhibition of the growth of *R. prowazekii* (33). Since tryptophan is not degraded in IFNγ-treated mouse cells (33), other mechanisms must also explain the inhibitory effects of IFNγ on the growth of other intracellular microorganisms in these cells. In IFNγ-treated human fibroblasts, tryptophan is severely depleted and cannot be detected; however, when the tryptophan pool is reconstituted in the IFNγ-treated cells, *R. prowazekii* does not grow (33). None of the other amino acids examined was dramatically lower in IFNγ-treated human fibroblasts (33). It is reasonable to assume that the depletion of tryptophan and another antirickettsial mechanism *both* play a role in inhibiting rickettsial growth in IFNγ-treated human fibroblasts. It thus appears that different mechanisms of inhibition by IFNγ operate in different types of cells and against different parasites.

A wide range of microorganisms (both procaryotic and eucaryotic) which have different sites of intracellular growth (cytoplasm, vacuoles, and phagolysosomes) are inhibited by IFNγ in host cells that are not specialized for the phagocytosis and destruction of microorganisms (27,32,36,42-44). It is not surprising that different mechanisms of inhibition by IFNγ would operate against different parasites, since the site of an intracellular parasite's growth would determine the parasite's exposure, and the parasite's metabolism would determine its sensitivity to a given inhibitory mechanism. It is interesting that in human cells, tryptophan supplementation reverses the IFNγ-induced inhibition of *T. gondii* and *C. psittaci*, both of which grow within vacuoles inside the host cell (36,42); however, the inhibition of the cytoplasmic parasite, *R. prowazekii*, is not reversed (33). Whether the rickettsia is exposed to an additional inhibitory mechanism because of its cytoplasmic location, or whether it is simply sensitive to another inhibitory mechanism to which *T. gondii* and *C. psittaci* are resistant is not known.

The significance of the toxicity of the combination of IFNγ treatment and rickettsial infection for host cells (35) (Section II.E.) and how it relates to the antirickettsial action of IFNγ is unknown. From the standpoint of the host, killing of the rickettsiae without damage to the host cell would be the most desirable outcome. If killing of infected host cells after exposure to IFNγ occurs in vivo, it could contribute to tissue damage in the infected host.

The mechanisms of host cell killing by IFNγ and *R. prowazekii* are unknown. In cultures of IFNγ-treated mouse macrophage-like cells, infection with *R. prowazekii* is followed by killing of many of the host cells within several hours (35). However, in cultures of IFNγ-treated mouse fibroblasts, killing of the cells after infection with *R. prowazekii* does not occur until much later (35) (Section II.E.). The ability of macrophage-like cells to undergo a respiratory burst is unlikely to be required for the killing of these cells by IFNγ and *R. prowazekii* because a macrophage-like cell line that is defective in this activity is also killed (Section III.E.). Perhaps the mechanisms of cell killing are similar in both fibroblasts and macrophage-like cells.

How high concentrations of lymphokines (or rIFNγ, in macrophage-like cells) suppress the ability of cells to become infected with *R. prowazekii* is also unknown (34,35). This type of activity, which was reported by others in studies with lymphokine-treated mouse macrophages and *R. tsutsugamushi* (17), has also been observed with other microorganisms (45,46).

These in vitro studies have demonstrated that IFNγ alters dramatically the interaction of *R. prowazekii* and host cells, making the host cells unsuitable for supporting rickettsial growth. In addition, these studies have defined experimental systems in which the mechanisms of action of IFNγ against *R. prowazekii* can be explored. IFNγ has the potential for acting on *R. prowazekii* growing within its host cells in vivo (endothelial cells and macrophages). In addition, IFNγ might function in host defense against rickettsial infections by modulating the activities of other cells of the immune system. Future studies must not only determine the mechanisms of action of IFNγ against *R. prowazekii* growing within its host cells, but must also assess the relative importance of this type of action by IFNγ in experimental rickettsial infections.

ACKNOWLEDGMENTS

This research was supported by U. S. Public Health Service grant AI-19659 from the National Institute of Allergy and Infectious Diseases.

The authors thank Genentech, Inc. (South San Francisco, CA) for providing the human and murine rIFN-γ used in this research.

REFERENCES

1. Weiss, E. 1982. The biology of rickettsiae. *Ann. Rev. Microbiol.* *36*:345–370.
2. Weiss, E., and J.W. Moulder. 1984. Genus I. Rickettsia. In *Bergey's Manual of Systematic Bacteriology*, Vol. 1, N.R. Krieg and J.G. Holt (Eds.). Williams and Wilkins, Baltimore, MD, pp. 688–698.
3. Wolbach, S.B., J.C. Todd, and F.W. Palfrey. 1922. *The Etiology and Pathology of Typhus*. Harvard University Press, Cambridge.
4. Gambrill, M.R., and C.L. Wisseman, Jr. 1973. Mechanisms of immunity in typhus infections. II. Multiplication of typhus rickettsiae in human macrophage cell cultures in the nonimmune system: Influence of virulence of rickettsial strains and of chloramphenicol. *Infect. Immun.* *8*:519–527.
5. Beaman, L., and C.L. Wisseman, Jr. 1976. Mechanisms of immunity in typhus infections. VI. Differential opsonizing and neutralizing action of human typhus rickettsia-specific cytophilic antibodies in cultures of human macrophages. *Infect. Immun.* *14*:1071–1076.
6. Yeomans, A., J.C. Snyder, and A.G. Gilliam. 1945. The effects of concentrated hyperimmune rabbit serum in louse borne typhus. *J. Am. Med. Assoc.* *129*:19–24.
7. Murphy, J.R., C.L. Wisseman, Jr., and P. Fiset. 1979. Mechanisms of immunity in typhus infection: Adoptive transfer of immunity to *Rickettsia mooseri*. *Infect. Immun.* *24*:387–393.
8. Murphy, J.R., C.L. Wisseman, Jr., and P. Fiset. 1980. Mechanisms of immunity in typhus infections: Analysis of immunity to *Rickettsia mooseri* infection of guinea pigs. *Infect. Immun.* *27*:730–738.
9. Crist, A.E., Jr., C.L. Wisseman, Jr., and J.R. Murphy. 1984. Characteristics of lymphoid cells that adoptively transfer immunity to *Rickettsia mooseri* infection in mice. *Infect. Immun.* *44*:55–60.
10. Shirai, A.P., J. Catanzaro, S.M. Phillips, and J.V. Osterman. 1976. Host defenses in experimental scrub typhus: Role of cellular immunity in heterologous protection. *Infect. Immun.* *14*:39–46.
11. Jerrells, T.R., B.A. Palmer, and J.G. MacMillan. 1984. Cellular mechanisms of innate and acquired immunity to *Rickettsia tsutsugamushi*. In *Microbiology–1984*, L. Leive and D. Schlessinger (Eds.). American Society for Microbiology, Washington, D.C., pp. 277–281.
12. Kenyon, R.H., and C.E. Pedersen, Jr. 1980. Immune responses to *Rickettsia akari* infection in congenitally athymic nude mice. *Infect. Immun.* *28*:310–313.
13. Kokorin, I.N., E.A. Kabanova, E.M. Shirokova, G.E. Abrosimova, N.N. Rybkina, and V.I. Pushkareva. 1982. Role of T lymphocytes in *Rickettsia conorii* infection. *Acta Virol.* *26*:91–97.
14. Rollwagen, F.M., A.J. Bakun, C.H. Dorsey, and G.A. Dasch. 1985. Mechanisms of immunity to infection with typhus rickettsiae: Infected fibroblasts bear rickettsial antigens on their surfaces. *Infect. Immun.* *50*:911–916.
15. Rollwagen, F.M., G.A. Dasch, and T.R. Jerrells. 1986. Mechanisms of im-

munity to rickettsial infection: Characterization of a cytotoxic effector cell. *J. Immunol. 136*:1418–1421.

16. Nacy, C.A., and J.V. Osterman. 1979. Host defenses in experimental scrub typhus: Role of normal and activated macrophages. *Infect. Immun. 26*: 744–750.

17. Nacy, C.A., and M.S. Meltzer. 1979. Macrophages in resistance to rickettsial infection: Macrophage activation in vitro for killing of *Rickettsia tsutsugamushi. J. Immunol. 123*:2544–2549.

18. Nacy, C.A., E.J. Leonard, and M.S. Meltzer. 1981. Macrophages in resistance to rickettsial infections: Characterization of lymphokines that induce rickettsiacidal activity in macrophages. *J. Immunol. 126*:204–207.

19. Winkler, H.H., and J. Turco. 1984. Role of lymphokine (γ interferon) in host defense against *Rickettsia prowazekii*. In *Microbiology–1984*, L. Leive and D. Schlessinger (Eds.). American Society for Microbiology, Washington, D.C., pp. 273–276.

20. Stewart, W.E., II, J.E. Blalock, D.C. Burke, C. Chany, J.K. Dunnick, E. Falcoff, R.M. Friedman, G.J. Galasso, W.K. Joklik, J.T. Vilcek, J.S. Youngner, and K.C. Zoon. 1980. Interferon nomenclature. *Nature 286*:110.

21. Hopps, H.E., S. Kohno, M. Kohno, and J.E. Smadel. 1964. Production of interferon in tissue cultures infected with *Rickettsia tsutsugamushi. Bact. Proc.* 115–116.

22. Kazàr, J. 1966. Interferon-like inhibitor in mouse sera induced by rickettsiae. *Acta Virol. 10*:277.

23. Kazàr, J., P.A. Krautwurst, and F.B. Gordon. 1971. Effect of interferon and interferon inducers on infections with a nonviral intracellular microorganism, *Rickettsia akari. Infect. Immun. 3*:819–824.

24. Vilcek, J., and R.I. Jahiel. 1970. Action of interferon and its inducers against nonviral infectious agents. *Arch. Intern. Med. 126*:69–77.

25. Stewart, W.E., II. 1981. *The Interferon System*, second edition, Springer-Verlag, New York, pp. 38–44 and 232–233.

26. Turco, J., and H.H. Winkler. 1983. Inhibition of the growth of *Rickettsia prowazekii* in cultured fibroblasts by lymphokines. *J. Exp. Med. 157*:974–986.

27. Turco, J., H.A. Thompson, and H.H. Winkler. 1984. Interferon-γ inhibits growth of *Coxiella burnetii* in mouse fibroblasts. *Infect. Immun. 45*:781–783.

28. Turco, J., and H.H. Winkler. 1983. Comparison of the properties of antirickettsial activity and interferon in mouse lymphokines. *Infect. Immun. 42*:27–32.

29. Jerrells, T.R., J. Turco, H.H. Winkler, and G.L. Spitalny. 1986. Neutralization of lymphokine-mediated antirickettsial activity of fibroblasts and macrophages with monoclonal antibody specific for murine interferon gamma. *Infect. Immun. 51*:355–359.

30. Spitalny, G.L., and E.A. Havell. 1984. Monoclonal antibody to murine gamma interferon inhibits lymphokine-induced antiviral and macrophage tumoricidal activities. *J. Exp. Med. 159*:1560–1565.

31. Wisseman, C.L., Jr., and A. Waddell. 1983. Interferonlike factors from antigen- and mitogen-stimulated human leukocytes with antirickettsial and cytolytic actions on *Rickettsia prowazekii* infected human endothelial cells, fibroblasts, and macrophages. *J. Exp. Med. 157*:1780-1793.

32. Turco, J., and H.H. Winkler. 1983. Cloned mouse interferon-γ inhibits the growth of *Rickettsia prowazekii* in cultured mouse fibroblasts. *J. Exp. Med. 158*:2159-2164.

33. Turco, J., and H.H. Winkler. 1986. Gamma interferon induced inhibition of the growth of *Rickettsia prowazekii* in fibroblasts cannot be explained by the degradation of tryptophan or other amino acids. *Infect. Immun. 53*: 38-46.

34. Turco, J., and H.H. Winkler. 1985. Lymphokines, interferon-γ and *Rickettsia prowazekii*. In *Rickettsiae and Rickettsial Diseases, Proceedings of the Third International Symposium*, J. Kazàr (Ed.). Publishing House of the Slovak Academy of Sciences, Bratislava, Czechoslovakia, pp. 211-218.

35. Turco, J., and H.H. Winkler. 1984. Effect of mouse lymphokines and cloned mouse interferon-γ on the interaction of *Rickettsia prowazekii* with mouse macrophage-like RAW264.7 cells. *Infect. Immun. 45*:303-308.

36. Pfefferkorn, E.R. 1984. Interferon γ blocks the growth of *Toxoplasma gondii* in human fibroblasts by inducing the host cells to degrade tryptophan. *Proc. Natl. Acad. Sci. (USA) 81*:908-912.

37. Turco, J., and H.H. Winkler. 1982. Differentiation between virulent and avirulent strains of *Rickettsia prowazekii* by macrophage-like cell lines. *Infect. Immun. 35*:783-791.

38. Nathan, C.F., H.W. Murray, M.E. Wiebe, and B.Y. Rubin. 1983. Identification of interferon-γ as the lymphokine that activates human macrophage oxidative metabolism and antimicrobial activity. *J. Exp. Med. 158*:670-689.

39. Damiani, G., C. Kiyotaki, W. Soeller, M. Sasada, J. Peisach, and B.R. Bloom. 1980. Macrophage variants in oxygen metabolism. *J. Exp. Med. 152*:808-822.

40. Kiyotaki, C., J. Peisach, and B.R. Bloom. 1984. Oxygen metabolism in cloned macrophage cell lines: Glucose dependence of superoxide production, metabolic and spectral analysis. *J. Immunol. 132*:857-866.

41. Austin, F.E., J. Turco, and H.H. Winkler. 1987. *Rickettsia prowazekii* requires host cell serine and glycine for growth. *Infect. Immun. 55*:240-244.

42. Byrne, G.I., L.K. Lehmann, and G.J. Landry. 1986. Induction of tryptophan catabolism is the mechanism for gamma-interferon-mediated inhibition of intracellular *Chlamydia psittaci* replication in T24 cells. *Infect. Immun. 53*:347-351.

43. de la Maza, L.M., E.M. Peterson, C.W. Fennie, and C.W. Czarniecki. 1985. The anti-chlamydial and anti-proliferative activities of recombinant murine interferon-γ are not dependent on tryptophan concentrations. *J. Immunol. 135*:4198-4200.

44. Sethi, K.K., Y. Omata, and H. Brandis. 1985. Contribution of immune interferon (IFN-γ) in lymphokine induced anti-toxoplasma activity: Studies with recombinant murine IFN-γ. *Immunobiology 170*:270-283.

45. Oster, C.N., and C.A. Nacy. 1984. Macrophage activation to kill *Leishmania tropica*: Kinetics of macrophage response to lymphokines that induce antimicrobial activities against amastigotes. *J. Immunol. 132*:1494–1500.

46. Horwitz, M.A., and S.C. Silverstein. 1981. Activated human monocytes inhibit the intracellular multiplication of Legionnaires' disease bacteria. *J. Exp. Med. 154*:1618–1635.

6

IFNγ as an Effector Molecule in Antirickettsial Immunity

Thomas R. Jerrells[*]/ Walter Reed Army Institute of Research, Washington, D.C.

I. IMMUNITY TO RICKETTSIAE

The rickettsiae are defined simply as obligate intracellular bacteria. As obligate intracellular parasites, they present a unique challenge to the immune system. In all rickettsial infections, it has been shown that both humoral and cellular immunity are associated with immunity to reinfection. Although it is not completely clarified, most investigators will agree that the major immune mechanism responsible for controlling rickettsial growth in vivo and clearing rickettsiae from an infected host is mediated by cellular mechanisms. In this section, I will present evidence for this statement, although the reader must be aware of the fact that evidence obtained using different species within the genus *Rickettsia* will be generalized, and this is not without a certain amount of danger. Where possible, differences in the rickettsiae will be discussed as they might affect the conclusions.

It has been argued that since antibody cannot enter viable cells, it is unlikely that antibody plays a direct effector role in rickettsial immunity. The fact that athymic (nu/nu) mice, which due to the lack of a thymus cannot generate a cellular immunity, are susceptible to overwhelming and ultimately lethal infections with spotted fever group rickettsiae (*Rickettsia conorii*) and a typhus group rickettsia (*R. typhi*) in spite of a marked IgM and IgG$_3$ antibody response, supports this conclusion (1-5). Other evidence supporting the notion that immunity is not directly mediated by antibody includes an inability to transfer passively protective immunity with immune sera (5,6) although some conflicting data have been reported (7), and the development of a local and systemic infec-

---

Present affiliation: University of Texas Medical Branch, Galveston, Texas

117

tion in monkeys following a secondary infection with *R. tsutsugamushi* in spite
of a rapid IgG anamnestic antibody response (8). Furthermore, susceptibility to
rickettsial infection has been shown to be associated with higher antibody titers
after infection of susceptible strains of mice with a lethal dose of rickettsiae (9).
These data suggest that once rickettsiae gain entrance to host cells, they are relatively protected from the effect of antibody.

On the other hand, protective immunity can be accomplished using spleen
cell suspensions from immune animals (2,6,10-12), and it has been shown that
the ability to transfer immunity can be eliminated by treatment of immune cells
with anti-Thy 1.2 antibody and complement, suggesting that passive immunity
is accomplished by the thymus-dependent lymphocyte or T cell (2). The Thy 1.2
cell population includes both the T cells that mediate "help" for antibody production to T-dependent antigens and T cells that are suppressor or cytotoxic
cells.

These two populations can be distinguished readily by surface markers such
as the L3T4 marker on T-helper cells (13) and the Lyt 2 marker on suppressor/
cytotoxic cells (14,15). In one study, it has been shown that the Lyt 2 cell population alone did not provide passive protection (2), and in recent experiments,
administration of anti L3T4 monoclonal antibody according to published
protocols (16) eliminated the ability of an animal to resist rickettsial infection
(Jerrells, T.R., unpublished data). Although the presence of cytotoxic T cells
specific for cells infected with rickettsiae has been reported recently (17) the
majority of data would suggest that the L3T4 T-helper cell is the major cell type
responsible for passive protection of naive animals as well as in the genetically
determined resistance of mice to infection that is dependent on the development
of a cellular immunity (1,2,9,18-20).

One important function of the L3T4 helper cell is the ability to produce
lymphokines in response to specific antigen stimulation of these cells. The
lymphokines that are produced include factors that interact with other T and B
cells (IL-2 and IL-4 [BSF-1]) to produce T-cell growth and B-cell activation to
provide antibody production (21). Previous studies using *Coxiella burnetii* (22)
or various species of *Rickettsia* (23-25) have suggested that one important
aspect of lymphokine biology is the ability to activate macrophages to kill
phagocytized rickettsiae. The molecule(s) that is responsible for this activation
has been collectively termed macrophage activating factors (MAF).

Another lymphokine that is being recognized for its importance in the cellular immune mechanisms is interferon gamma (IFNγ). This molecule, like all
lymphokines, is produced by the T lymphocyte and the majority, although not
all, of the IFNγ is produced by the L3T4 helper lymphocyte. It has been shown
that IFNγ is involved in effector mechanisms including macrophage activation
to mediate tumor cell killing, antimicrobial activity, and more recently, antirickettsial activity (26-28). This molecule also is involved in the induction of the

immune response by inducing macrophages and other antigen-presenting cells to express class II major histocompatability complex (MHC) molecules (29). The class II MHC antigen complex in the mouse is termed the Ia antigen and it is a necessary recognition element involved in T-cell receptor recognition of antigen after processing and presentation by the antigen-presenting cell (30).

In the following sections, data will be presented showing the importance of IFN in immunity to rickettsial infection of experimental animals at both the effector level and the level of induction of the immune response to rickettsial antigens, Studies will be reported which describe mechanisms of down regulation of IFNγ effects and the effect of this down regulation on the outcome of the infection.

II. IN VIVO PRODUCTION OF INTERFERON-GAMMA BY MICE IMMUNE TO RICKETTSIA

Early studies designed to investigate an association of IFNγ and antirickettsial immunity used a murine model system of *R. tsutsugamushi* infection where a subcutaneous infection produces a long-lived immunity to rechallenge (20,31, 32). In the original studies of IFNγ in this laboratory, mice that were immune to rechallenge with rickettsia (28 days after the immunizing infection) were administered various amounts of irradiated rickettsiae ranging from 10 μg to 1 mg and sera collected at time intervals from 1 to 4 hr after antigen administration (33). It was found that a significant amount of antiviral activity, as assayed by inhibition of cytopathic effects of vesicular stomatitis virus, was demonstrable in the sera of immune mice but not in the sera obtained from nonimmunized animals administered equivalent amounts of antigen. The amount of antiviral activity was dependent on the dose of antigen administered and was consistently found to reach peak levels 4 hr after antigen administration (33).

Based on the kinetics of production, sensitivity to treatment of pH 2.0, and neutralization with a polyclonal antiserum with specificity for IFNγ we concluded that the characteristics of the antiviral activity in our model system were consistent with IFNγ. This idea was further suggested by the finding that IFNγ activity was absent from nu/nu mice infected with the same rickettsia and treated with antibiotics, which has been shown previously to prevent lethal rickettsial infections but not the development of the immune response (34).

To determine the relationship of IFNγ production and the development of protective immunity we immunized groups of mice with *R. tsutsugamushi* subcutaneously as before, and at various intervals after immunization, animals were either challenged with a potentially lethal dose of rickettsiae or administered rickettsial antigen intravenously and sera was examined for IFNγ. These studies demonstrated that a clear association existed between the ability to produce IFNγ in response to rickettsial antigen and the ability to resist an otherwise lethal infection (33,35).

To determine if IFNγ was involved in rickettsial clearance from immune mice, sera were collected from immune mice at various times after challenge with a viable inoculum of rickettsiae. It was found that a bimodal peak of antiviral activity was demonstrable in sera from the challenged animals: an initial peak that was detectable relatively early after challenge (3–5 days) and a peak that occurred at a time when rickettsial clearance had been reported previously to be occurring (20). The early peak is thought to be due to stimulation of immune cells by the expanding rickettsial infection and the later peak at 14 days after infection probably represents the ongoing immune response.

Another important aspect of IFNγ in terms of the immune response is the induction of class II MHC antigens on antigen-presenting cells that are required for T-helper cell recognition of antigen on the presenting cell. In murine systems, these surface proteins are termed I region-associated (Ia) antigens.

The importance of class II antigen-bearing cells in rickettsial immunity is suggested in studies examining the nature of the macrophage inflammation resulting from infection of strains of mice genetically resistant or susceptible to infection with *R. tsutsugamushi*. It was shown that an intraperitoneal infection of susceptible mice resulted in a large inflammatory response containing a relatively large proportion of macrophages, in contrast to the infection in resistant animals that was characterized as having a significantly lesser inflammatory response (19). Further examination of the inflammatory exudate revealed that the response of the resistant animals was notable in the abundance of Ia antigens in contrast to the essentially Ia antigen-negative macrophage response of the susceptible mice (36).

Resistance to rickettsial infection in this system has been shown to be due in part to the ability of these animals to generate a cellular immune response as evidenced by susceptibility of nu/nu animals on a resistant background (1,2) and susceptibility of resistant mice after adult thymectomy, irradiation, and bone marrow reconstitution (Jerrells, T.R., unpublished data). These data would suggest that one important aspect of genetic resistance to rickettsial infection is the ability to produce IFNγ, and consequently, Ia antigen induction on macrophages so that antigen presentation can occur.

III. IN VITRO PRODUCTION OF INTERFERON-GAMMA

At the time the above experiments were performed, it was not clear what the source of IFNγ was, although it was thought that IFNγ was produced by the T cell. To address this question in our model system of rickettsial immunity, we examined the ability of spleen cells to produce IFNγ in vitro after stimulation with specific antigen. Using this approach, it was shown that cells from immune animals produced high levels of IFN into culture supernatants with kinetics and biochemical criteria compatible with IFNγ (35). By both negative (depletion with anti-Thy 1.2 and complement) and positive (nylon-wool nonadherent cells) selection methodologies, it was established that at least the major source of IFNγ was the T cell. The antigenic specificity of IFNγ production by immune cells was essentially the same as the specificity established using proliferation as the indicator of antigen recognition (32).

In order to investigate the relative importance of T-cell IFNγ production at the localized site of infection compared to IFNγ production after systemic spread of the infection, we compared the response of draining lymph node lymphocytes and the response of spleen cells to antigen. It was found that the lymph node lymphocytes from nodes draining an infection originating in the foot pad responded relatively early, and the peak activity of IFNγ production in these cultures occurred from 10 to 14 days after infection. The spleen cell response was evident later and peaked about 28 days after infection and approximated the development of resistance to reinfection. Lymph node cells were unresponsive at a time when the spleen cell response was at its peak, and it was found that the spleen cell response was long-lived. These data suggest that local immune responses occur, but are not able to prevent the spread of rickettsiae, and only after systemic immunization, as evidenced by spleen cell responses to antigen, will rickettsiae be cleared from the host. It is interesting to note that only after the development of systemic immunity will animals resist challenge, suggesting the necessity for rickettsial replication to generate a long-lived immune response.

Experiments were designed to test these findings from the murine model in a larger animal model system that more closely approximates scrub typhus in humans. To accomplish this, we infected cynomolgus monkeys, which have been shown to be an acceptable primate model of scrub typhus (37), with *R. tsutsugamushi* and followed the development of the cell-mediated immune response. One measure that we chose to use was serum levels of IFNγ and IFNγ production by isolated peripheral blood mononuclear cells. The data generated in this study closely approximated earlier findings in the murine model, namely that IFNγ was detectable in the sera and recovery from infection was associated with the ability to produce IFNγ. Furthermore, secondary immune responses to this rickettsiae were shown to be associated with the rapid development of the ability to produce IFNγ in cultured peripheral blood mononuclear cells (8).

IV. MACROPHAGE ACTIVATION BY INTERFERON-GAMMA

Previous work by Nacy and Meltzer showed that macrophage activation for anti-rickettsial activity was important not only in vitro but also in vivo (38), and that this activation was accomplished by immunological mechanisms. Based on these data and the data that suggested IFNγ to be an important mediator of macrophage activation, studies were designed to examine the role of IFNγ in the ob-

served macrophage activation to an antirickettsial state. These studies were greatly aided by the development by Spitalny and Havell (27) of a monoclonal antibody produced in a rat that had neutralizing specificity for murine IFNγ and the
development of a recombinant source of IFNγ.

Initially, we used a recombinant murine IFNγ preparation, kindly supplied
by Dr. Patrick Gray, to treat macrophages prior to infection with rickettsiae. It
was found that the recombinant product was as active in macrophage activation
as was a lymphokine preparation adjusted to equivalent concentrations of IFN.
It was further suggested that IFNγ was active in macrophage activation when it
was shown that treatment of macrophages with a lymphokine preparation that
had been passed over a column containing bound monoclonal antimurine IFNγ
had no activating capacity. The addition of soluble antibody directly to cultures
of macrophages and lymphokine also resulted in a loss of activation, but it required considerably more monoclonal antibody for neutralization, and complete
neutralization of macrophage activation was not always achieved (26).

It is unclear why differences exist in the ability of bound antibody and
soluble antibody to neutralize IFNγ-mediated macrophage activation, although it
is possible that the macrophage receptor for IFNγ has a greater affinity for IFNγ
than the antibody or that IFN-antibody complex is capable of activating macrophages. Removal of IFNγ activity by passage over a column of bound antibody
would completely remove IFNγ from a preparation and eliminate these
problems. It also must be remembered that the activity of monoclonal antibody
is measured by neutralization of antiviral activity, which is a relatively insensitive measure of IFNγ activity. Perhaps macrophage activation or Ia-antigen induction is a more sensitive parameter of IFNγ activity than antiviral activity as
has been suggested by Vogel et al. (39). Nevertheless, it is clear from this study
that the major, if not the exclusive lymphokine that activates macrophages for
antirickettsial activity in vitro is IFNγ.

In a recent study, the monoclonal antimurine IFNγ antibody that was used
in the previous studies was administered to animals and the ability of these
treated animals to resist rickettsial infection was evaluated (40). To provide a
source of concentrated antibody, the hybridoma was grown in nu/nu mice and
the ascites collected. This material was found to contain sufficiently high level of
anti-IFNγ activity (4.2×10^6 neutralizing units/ml) that was thought to allow
depletion of activity in vivo. The antibody was injected into mice that are
genetically resistant to *R. conorii* (9) on days 0,1,3,5, and 7 relative to infection
with *R. conorii* and evaluated for effects of the infection. It was found that injection of the antibody alone had no demonstrable effect on the mice; however,
if antibody-treated animals were infected with rickettsiae, all the treated and infected animals showed signs of infection including ruffled fur, apathy, anorexia,
etc. Additionally, approximately 50% of the treated infected group died 10-12
days after infection. None of the animals treated with monoclonal anti-IFNγ or

with an irrelevant monoclonal antibody (anti-vero cell) alone showed signs of infection or died.

Using a technique of quantitation of rickettsiae in tissue sections by direct immunofluorescence (41) it was shown that mice treated with monoclonal anti-IFNγ had significantly higher numbers of rickettsiae in the spleen, liver, and kidney than mice treated with anti-vero cell antibody. Sera from the groups were assayed for IFNγ levels and it was found that the sham-treated animals produced demonstrable levels of IFNγ at the appropriate time after infection, but the animals treated with anti-IFNγ did not produce circulating levels of IFNγ. The anti-IFNγ-treated animals were shown to have detectable levels of neutralizing anti-IFNγ antibody, showing that sufficient monoclonal antibody was given.

Interestingly, some animals that were treated with the monoclonal anti-IFNγ did survive and clear the rickettsiae from the spleen, liver, and kidney. It was shown that administration of monoclonal anti-IFNγ past 7 days after infection did not affect the ability of these animals to survive the infection. This would suggest that an IFNγ-independent antirickettsial mechanism exists in these mice as previously suggested for R. tsutsugamushi (20) but no data have been generated to further elucidate this observation.

In related studies, it has been shown that treatment of resistant mice with anti-IFNγ monoclonal antibody resulted in the elimination of the Ia$^+$ macrophage influx associated with infection with the appropriate strain of R. tsutsugamushi (42). In the same study, it was shown that elimination of the L3T4 T-cell subset by injection of monoclonal anti-L3T4 (16) was associated with an inability to resist infection. This treatment would remove the T cells that produce IFNγ but it must be noted that ability to produce antibody to the T-dependent scrub typhus rickettsiae (1,2) also would be impaired. Further studies are required to determine the role of IFNγ-independent mechanism in antirickettsial immunity including the role of antibody. In summary, IFNγ has been shown to be a major immunological mechanism operative in antirickettsial immunity using in vitro, and importantly, in vivo studies.

V. REGULATION OF INTERFERON-GAMMA ACTION

Depending on the strain of mouse and the strain of R. tsutsugamushi, an intraperitoneal infection will produce either a systemic lethal infection or a chronic immunizing infection (19,20,31–33,35,36). This model system has been used extensively in this laboratory to examine the immunoregulatory effects resulting from each type of infection (43). It was found that an acute lethal infection of C3H/He mice was associated with an early inability of spleen cells to produce IFNγ. However, if spleen cells or mesenteric lymph node cells that were obtained from mice late in the infection were stimulated in vitro with antigen, it was found that these cells produced significantly more IFNγ than cells from resistant mice at the same time after infection. In a more direct comparison, it was shown that stimulation of these two cell populations from lethally infected mice or uninfected mice with concanavalin A, a nonspecific T-cell stimulus, also resulted in a hyperproduction of IFNγ by the cells from the lethally infected animals (Palmer, B.A. and T.R. Jerrells, unpublished data). To examine if there was a defect in the ability of IFNγ to reach the primary site of the infection, we examined peritoneal washouts and ascites obtained from mice that had been infected intraperitoneally for IFNγ levels. Surprisingly, we found that the ascites obtained from animals late in a lethal infection contained high levels of IFNγ but the inflammatory macrophage population from the same peritoneal washout was essentially negative for Ia antigen expression (44). In contrast, as discussed earlier, the macrophages from resistant mice were found to be mostly Ia positive at this time (36).

The kinetics of the Ia antigen-bearing macrophage influx in susceptible mice showed that the Ia antigen expression on inflammatory macrophages occurred but this relatively early peak of Ia$^+$ macrophages declined rapidly and apparently was being inhibited (36). Taken together, these data suggested that somehow the effect of IFNγ, at least at the level of Ia antigen expression, was being suppressed by mechanisms resulting from a progressive rickettsial infection.

Work in other laboratories has shown that several factors can inhibit Ia antigen expression on macrophages, including corticosteroids and prostaglandins (45–47). In initial studies to define mechanisms of suppression of Ia antigen expression in our system, we examined the ascites from lethally infected mice for levels of prostaglandin E_2 and corticosterone using sensitive radioimmunoassays. These studies showed that the progressive infection of susceptible mice was associated with increasing levels of corticosterone in the ascites and this corticosteroid was virtually absent in the resistant mice. The levels of PGE_2 also increased in lethal infections and it was further shown that inflammatory macrophages from lethally infected mice produced from 2 to 5 times more PGE_2 than an equivalent number of cells from resistant mice. Interestingly, it was found that the ascites would induce Ia antigens on the macrophage-like cell line J774 in spite of the PGE_2 and steroid levels in this material. When macrophages were removed from the peritoneal cavity of susceptible mice, washed, and incubated in IFNγ-containing lymphokine they still failed to express Ia antigen relative to expression by macrophages induced with thioglycollate, suggesting that factors produced by the macrophages and acting in close proximity may be important in the noted suppression.

In more recent studies, it was found that incubation of macrophages from susceptible animals with indomethacin, a prostaglandin inhibitor, resulted in partial but not complete restoration of Ia antigen expression (Jerrells, T.R., unpublished data). These data would suggest that the PGE_2 production by the inflammatory macrophage, perhaps in response to phagocytic stimuli provided by the phagocytosis of rickettsiae in the peritoneal cavity, was responsible for partial inhibition of Ia antigen expression. The failure to reverse this completely with indomethacin would suggest other factors were being produced by the inflammatory cell population.

Another potent inhibitor of Ia antigen induction by IFNγ has been shown to be α/β-type IFN (48). Relatively high levels of this type of IFN are present in the ascites from lethally infected mice (44) and, as discussed earlier, the ascites also contained high levels of IFNγ. It was found that the inflammatory cell population obtained from lethally infected mice produced α/β-type IFN for the most part in culture and relatively little IFNγ (44). Studies are in progress to identify the cell type that is producing the α/β-type IFN and also it is possible that factors produced by cells within the large influx of neutrophils late in the infection are mediating suppression of IFNγ action on the macrophages.

In early studies, Nacy and Groves (49) suggested that one possible reason for genetic susceptibility to scrub typhus rickettsiae was the inability of inflammatory macrophages to be activated by lymphokines to inhibit the replication of the rickettsiae. From the current studies, it would seem that the inability of the macrophages to be activated may largely result from inhibition of IFNγ action by corticosteroids and factors such as PGE_2 and α/β-type IFN produced by the inflammatory macrophages themselves and perhaps by the polymorphonuclear neutrophil. Further studies in this area will provide insight into regulatory mechanisms active in progressive rickettsial infections as well as basic mechanisms involved in the down regulation of IFNγ action on the macrophage.

VI. SUMMARY

The experimental studies described in this chapter have described the effect of IFNγ on immunological mechanisms involved in the clearance of rickettsiae from an infected host. In another chapter, a large amount of data is summarized showing the antirickettsial effect of IFNγ on cells outside of the immune response. It is very likely that the total effect of IFNγ includes the effect of this molecule not only on macrophages and other immunological cells but also on the endothelium and perhaps the epithelium at the initial site of rickettsial infection by an arthropod.

The reversal of resistance to rickettsial infection by depletion of IFNγ using a monoclonal antibody does not provide insight into the relative importance of these possible sites of IFNγ action, but has clearly established this lymphokine as a critical factor involved in antirickettsial immunity. These data indirectly define also the relative importance of the T-helper cell type that produces the majority of IFNγ. This molecule is involved in the activation of macrophages to mediate antirickettsial activity and this perhaps has in vivo correlation with clearance of rickettsiae from the blood by the cells of the reticuloendothelial system in the liver and spleen.

The afferent aspects of cellular immunity to rickettsial antigens also are critically dependent on IFNγ in that this molecule is involved in the induction of class II MHC antigens required for T-cell recognition of processed antigen. This aspect seems to be involved in the noted susceptibility to certain strains of inbred mice, since the lack of an Ia antigen-bearing macrophage population is associated with uncontrolled replication of rickettsiae.

In this system it has been shown that factors, including corticosteroids and prostaglandins and perhaps α/β-type IFN, are produced and down regulate the effects of IFNγ, and the overproduction of these factors is at least one contributing factor in the inability to control the infection. It also is clear that a great deal of further work is required to define completely the role of this and other lymphokines in antirickettsial immunity. The role of IFNγ in the development of the recently described cytotoxic T-cell activity is an important area of research, as is the definition of the relevance of the reported cytotoxic effect of IFNγ on infected macrophages and macrophage-like cells (50). If this IFNγ-induced cytolysis occurs in vivo there would be important ramifications on host defenses and perhaps the development of chronic infections with members of the genus *Rickettsia*.

REFERENCES

1. Jerrells, T.R., and C.S. Eisemann. 1983. Role of T-lymphocytes in production of antibody to antigens of *Rickettsia tsutsugamushi* and other *Rickettsia species. Infect. Immun.* 41:666–674.
2. Jerrells, T.R., B.A. Palmer, and J.G. MacMillan. 1984. Cellular mechanisms of innate and acquired immunity to *Rickettsia tsutsugamushi. Microbiology 1984*, 277–281.
3. Murata, M., and A. Kawamura, Jr. 1977. Restoration of the infectivity of *Rickettsia tsutsugamushi* to susceptible animals by passage in athymic nude mice. *Japan J. Exp. Med.* 47:385–391.
4. Kenyon, R.H., and C.E. Pedersen, Jr. 1980. Immune responses to *Rickettsia akari* in congenitally athymic nude mice. *Infect. Immun.* 28:310–313.
5. Humphres, R.C., and D.J. Hinrichs. 1981. Role for antibody in *Coxiella burnetii* infection. *Infect. Immun.* 31:641–645.
6. Shirai, A., P.J. Catanzaro, S.M. Phillips, and J.V. Osterman. 1976. Host defenses in experimental scrub typhus: role of cellular immunity in heterologous protection. *Infect. Immun.* 14:39–46.
7. Robinson, D.M., and D.L. Huxsoll. 1975. Protection against scrub typhus infection engendered by the passive transfer of immune sera. *Southeast Asian J. Trop. Med. Pub. Health* 6:477–482.
8. MacMillan, J.G., R.M. Rice, and T.R. Jerrells. 1985. Development of antigen specific cell-mediated immune responses after infection of cynomolgus

monkeys (*Macaca fasicularis*) with *Rickettsia tsutsugamushi. J. Infect. Dis. 152*:739–749.

9. Eisemann, C.S., M.J. Nypaver, and J.V. Osterman. 1984. Susceptibility of inbred mice to rickettsiae of the spotted fever group. *Infect. Immun. 43*: 143–148.

10. Catanzaro, P.J., A. Shirai, L.D. Agniel, Jr., and J.V. Osterman. 1977. Host defenses in experimental scrub typhus: role of spleen and peritoneal exudate lymphocytes in cellular immunity. *Infect. Immun. 18*:118–123.

11. Murphy, J.R., C.L. Wisseman, Jr., and P. Fiset. 1980. Mechanisms of immunity in typhus infection: analysis of immunity to *Rickettsia mooseri* infection of guinea pigs. *Infect. Immun. 27*:730–738.

12. Crist, A.E., Jr., C.L. Wisseman, Jr., and J.R. Murphy. 1984. Characteristics of lymphoid cells that adoptively transfer immunity to *Rickettsia mooseri* infection in mice. *Infect. Immun. 44*:55–60.

13. Dialynas, D.P., Z.S. Quan, K.A. Wall, A. Pierres, J. Quintans, M.R. Loken, M. Pierres, and F.W. Fitch. 1983. Characterization of the murine T cell surface molecule, designated L3T4, identified by monoclonal antibody GK1.5: similarity of L3T4 to the human Leu-3/T4 molecule. *J. Immunol. 131*:2445–2451.

14. Cantor, H., and E.A. Boyse. 1975. Functional subclasses of T lymphocytes bearing different Ly antigens. I. The generation of functionally distinct T cell subclasses is a differentiative process independent of antigen. *J. Exp. Med. 141*:1376–1389.

15. Cantor, H., and E.A. Boyse. 1975. Functional subclasses of T lymphocytes bearing different Ly antigens. II. Cooperation between subclasses of Ly$^+$ cells in the generation of killer activity. *J. Exp. Med. 141*:1390–1399.

16. Goronzy, J., C.M. Weyand, and C.G. Fathman. 1986. Long-term humoral unresponsiveness in vivo induced by treatment with monoclonal antibody against L3T4. *J. Exp. Med. 164*:911–925.

17. Rollwagen, F.M., G.A. Dasch, and T.R. Jerrells. 1986. Mechanisms of immunity to rickettsial infection: characterization of a cytotoxic effector cell. *J. Immunol. 136*:1418–1421.

18. Groves, M.G., and J.V. Osterman. 1978. Host defenses in experimental scrub typhus: genetics of natural resistance to infection. *Infect. Immun. 19*:583–588.

19. Jerrells, T.R., and J.V. Osterman. 1981. Host defenses in experimental scrub typhus: inflammatory response of congenic C3H mice differing at the *Ric* gene. *Infect. Immun. 31*:1014–1022.

20. Jerrells, T.R., and J.V. Osterman. 1982. Role of macrophages in innate and acquired host resistance to experimental scrub typhus infection of inbred mice. *Infect. Immun. 37*:1066–1073.

21. Singer, A., and R.J. Hodes. 1983. Mechanisms of T cell-B cell interaction. *Annu. Rev. Immunol. 1*:211–241.

22. Hinrichs, D.J., and T.R. Jerrells. 1976. *In vitro* evaluation of immunity to *Coxiella burnetii. J. Immunol. 117*:996–1003.

23. Nacy, C.A., and J.V. Osterman. 1979. Host defenses in experimental scrub

typhus: role of normal and activated macrophages. *Infect. Immun. 26*: 744–750.

24. Nacy, C.A., and M.S. Meltzer. 1979. Macrophages in resistance to rickettsial infections: macrophage activation *in vitro* for killing *Rickettsia tsutsugamushi. J. Immunol. 123*:2544–2549.

25. Nacy, C.A., E.J. Leonard, and M.S. Meltzer. 1981. Macrophages in resistance to rickettsial infections: characterization of lymphokines that induce antirickettsial activity in macrophages. *J. Immunol. 126*:204–207.

26. Jerrells, T.R., J. Turco, H.H. Winkler, and G.L. Spitalny. 1986. Neutralization of lymphokine-mediated antirickettsial activity of fibroblasts and macrophages with monoclonal antibody specific for murine interferon gamma. *Infect. Immun. 51*:355–359.

27. Spitalny, G.L., and E.A. Havell. 1984. Monoclonal antibody to murine gamma interferon inhibits lymphokine-induced antiviral and macrophage tumoricidal activities. *J. Exp. Med. 159*:1560–1565.

28. Wisseman, C.L., Jr., and A. Waddell. 1983. Interferon-like factors from antigen- and mitogen-stimulated human lymphocytes with antirickettsial and cytolytic action on *Rickettsia prowazekii* infected human endothelial cells, fibroblasts, and macrophages. *J. Exp. Med. 157*:1780–1793.

29. Steeg, P.S., R.N. Moore, H.M. Johnson, and J.J. Oppenheim. 1982. Regulation of murine macrophage Ia antigen expression by a lymphokine with immune interferon activity. *J. Exp. Med. 156*:1780–1793.

30. Unanue, E.R. 1984. Antigen-presenting function of the macrophage. *Annu. Rev. Immunol. 2*:395–428.

31. Jerrells, T.R., and J.V. Osterman. 1982. Host defenses in experimental scrub typhus: Delayed-type hypersensitivity responses of inbred mice. *Infect. Immun. 35*:117–123.

32. Jerrells, T.R., and J.V. Osterman. 1983. Development of specific and cross-reactive lymphocyte proliferative responses during chronic immunizing infections with *Rickettsia tsutsugamushi. Infect. Immun. 40*:147–156.

33. Palmer, B.A., F.M. Hetrick, and T.R. Jerrells. 1984. Production of gamma (γ) interferon in mice immune to *Rickettsia tsutsugamushi. Infect. Immun. 43*:59–65.

34. Shirai, A., P.J. Catanzaro, S.M. Phillips, G.H.G. Eisenberg, Jr., and J.V. Osterman. 1977. Host defenses in experimental scrub typhus: effect of chloramphenicol. *Infect. Immun. 18*:324–329.

35. Palmer, B.A., F.M. Hetrick, and T.R. Jerrells. 1984. Gamma interferon production in response to homologous and heterologous strain antigens in mice chronically infected with *Rickettsia tsutsugamushi. Infect. Immun. 46*: 237–244.

36. Jerrells, T.R. 1983. Association of an inflammatory I region-associated antigen-positive macrophage influx and genetic resistance of inbred mice to *Rickettsia tsutsugamushi. Infect. Immun. 42*:549–557.

37. Ridgway, R.L., S.C. Oaks, and D.D. LaBarre. 1986. Laboratory animal models for human scrub typhus. *Lab. Animal Sci. 36*:481–485.

38. Nacy, C.A., and M.S. Meltzer. 1984. Macrophages in resistance to rickettsial infections: protection against lethal *Rickettsia tsutsugamushi* infections by treatment of mice with macrophage-activating agents. *J. Leukocyte Biol.* *35*:385–396.

39. Vogel, S.N., E.A. Havell, and G.L. Spitalny. 1986. Monoclonal antibody-mediated inhibition of interferon-γ-induced macrophage antiviral resistance and surface antigen expression. *J. Immunol.* *136*:2917–2923.

40. Li, H., T.R. Jerrells, G.L. Spitalny, and D.H. Walker. 1987. Interferon-gamma as a crucial host defense against *Rickettsia conorii in vivo. Infect. Immun.* *55*:1252–1255.

41. Montenegro, M.R., S. Mansueto, B.C. Hegarty, and D.H. Walker. 1983. The histology of "taches noires" of boutonneuse fever and demonstration of *Rickettsia conorii* in them by immunofluorescence. *Virchows Arch. (Pathol. Anat.) 400*:309–317.

42. Jerrells, T.R., H. Li, and D.H. Walker. 1987. In vivo and in vitro role of gamma interferon in immune clearance of *Rickettsia species.* In *Proceedings of the Symposium on Host Defenses and Immunomodulation to Intracellular Pathogens*, T.K. Eisenstein, W.E. Bullock, and N. Hanna (Eds.). Plenum Press, New York, in press.

43. Jerrells, T.R. 1985. Immunosuppression associated with the development of chronic infections with *Rickettsia tsutsugamushi*: association of adherent suppressor cell activity and macrophage activation. *Infect. Immun.* *50*:175–180.

44. Jerrells, T.R., and C.J. Hickman. 1987. Down regulation of Ia antigen expression on inflammatory macrophages by factors produced during acute infections with *R. tsutsugamushi.* In *Immune Regulation by Characterized Polypeptides* (UCLA Symposia on Molecular and Cellular Biology), G. Goldstein, J.R. Bach, and H. Wigzell (Eds.). Alan R. Liss, Inc., New York, Vol. 41.

45. Snyder, D.S., D.I. Beller, and E.R. Unanue. 1982. Prostaglandins modulate macrophage Ia expression. *Nature 299*:163–165.

46. Snyder, D.S., and E.R. Unanue. 1982. Corticosteroids inhibit murine macrophage Ia expression and interleukin I production. *J. Immunol. 129*:1803–1805.

47. Warren, K.M., and S.N. Vogel. 1985. Opposing effects of glucocorticoids on interferon-γ-induced macrophage Fc receptor and Ia antigen expression. *J. Immunol. 134*:2462–2469.

48. Ling, P.D., M.K. Warren, and S.N. Vogel. 1985. Antagonistic effect of interferon-β on the interferon-γ-induced expression of Ia antigen in murine macrophages. *J. Immunol. 135*:1851–1863.

49. Nacy, C.A., and M.G. Groves. 1981. Macrophages in resistance to rickettsial infections: early host defense mechanisms in experimental scrub typhus. *Infect. Immun. 31*:1239–1250.

50. Turco, J., and H.H. Winkler. 1984. Effect of mouse lymphokines and cloned mouse interferon-γ on the interaction of *Rickettsia prowazekii* with mouse macrophage-like RAW264.7 cells. *Infect. Immun. 45*:303–308.

7

Toxoplasma

Carolyn M. Black and Jack S. Remington / Palo Alto Medical Foundation, Palo Alto, and Stanford University Medical Center, Stanford, California

Robert E. McCabe[*]/ Martinez Veterans Administration Medical Center, Martinez, California

I. INTRODUCTION

Toxoplasma gondii is a protozoan parasite capable of invading virtually every cell in the body. Individuals who are immunologically normal are relatively resistant to *T. gondii* in that the organism does not cause death or even severe disease in such individuals. In contrast, in immunocompromised individuals, the incidence of life-threatening toxoplasmosis has risen sharply in recent years—a reflection of the AIDS epidemic and increased use of immunosuppressive drugs. The ability of *T. gondii* to grow and multiply within host cells has provided an excellent model for assessing the role that cell-mediated immunity plays in the development of resistance to this parasite. The effector arm of cell-mediated immune responses to *T. gondii* appears to be the activated macrophage.

More than twenty years have passed since it was first hypothesized that interferon (IFN) may play a major role in resistance to *T. gondii* (1,2). These early studies demonstrated that infection with a virulent strain of *T. gondii* induces high levels of antiviral activity in the serum and peritoneal fluid of mice within 24 hours. Recently, it was reported that *T. gondii* induces production of IFNγ in vivo (3)—a gratifying corroboration of studies performed in the years before identification of the type of IFN was possible (1,2). Investigation of the function of interferon in *T. gondii* and other infections has benefitted from the identification of the three different types of interferon (α, β, γ), the discovery

Present affiliation: University of California, Davis, California

that IFNγ is the primary macrophage-activating factor, and the development of recombinant DNA technology. Along with identification of the macrophage as a major effector cell in resistance to toxoplasma infection, these advances have resulted in a resurgence of interest in the role of interferon in resistance to *T. gondii* infection. Recently, these studies have focused on the areas of the effect of IFNγ-activated macrophages on toxoplasma both in vitro and in vivo, inhibition of *T. gondii* by IFNγ-treated somatic cells, and the therapeutic value of IFNγ administered in vivo in protection against *T. gondii* infection.

The mechanism whereby replication of *T. gondii* is inhibited by IFNγ-treated somatic cells has been discovered and elegantly described by Pfefferkorn and colleagues (4-6), and is discussed in a chapter of this book. This mechanism is dependent on the induction in the IFNγ-treated cell of indoleamine 2,3-dioxygenase that converts tryptophan to N-formylkynurenine, thereby inhibiting replication of the parasites through tryptophan starvation (6). Of interest in this regard is the application of IFNγ to the in vitro cultivation of toxoplasma cysts (7).

A.   Studies of Macrophage Activation

The importance of lymphokines in inducing antitoxoplasma activity in macrophages has been known for over a decade (8-10). By 1979, evidence was increasing that IFNγ accounts for most if not all the macrophage-activating activity of lymphokine preparations in assays of macrophage activity against *T. gondii* (11). With the advent of technology for preparation of highly purified and recombinant proteins, Nathan and colleagues demonstrated in 1983 that the lymphokine responsible for activation of human macrophage oxidative metabolic and antitoxoplasma activities is IFNγ (12). In 1984–1985, studies which compared α, β, and γ interferons in activation of antitoxoplasma activity in human monocyte-derived macrophages (13) and human tissue macrophages (14) were published. Following these breakthroughs, a plethora of data appeared in the literature which argued the relative importance of oxygen-dependent versus -independent mechanisms of killing by activated macrophages. In the case of human monocytes and monocyte-derived macrophages, Murray and his colleagues clearly showed that both types of activity can be enhanced by IFNγ, but that the primary killing mechanism used by these cells is oxygen dependent (15). Murray's group later extended the cell types which are known to be activated in vitro by IFNγ against *T. gondii* to include mouse peritoneal macrophages (16) and human alveolar macrophages from AIDS patients (17).

The remainder of this chapter will focus chronologically on work which has been performed in our own laboratory concerning in vitro and in vivo activation of macrophages by IFNγ, therapeutic effects of in vivo administration of IFNγ, and preliminary data on in vivo synergism of IFNγ with other agents against *T. gondii*.

II. PROTECTIVE EFFECT OF IFNγ ADMINISTERED IN VIVO

With the knowledge that macrophages are important in host defense against intracellular pathogens and that recombinant IFNγ (rIFNγ) activates anti-protozoal activity of human monocyte-derived macrophages in vitro (12), we set out to determine if in vivo administration of rIFNγ would affect mortality of mice infected with *T. gondii* (18). Swiss-Webster female mice were infected by intravenous injection with *T. gondii* on day 0. As shown in Figure 1, 5×10^3 units of murine rIFNγ administered intraperitoneally every other day until control mice started to die completely protected mice against death, whereas 50% of control mice died within 17 days (p = .016). Significant prolongation of time to death was also observed in mice that received injections every other day of 5×10^2, and 5×10^1 units of murine IFNγ (18). Lower doses of IFNγ were not effective in these experiments. In a separate experiment, only two injections of 5×10^4 or 5×10^3 units of rIFNγ were administered per group at the following times before, during, and after *T. gondii* infection: days –1 and 0, days 0 and 1, and days 1 and 2. All of the treated groups showed significant prolongation of time to death, but treatment on days 0 and 1 produced the best overall results (Fig. 2a,b), suggesting the importance of inhibiting *T. gondii* very early in the

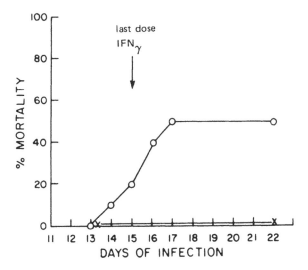

Figure 1   Effect of murine IFNγ on mortality of mice infected with *T. gondii*. Mice were infected by intravenous injection on day 0. IFNγ was administered intraperitoneally every other day beginning 48 hours before infection. (X) 5×10^3 units per dose IFNγ. (O) Phosphate-buffered saline. Eight to ten mice were in each experimental group. (From Ref. 18.)

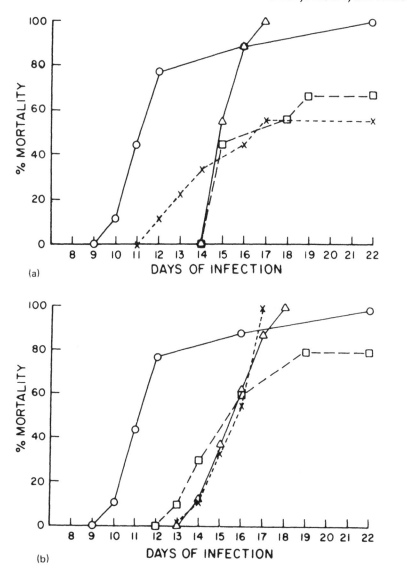

Figure 2 Effect of murine IFNγ on mortality of mice infected with *T. gondii*.
Mice were infected as described in Fig. 1. (a) Mice received two doses of 5 × 10⁴
units of IFNγ intraperitoneally on days (X) – 1 and 0; (□) 0 and 1; (△) 1 and 2;
(○) controls received PBS. (b) Identical to (A) except that mice received two
doses of 5 × 10³ units of IFNγ. Eight to ten mice were in each experimental
group. (From Ref. 18.)

infection. It is not clear why injections on days -1 and 0 did not produce better results than injections on days 0 and 1 [slower time to death at the higher dose of IFNγ (Fig. 2a), and lower mortality at the lower dose of IFNγ (Fig. 2b)]. One possible explanation is that the injections of IFNγ given prior to *T. gondii* infection, on day -1, may have activated a subpopulation of macrophages with suppressor activity, thereby producing partial inhibition of the immune response to *T. gondii*, whereas this suppressor macrophage population may not become activated to the same extent when IFNγ is administered at the time of, or after infection.

Serum toxoplasma antibody titers were determined by the Sabin-Feldman dye test on days 7 and 11 of infection. Mice that received IFNγ and that showed prolonged time to death had significantly higher titers than control mice, whereas IFNγ-treated but unprotected groups of mice had titers similar to those of the control mice (18). These studies demonstrated that murine IFNγ has significant activity against *T. gondii* in vivo, and that this activity appears to be associated with an enhanced antibody response to the parasite.

Currently, it is not known whether the protective effect of IFNγ administered in vivo is mediated by activation of antiprotozoal activity of macrophages, by tryptophan starvation within somatic cells, or by any other of the many immunomodulating effects of IFNγ in vivo. Most likely, it is due to a combination of these factors. In this regard, the inhibition of *T. gondii* by IFNγ-activated macrophages probably does not involve tryptophan starvation since macrophages appear to lack the enzyme indoleamine dioxygenase (E.R. Pfefferkorn, personal communication).

III. ACTIVATION OF MACROPHAGES IN VIVO AND IN VITRO

A. Peritoneal Macrophages

To determine whether in vivo administration of IFNγ can activate macrophages to inhibit replication of *T. gondii*, we injected IFNγ intraperitoneally into mice at various doses, both below and as high as those known to protect mice (Fig. 1). Peritoneal macrophages were harvested 24 hours later, adhered to glass in monolayers, and infected with *T. gondii* (19). Macrophages harvested from mice injected with as few as 5 units of IFNγ inhibited replication of *T. gondii* (Fig. 3). These experiments, published as part of the protection study described above (18), provided the first demonstration of in vivo activation of macrophages by IFNγ against an intracellular pathogen. The results were corroborated in a study of in vivo and in vitro activation of mouse peritoneal macrophages performed by Murray et al. (16).

Another set of experiments were performed using antisera to IFNγ and to IFNα/β to determine if the effects observed in the preceding experiments can be

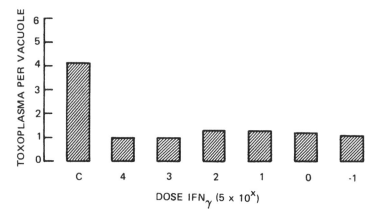

Figure 3 Effect of in vivo administration of IFNγ on replication of *T. gondii* within vacuoles of peritoneal macrophages. Mice received a single dose of rIFN-γ at the concentrations shown. Macrophages were harvested 24 hours later and challenged with *T. gondii* in vitro. Control represents macrophages from mice injected with phosphate-buffered saline.

attributed to IFNγ. Mice were injected with IFNγ, IFNγ plus anti-IFNγ, IFNγ plus anti-IFNα/β, anti-IFNγ antiserum, anti-IFNα/β antiserum, and phosphate-buffered saline as control. These preparations were made by incubating 5×10^2 units of IFNγ alone and with 5×10^2 units of neutralizing activity of each antiserum at 37°C for 2 hr, and then injected intraperitoneally into mice. Specific neutralizing activities of anti-IFNγ and α/β antiserum were determined by the supplier of IFNγ and anti-IFNγ antiserum (Genentech Inc.), and anti-IFNα/β antiserum (Lee Biochemicals). Macrophages were harvested 24 hours later and challenged with *T. gondii*. The results revealed that, after 18 hours of incubation, the percentage of infected control macrophages was 55, whereas the percentage of infected macrophages from IFNγ-treated mice was 4 (Fig. 4). Anti-IFNγ antiserum completely abrogated the effect of IFNγ ($p < .025$ as compared to IFNγ alone and not significantly different from the control value). Anti-IFNα/β antiserum only partially blocked the effect of IFNγ. Similar results were observed with respect to the percentage of macrophages infected with *T. gondii*, as shown in Figure 4, and with the number of toxoplasma per vacuole (data not shown).

B. Alveolar Macrophages

Most previous studies of macrophage activation have been performed with peritoneal macrophages, whose role in naturally occurring infections is unclear,

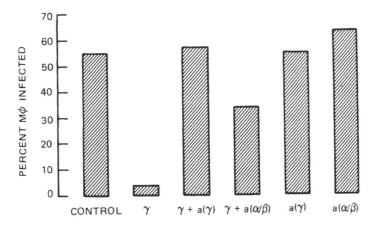

Figure 4 Neutralization of IFNγ-induced activation of peritoneal macrophages by antisera to IFNγ and IFNα/β in vivo. 5×10^3 units of IFNγ were incubated alone (γ) or with 5×10^3 units of neutralizing activity of anti-IFNγ [γ + a (γ)] and anti-IFNα/β [γ + a (α/β)] antisera at 37°C for 2 hours, then injected into mice. Control groups were injected with anti-IFNγ antiserum [a (γ)]; anti-IFNα/β antiserum [a (α/β)]; and phosphate-buffered saline (control). Macrophages were harvested 24 hours later and challenged with *T. gondii* in vitro.

or with macrophages derived artificially from peripheral blood monocytes by prolonged culture in vitro. The activation of alveolar macrophages, on the other hand, is poorly understood despite their importance in protection from pulmonary infection. Previous studies indicate that of immunosuppressed patients who have died of toxoplasmosis, especially patients with AIDS, the lungs were infected with *T. gondii* in 20 to 100% of cases (20). Since several functional capacities of peritoneal and alveolar macrophages have been demonstrated to differ even when obtained from the same animal (21-23), we considered it to be important to determine if in vivo administration of IFNγ can activate the antitoxoplasma activity of resident murine alveolar macrophages.

Preliminary to performing the in vivo experiments, we tested the ability of IFNγ to activate alveolar macrophages in vitro, using peritoneal macrophages as a control. After 24 hours of incubation with concentrations of either 0.1, 0.5, 1, 10, or 1000 units/ml murine rIFNγ, alveolar macrophages markedly inhibited replication of *T. gondii* (24). A dose-dependent response was observed for alveolar macrophages in the range of 0.1-1 unit/ml rIFNγ, with the maximum antiprotozoal effect essentially demonstrable at 1 unit/ml (Fig. 5; solid bars). In peritoneal macrophage monolayers, concentrations ≥ 1 unit/ml rIFNγ inhibited replication of *T. gondii* in a dose-dependent fashion (Fig. 5; hatched bars). The concentration of rIFNγ which induced maximal inhibition of *T. gondii* was

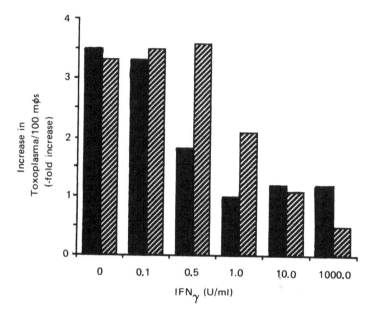

Figure 5 Effect of incubation of murine macrophages for 24 hours with rIFNγ in vitro on intracellular multiplication of *T. gondii*. (■) Alveolar macrophages. (▨) Peritoneal macrophages. Data are representative of 2 experiments performed in duplicate. (From Ref. 24.)

lower for alveolar macrophages than for peritoneal macrophages, suggesting a greater sensitivity of alveolar macrophages to the in vitro activity of rIFNγ.

Alveolar and peritoneal macrophages from mice treated in vivo with rIFNγ by either the intravenous or intraperitoneal route markedly inhibited intracellular replication of *T. gondii* over 18 hours of incubation in vitro (Table 1). In contrast, alveolar and peritoneal macrophages from mice injected by either route with excipient control solution showed a threefold or greater increase in numbers of parasites per 100 macrophages. Occasional alveolar macrophages which lacked antiprotozoal activity and permitted replication of *T. gondii* within vacuoles were noted in monolayers from the rIFNγ-treated mice; this was not the case for peritoneal macrophages from these mice. No important differences in results were noted between alveolar and peritoneal macrophages for the two routes of administration of rIFNγ.

Monolayers of alveolar and peritoneal macrophages inhibited intracellular multiplication of *T. gondii* in a dose-dependent fashion in mice injected intravenously with 10^3–10^5 units rIFNγ/mouse (Fig. 6a,b). This dose-dependent effect was most apparent in macrophages harvested at 24 hours after injection of

Table 1 Effect of In Vivo Administration of rIFNγ on Survival and Replication of T. gondii in Alveolar and Peritoneal Macrophages

Mice treated with[a]	% Cells infected		Toxoplasma/100 cells		Toxoplasma/vacuole	
	0 hr	18 hr	0 hr	18 hr	0 hr	18 hr
Alveolar						
i.v. Control	29.0	28.4	45	134	1.0	2.4
rIFNγ	31.2	12.1	49	53	1.1	3.7
i.p. Control	27.4	36.0	29	226	1.0	5.6
rIFNγ	42.0	11.8	57	66	1.0	5.6
Peritoneal						
i.v. Control	24.7	17.9	37	84	1.0	3.7
rIFNγ	57.1	25.4	112	36	1.0	1.1
i.p. Control	49.5	43.1	73	284	1.0	5.8
rIFNγ	66.3	26.4	110	120	1.0	3.3

[a]BALB/c mice were injected intravenously (i.v.) or intraperitoneally (i.p.) with 0.2 ml PBS containing 1 μg/ml bovine serum albumin alone or with 10^4 U rIFNγ. Macrophages were harvested at 24 hr after administration of rIFNγ. Data represent 1 of 2 experiments in which results were similar. (From Ref. 24.)

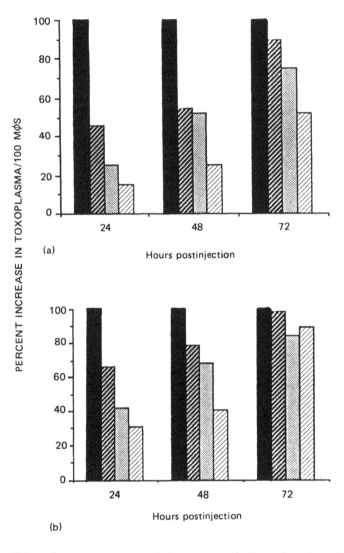

(a)

Hours postinjection

(b)

Hours postinjection

Figure 6 Dose response and time course of effect of in vivo administration of rIFNγ on intracellular multiplication of *T. gondii* within murine macrophages. (a) Alveolar macrophages. (b) Peritoneal macrophages. Mice received a single intravenous injection of (▨) 10^5 ; (▨) 10^4 ; (▨) 10^3 units of rIFN-γ; (■) PBS control. Thereafter, macrophages were harvested and assayed at the times shown. Data shown are from three experiments and are representative of duplicate monolayers. (From Ref. 24.)

rIFNγ. When harvested and assayed at 24 hr intervals, the antiprotozoal activity of alveolar and peritoneal macrophages from mice that received a single injection of rIFNγ diminished over the course of 72 hr (Fig. 6a,b). In the optimal case in which mice were injected with 10^5 units rIFNγ/mouse, the alveolar macrophages harvested and assayed at 48 hr after injection showed a 12% loss and the peritoneal macrophages a 15% loss in antiprotozoal activity from that observed at 24 hr after injection. At 72 hr postinjection of the same dosage of rIFNγ, the alveolar macrophages showed a 44% loss and the peritoneal macrophages an 83% loss in activity from the level observed at 24 hours. Similar losses in activity over time were observed with alveolar and peritoneal macrophages from mice injected with 10^3 and 10^4 units rIFNγ (Fig. 6a,b).

As indicated above, parasites which survive within vacuoles of monolayers of activated macrophages were noted in alveolar but not peritoneal populations. Thus, it appears from these results that administration of rIFNγ in vivo and in vitro activates the antimicrobial activity of peritoneal macrophages against $T.$ $gondii$ to a greater degree than it does for alveolar macrophages from the same mice (24). It is not known whether this difference is a result of greater heterogeneity of activation in the alveolar macrophage populations, or to lesser stability of the activated state during in vitro culture of alveolar macrophages as compared to peritoneal macrophages. In what may at first appear to be paradoxical, however, is the finding that 0.5 units/ml rIFNγ was the minimum dose sufficient to activate the antiprotozoal activity of alveolar macrophages but not peritoneal macrophages in vitro. Similarly, when different doses of rIFNγ were administered in vivo, smaller increases were observed in the number of toxoplasma per 100 alveolar macrophages than for peritoneal macrophages from the same mice, although the parasites that survived went on to multiply only in the monolayers of alveolar macrophages. Of interest in this regard is the observation that, of the doses tested (0.1, 0.5, 1, 10, 100, 1000), the concentration of rIFNγ which induced maximal inhibition of $T.$ $gondii$ in vitro was also lower for alveolar (1 unit/ml) than for peritoneal macrophages (1000 units/ml). Taken together, these results suggest that there may be a dissociation between the sensitivity of macrophages from different anatomic sites to rIFNγ-induced activation (alveolar > peritoneal) and the degree of activation for intracellular inhibition or killing induced by rIFNγ (peritoneal > alveolar). Whether these results are due to a quantitative or qualitative difference in rIFNγ receptors or to some other mechanism of rIFNγ-induced activation between alveolar and peritoneal macrophages is unclear. In addition, the alveolar and peritoneal macrophage populations differed in the duration of their activation that was induced by a single injection of rIFNγ in vivo (alveolar > peritoneal). This finding suggests that alveolar macrophages may intrinsically be capable of a longer-lived activation state than peritoneal macrophages. The tissue level and rate of degradation of intravenously administered rIFNγ will also certainly differ for different anatomical sites.

The concentration of rIFNγ necessary to achieve activation of alveolar macrophages in vivo was comparable to that previously shown to protect mice from lethal doses of *T. gondii* (18). Our results indicate that IFNγ can gain access to the lungs from peripheral sites of administration while retaining biological activity, and provide a rationale for the evaluation of rIFNγ in therapy of pulmonary infections. This approach may be particularly useful in patients with an impaired ability to produce IFNγ, including patients with AIDS (25).

IV. COMBINATION OF INTERFERON AND OTHER AGENTS IN VIVO

A. IFNγ and Roxithromycin

Since in vivo administration of IFNγ is not completely protective in all models of *T. gondii* infection (26) and since some patients are particularly prone to adverse reactions associated with sulfonamides, there currently exists a need for improved therapies of toxoplasmosis. Roxithromycin is a macrolide antibiotic that provides protection in mice infected intraperitoneally with *T. gondii* (27) and in a murine model of toxoplasmic encephalitis developed in our laboratory (28). In preliminary studies in the encephalitis model, neither roxithromycin nor IFNγ were completely protective (26). We thus considered it of interest to investigate the efficacy of these two agents in combination in the murine model of toxoplasmic encephalitis.

Mice were infected with *T. gondii* by intracerebral injection (28). Roxithromycin was fed ad libitum in mouse feed at doses calculated on a basis of approximate feed intake per day per mouse. IFNγ was administered intravenously every other day for three doses beginning 6 hours after infection. Roxithromycin alone at a dose of 25 mg/mouse/day begun 2 days prior to infection had no effect on mortality or median time to death (10.5 days) (26). IFNγ alone at a dose of 1×10^6 units significantly decreased mortality on day 8 and prolonged time to death (median: 10 vs. 8 days) but failed to alter mortality (100%) at 30 days. When IFNγ at 5×10^4 units/mouse, and roxithromycin at 25 mg/mouse/day begun 2 hours before infection were both administered, there was a remarkable synergistic effect (Fig. 7). Survival was increased from 0 to 100% (p = .002).

The mechanism of the synergistic protection observed is unknown and may be complex. We consider it likely that the synergy which we observed in this model will occur when IFNγ is combined with other antibiotics which have activity against *T. gondii* in vivo.

B. rIFNγ and Interleukin-2 (IL-2)

To date, IL-2 represents the only lymphokine other than IFNγ that has been shown to have a protective effect in vivo in models of acute *T. gondii* infection

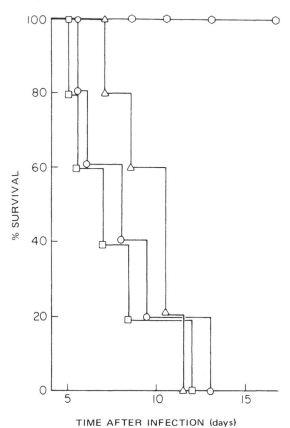

Figure 7 Synergistic effect of IFNγ and roxithromycin (RU965) in protection of mice infected with *T. gondii*. Mice were infected by intracerebral injection on day 0. Mice received (△) 5 X 10⁴ units IFNγ intravenously; (□) 25 mg/day roxithromycin in feed; (○) 25 mg/day roxithromycin + 5 X 10⁴ units IFNγ; (○) phosphate-buffered saline. There were 10 mice in each experimental group. (From Ref. 26.)

(29). IL-2 has multiple effects on the immune response and which one or more of these is operative in the protection of the *T. gondii* model has not been defined. In view of these findings, it was considered of interest to determine the effect of combinations of IFNγ with IL-2 on mice infected with *T. gondii*. For these experiments, MRL/MPJ/lpr strain mice (NIH) were used which display ineffective helper T-cell function (L3T4⁻, dull Lyt 1⁺2⁻) (30). This defect leads to markedly depressed in vivo and in vitro immune responses, a situation encountered in AIDS. In spite of massive lymphoproliferation in vivo, MRL/lpr mice

spleen and lymph node cells display markedly impaired IL 2 production and proliferate poorly in response to mitogens (30). Thus, the MRL/lpr strain provides
an attractive opportunity to develop a possible model of toxoplasmosis which
does not require drugs to cause the immunosuppression found associated with
patients with AIDS and toxoplasmosis.

The results of preliminary studies performed in our laboratory suggest that
the lpr gene results in exquisite susceptibility of these mice to morbidity and
mortality with a strain and dose of *T. gondii* that has no clinically apparent effect on normal mice. In a representative experiment, untreated MRL/lpr mice infected intraperitoneally with with an avirulent strain of *T. gondii* died within 2
weeks, whereas no deaths occurred in the MRL/lpr[++] control mice. When the
number of cysts present in the brains of lpr mice treated with sulfadiazine (to
allow for development of chronic infection) was compared to corresponding
control mice, lpr mouse brains contained an average of 10-20 fold more cysts
(S.D. Sharma, manuscript in preparation).

To test the effect of therapy with combinations of lymphokines on the
MRL/lpr strain, mice were injected with 5×10^3 units rIFNγ, 100 units rIL-2,

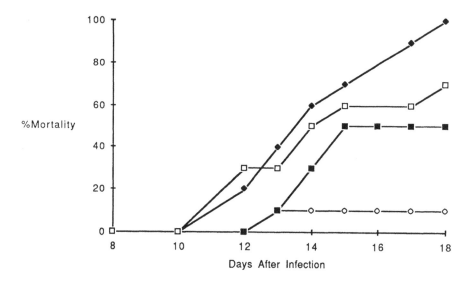

Figure 8 Effect of in vivo administration of rIFNγ, rIL-2, and rIFNγ plus rIL-2
on mortality of MRL/lpr mice infected intraperitoneally on day 0 with *T. gondii*.
Mice were injected with (■) 5×10^3 units of rIFNγ, (◊) 100 units rIL-2, (□)
rIFNγ and rIL-2, and (♦) phosphate buffered saline on days −2, +2, +4, +6, and
+8. There were 10 mice in each experimental group. (S.D. Sharma, manuscript in
preparation.)

rIFNγ and rIL-2, or phosphate-buffered saline on day -2, +2, +4, +6, and +8. On day 0, mice were infected with 10 cysts of the avirulent Me49 strain of *T. gondii*. Mice that received IL-2 alone were the most protected; this group showed 90% survival as compared to 100% mortality of the untreated control mice at 18 days after infection (Fig. 8). Mice that received IFNγ alone were also protected; this group showed 50% mortality at 18 days after infection. In contrast to these results, mice that received both IFNγ and IL-2 showed poorer survival than mice treated with either lymphokine alone, although the mortality in this group was significantly less than the control (Fig. 8; $p < .05$).

These findings suggest that IL-2 treatment may ameliorate the infection of MRL mice with *T. gondii* by partially reversing the T-cell functional defect. IFNγ treatment of these mice also improved survival; this result may be due to IFNγ-induced activation of macrophages, although this was not tested. At present, it is not clear why combinations of IFNγ and IL-2 fail to protect MRL mice as well as either lymphokine alone. One possible explanation is that the combination of lymphokines at the concentrations and conditions used resulted in increased toxicity for MRL strain mice. It remains to be tested whether different concentrations and/or conditions of therapy would avoid such toxicity and allow for synergism between IFNγ and IL-2.

V. SUMMARY

Recombinant IFNγ activates antiprotozoal activity of alveolar and peritoneal macrophages in vitro and in vivo. The duration of activation of macrophages after a single dose of rIFNγ in vivo is at least 3 days. When recombinant IFNγ was administered at a dose which prolongs time to death in mice infected intraperitoneally with *T. gondii*, and was used in combination with the macrolide antibiotic roxithromycin, at a dose which protected only 50% of the mice, the combination was synergistic and provided complete protection against a lethal intracerebral challenge of *T. gondii*. Preliminary results of experiments in genetically immunodeficient mice (MRL/lpr strain) revealed that the combination of rIFNγ with IL-2 is not as effective in protection against *T. gondii* as either lymphokine alone.

Further investigation of the chemical nature of IFNγ-induced antiprotozoal activity of macrophages is needed. It is possible that study of the inherent differences in sensitivity to IFNγ-induced activation and the degree of activation produced in macrophage populations from different anatomic sites may help to elucidate this mechanism. In addition, the effect of in vivo administration of IFNγ, in combination with other drugs and lymphokines such as IL-2, on the immune response to microbial agents needs further study. This would help in the design of optimal therapies that result in synergistically protective responses and eliminate possible toxic combinations.

ACKNOWLEDGMENTS

This work was supported by National Institute of Health Grant A104717, a MacArthur Foundation Grant in Molecular Parasitology, and a grant from Hoechst-Roussel Pharmaceuticals, Inc.

The authors wish to thank Drs. Christine Czarniecki and H. Michael Shepard of Genentech, Inc., for supplying recombinant IFNγ and for their helpful discussions.

REFERENCES

1. Freshman, M.M., T.C. Merigan, J.S. Remington, and I.E. Brownlee. 1966. In vitro and in vivo antiviral action of an interferon-like substance induced by *Toxoplasma gondii*. *Proc. Soc. Exp. Biol. Med. 123*:862–866.

2. Rytel, M.W., and T.C. Jones. 1966. Induction of interferon in mice infected with *Toxoplasma gondii. Proc. Soc. Exp. Biol. Med. 123*:859–862.

3. Jones, T.C., S. Alkan, and P. Erb. 1986. Spleen and lymph node cell populations, in vitro cell proliferation and interferon-γ production during the primary immune response to *Toxoplasma gondii. Paras. Immunol. 8*:619–631.

4. Pfefferkorn, E.R. 1984. Interferon-γ blocks the growth of *Toxoplasma gondii* in human fibroblasts by inducing the host cells to degrade tryptophan. *Proc. Natl. Acad. Sci. (USA) 81*:908–912.

5. Pfefferkorn, E.R., and P.M. Guyre. 1984. Inhibition of growth of *Toxoplasma gondii* in cultured fibroblasts by human recombinant gamma interferon. *Infect. Immun. 44*:211–216.

6. Pfefferkorn, E.R., M. Eckel, and S. Rebhun. 1986. Interferon-γ suppresses the growth of *Toxoplasma gondii* in human fibroblasts through starvation for tryptophan. *Mol. Biochem. Parasitol. 20*:215–224.

7. Jones, T.C., K.A. Bienz, and P. Erb. 1986. In vitro cultivation of *Toxoplasma gondii* cysts in astrocytes in the presence of gamma interferon. *Infect. Immun. 51*:147–156.

8. Krahenbuhl, J.L., and J.S. Remington. 1971. In vitro induction of nonspecific resistance in macrophages by specifically sensitized lymphocytes. *Infect. Immun. 4*:337–343.

9. Anderson, S.E., Jr., and J.S. Remington. 1974. Effect of normal and activated human macrophages on *Toxoplasma gondii. J. Exp. Med. 139*:1154–1174.

10. Jones, T.C., L. Len, and J.G. Hirsch. 1975. Assessment in vitro of immunity against *Toxoplasma gondii. J. Exp. Med. 141*:466–482.

11. Shirahata, T., and K. Shimizu. 1979. Some physicochemical characteristics of an immune lymphocyte product which inhibits the multiplication of toxoplasma within mouse macrophages. *Microbiol. Immunol. 23*:17–30.

12. Nathan, C.F., H.W. Murray, M.E. Wiebe, and B.Y. Rubin. 1983. Identification of interferon as the lymphokine that activates human macrophage oxidative metabolism and antimicrobial activity. *J. Exp. Med. 158*:670–689.

13. Nathan, C.F., T.J. Prendergast, M.E. Wiebe, E.R. Stanley, E. Platzer, H.G. Remold, K. Welte, B.Y. Rubin, and H.W. Murray. 1984. Activation of human macrophages: comparison of other cytokines with interferon-γ. *J. Exp. Med. 160*:600–605.

14. Wilson, C.B., and J. Westall. 1985. Activation of neonatal and adult human macrophages by alpha, beta, and gamma interferons. *Infect. Immun. 49*: 351–356.

15. Murray, H.W., B.Y. Rubin, S.M. Carriero, A.M. Harris, and E.A. Jaffee. 1985. Human mononuclear phagocyte antiprotozoal mechanisms: oxygen-dependent vs oxygen-independent activity against intracellular *Toxoplasma gondii. J. Immunol. 134*:1982–1988.

16. Murray, H.W., G.L. Spitalny, and C.F. Nathan. 1985. Activation of mouse peritoneal macrophages in vitro and in vivo by interferon-γ. *J. Immunol. 134*:1619–1622.

17. Murray, H.W., R.A. Gellene, D.M. Libby, C.D. Rothermel, and B.Y. Rubin. 1985. Activation of tissue macrophages from AIDS patients: in vitro response of AIDS alveolar macrophages to lymphokines and interferon-γ. *J. Immunol. 135*:2374–2377.

18. McCabe, R.E., B.J. Luft, and J.S. Remington. 1984. Effect of murine interferon gamma on murine toxoplasmosis. *J. Infect. Dis. 150*:961–962.

19. Wilson, C.B., V. Tsai, and J.S. Remington. 1980. Failure to trigger the oxidative metabolic burst by normal macrophages. Possible mechanism for survival of intracellular pathogens. *J. Exp. Med. 151*:328–346.

20. Catterall, J.R., J.M. Hofflin, and J.S. Remington. 1986. Pulmonary perspective: pulmonary toxoplasmosis. *Am. Rev. Respir. Dis. 133*:704–705.

21. Black, C.M., B.L. Beaman, R.M. Donovan, and E. Goldstein. 1985. Intracellular acid phosphatase content and the ability of different macrophage populations to kill *Nocardia asteroides. Infect. Immun. 47*:375–383.

22. Collins, F.M., C.J. Niederbuhl, and S.G. Campbell. 1983. Bactericidal activity of alveolar and peritoneal macrophages exposed in vitro to three strains of *Pasteurella multocida. Infect. Immun. 39*:779–784.

23. Lehrer, R.I., L.G. Ferrari, J. Patterson-Delafield, and T. Sorrell. 1980. Fungicidal activity of rabbit alveolar and peritoneal macrophages against *Candida albicans. Infect. Immun. 28*:1001–1008.

24. Black, C.M., J.R. Catterall, and J.S. Remington. 1987. In vivo and in vitro activation of alveolar macrophages by recombinant gamma-interferon. *J. Immunol. 138*:491–495.

25. Murray, H.W., B.Y. Rubin, H. Masur, and R.B. Roberts. 1984. Impaired production of lymphokines and immune (gamma) interferon in the acquired immunodeficiency syndrome. *N. Engl. J. Med. 310*:883–887.

26. Hofflin, J.M., and J.S. Remington. 1987. In vivo synergism of roxithromycin and interferon against *Toxoplasma gondii. Antimicrob. Agents Chemother. 31*:346–348.

27. Luft, B.J., J.M. Hofflin, J. Chan, and J.S. Remington. 1986. The activity of RU28965, a macrolide, in the treatment of toxoplasmic encephalitis. 26th Interscience Conference on Antimicrobial Agents and Chemotherapy, Abstract #1105, September 1986, New Orleans.

28. Hofflin, J.M., F.K. Conley, and J.S. Remington. 1987. Murine model of intracerebral toxoplasmosis. *J. Infect. Dis. 155*:550–557.

29. Sharma, S.D., J.M. Hofflin, and J.S. Remington. 1985. In vivo recombinant interleukin 2 administration ehhances survival against a lethal challenge with *Toxoplasma gondii. J. Immunol. 135*:4160–4163.

30. Rosenberg, V., A.D. Steinberg, and T.J. Santoro. 1984. The basis of auto-immunity in MRL-Lpr/lpr mice: a role for self Ia-reactive T cells. *Immunol. Today 5*:64–67.

8

Toxoplasma gondii

E. R. Pfefferkorn / Dartmouth Medical School, Hanover, New Hampshire

I. BACKGROUND

Toxoplasma gondii is an obligate intracellular protozoan parasite that grows within a vacuole in the cytoplasm of its host cell. Although overt disease is comparatively rare, infection is common in most mammals, including humans. Severe human disease is usually seen only in transplacental infections of the developing fetus and in primary or recrudescent infections of immunosuppressed patients. *Toxoplasma gondii* has a complex life cycle in which members of the cat family serve as the definitive host in which male and female gametes fuse to form a zygote. In all other animals, *T. gondii* multiplies only asexually. The rapidly multiplying asexual stage, the tachyzoite, grows in a wide variety of cultured cells, thus offering the opportunity to study the actions of interferons (IFNs) under in vitro conditions in which the antiparasitic mechanism can be readily examined. Various methods are available for measurement of the in vitro growth of *T. gondii*. Direct microscopic enumeration is reliable but tedious. It has been largely replaced in our laboratory by a plaque assay (1) that is analogous to that used with animal viruses or by the incorporation of radioactive uracil, a metabolite that is used by the parasite for DNA and RNA synthesis, but cannot be used by the host cell.

The superficial similarity of *T. gondii* and viruses as obligate intracellular parasites led to various attempts to detect an antitoxoplasma effect of IFN. Thus mouse IFNs have been reported both to block (2) and not to affect (3) the growth of *T. gondii* in homologous host cells. Since these experiments were done with IFN prepared in vivo and were done before the distinctions among the various types of IFN were completely realized, their interpretation is difficult. More recently, a plaque assay was used to show that human IFNα and β had no effect on the growth of *T. gondii* in human fibroblasts (4).

II. CHARACTERISTICS OF THE IN VITRO INHIBITION OF *T. GONDII* BY IFNγ

The availability of highly purified human recombinant IFNα, β, and γ from Genentech, Inc., encouraged us to renew the study of their antitoxoplasma action. All experiments with human IFNs included an antiviral titration of the IFN dilutions using encephalomyocarditis (EMC) virus as the challenge and a parallel antiviral titration of an appropriate IFN reference standard from the National Institutes of Health. (All data in this chapter report human IFN titers in reference units. In some earlier publications (5,6) IFN titers furnished by Genentech, Inc., were used. At most, the discrepancy between reference units and those of Genentech, Inc., were twofold, with the titer in reference units being lower.)

Our initial experiments on the antitoxoplasma activity of human IFNs used a plaque reduction assay in human fibroblasts analogous to that used to titrate the antiviral activity of IFN. These experiments showed potent antitoxoplasma activity for IFNγ: 4 units/ml completely suppressed the formation of plaques. In contrast, human recombinant IFNα and IFNβ had no antitoxoplasma activity at the highest concentration tested, 4000 units/ml (unpublished results). Results from plaque assays showed that IFNγ had no effect on the viability of extracellular parasites (6) and that the effect on intracellular parasites was largely to suppress their growth rather than killing them (7). Additional experiments that employed the incorporation of [^{3}H]uracil to measure the intracellular growth of *T. gondii* revealed that the antitoxoplasma action had many of the characteristics of the antiviral action of IFN. As shown in Table 1, when human fibroblasts were treated with INFγ for one day and then challenged with *T. gondii*, inhibition of parasite growth was seen one day after infection and was more evident when measured two days after infection. However, when INFγ was added at the same time as *T. gondii*, no significant effect on parasite growth was seen during the first day of infection (Table 2). Under these conditions, growth was inhibited during the second day. Thus some time was required for IFNγ to establish an "antitoxoplasma state" in human fibroblast cultures. Once this "antitoxoplasma state" was established, the free IFNγ could be washed out of the cultures and, as shown in Table 3, the growth of *T. gondii* was still suppressed. The establishment of an "antitoxoplasma state" required synthesis of host cell mRNA and, presumably, protein synthesis. This requirement for RNA synthesis was demonstrated by inhibiting host cell mRNA with α-amanitin at a concentration that had no effect on the growth of the parasite (Table 4).

Table 1 Inhibition of the Growth of *T. gondii* by IFNγ Added to the Human Fibroblast Host Cells 24 Hours Before Infection

IFNγ (units/ml)	Percent of control [³H] uracil incorporation[a] at following times after infection	
	24 hr	48 hr
32	18 ± 6	3 ± 0.2
16	20 ± 3	5 ± 0.5
8	23 ± 4	6 ± 1
4	78 ± 30	27 ± 8

[a]Mean and standard deviation of quadruplicate samples.
Source: From Ref. 6, with the IFNγ titers converted to reference units.

Table 2 Inhibition of the Growth of *T. gondii* by IFNγ Added at the Time of Infection of the Human Fibroblast Host Cells

IFNγ (units/ml)	Percent of control [³H] uracil incorporation[a] at following time after infection	
	24 hr	48 hr
64	91 ± 4[b]	12 ± 2
32	82 ± 3	9 ± 1
16	102 ± 3	11 ± 1
8	107 ± 3	16 ± 3

[a]Mean and standard deviation of quadruplicate samples.
Source: From Ref. 6, with the IFNγ titers converted to reference units.

Table 3 Inhibition of the Growth of *T. gondii* when IFNγ Was Removed

IFNγ (units/ml)	Percent of control [³H] uracil incorporation[a]	
	IFNγ washed out at the time of infection	IFNγ present throughout
64	23 ± 2[b]	16 ± 2
32	23 ± 2	15 ± 3
16	40 ± 1	20 ± 1
8	117 ± 18	27 ± 3
4	107 ± 18	104 ± 6

[a]Measured 24 hours after infection.
[b]Mean and standard deviation of quadruplicate samples.
Source: From Ref. 6, with the IFNγ titers converted to reference units.

Table 4 Reversal of the Antitoxoplasma Effect of IFNγ by Treatment of Human Fibroblast Cultures with α-Amanitin

	Percent of control parasite growth[a]	
	without α-amanitin	with α-amanitin, 1 μg/ml[b]
without IFNγ	100	97
with IFNγ, 16 units/ml[b]	17	93

[a]Measured one day after infection.

[b]Added one day before infection.

Source: From Ref. 8, with the IFNγ titers converted to reference units.

III. DEGRADATION OF TRYPTOPHAN

Despite the superficial similarities between the antiviral and antitoxoplasma effects of IFNγ, it seemed unlikely that the same fundamental mechanisms could be involved. The known antiviral mechanisms of IFN involve the synthesis of two host cell enzymes (reviewed in Ref. 9). One is $2',5'$-A polymerase, the product of which activates a host cell endonuclease that attacks mRNA. The other is a protein kinase that phosphorylates and hence inactivates initiation factor 2 of the host cell. Thus these two enzymes induced by IFNs ultimately act to inhibit viral protein synthesis in the cytoplasm of infected cells. Since *T. gondii* is insulated from the cytoplasm of its host cell by the membrane of its vacuole and by its own plasma membrane, enzymes induced in the host cell by IFNγ should have limited access to the parasite. Furthermore, if these enzymes played a significant antitoxoplasma role, then IFNα and β, which also induce these enzymes, should block the growth of *T. gondii*. But, as noted above, these two IFNs had no antitoxoplasma activity in our in vitro system. While seeking an alternative mechanism for IFNγ, we happened upon the observation that its antitoxoplasma activity was highly dependent upon the tryptophan concentration in the medium. Figure 1 shows the results of a titration of IFNγ in four media that differed only in tryptophan content. As the initial tryptophan concentration of the medium was increased, progressively more IFNγ was required to give significant inhibition of *T. gondii*.

The simplest explanation for dependence of the antitoxoplasma titer of IFNγ upon tryptophan concentration was that this amino acid was destroyed in the treated cultures. In our search for tryptophan degradation, we first examined the media of uninfected human fibroblast cultures treated with various concentrations of IFNγ. The fate of tryptophan was followed by adding [^{14}C] tryptophan to the medium and determining, at intervals, the radioactivity that cochromatographed with authentic tryptophan. Figure 2 documents the progres-

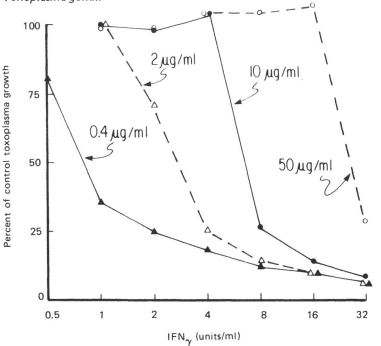

Figure 1 Titration of the antitoxoplasma activity of IFNγ in media with four tryptophan concentrations. Quadruplicate cultures in 24-well trays were treated with twofold dilutions of IFNγ in four media that differed only in tryptophan content. After 24 hr, all cultures were infected. At 48 hr after infection the parasite growth was measured by [³H]uracil incorporation. (Redrawn from Ref. 5 with the IFNγ concentrations converted to reference units.)

sive disappearance of tryptophan from the media of cultures treated with IFNγ. The rate of disappearance was a function of the IFNγ concentration in the medium. The final plateau seen in Figure 2 in which about 5% of the initial radioactivity continued to comigrate with tryptophan is an artifact. We found that most of the residual radioactivity that comigrated with tryptophan was actually D-[¹⁴C] tryptophan that was a contaminant of the L-[¹⁴C] tryptophan used to label the medium (7). Thus nearly all of the L-tryptophan in the media of cultures treated for 2 days with 4 or more units/ml of IFNγ was degraded.

We further examined the medium of cultures treated with 32 units/ml of IFNγ to identify the degradation products. Most of these products had to be in the medium because there was no significant loss of radioactivity from the medium when [¹⁴C] tryptophan was added at the same time as the IFNγ. Two-dimensional chromatography-electrophoresis followed by autoradiography

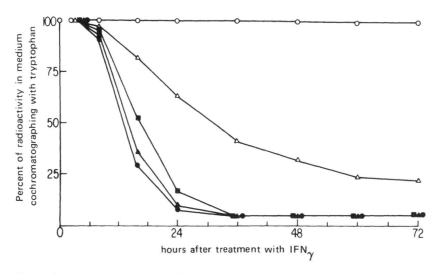

Figure 2 The disappearance of tryptophan from the medium of confluent
cultures treated with various concentrations of IFNγ. Duplicate cultures were incubated with zero (○——○), 1 (△——△), 4 (■——■), 16 (▲——▲), and 64 (●——●)
units/ml IFNγ in medium that contained [^{14}C] tryptophan. Samples of medium
were obtained at intervals and analyzed by chromatography. (From Ref. 7.)

showed that most of [^{14}C] tryptophan was degraded to two known tryptophan
metabolites N-formylkynurenine and kynurenine (5). In addition to these major
metabolites, approximately 5% of the radioactivity supplied as [^{14}C] tryptophan
was found distributed among four spots located autoradiographically but not, as
yet, identified chemically (5). The identification of the major tryptophan metabolites in the medium of human fibroblasts treated with human recombinant
IFNγ immediately suggested the well-known pathway for the degradation of
tryptophan shown in Figure 3.

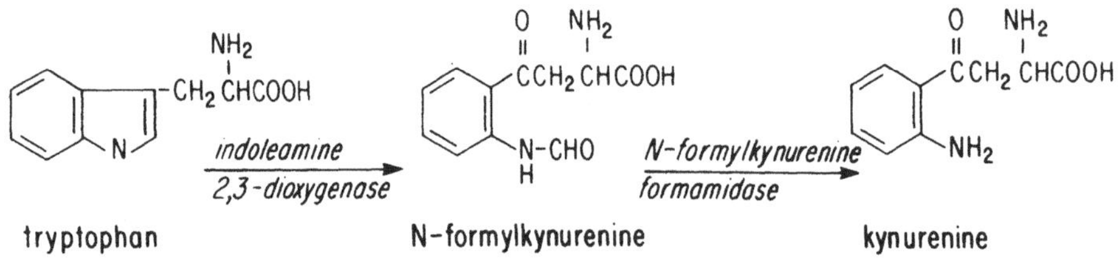

Figure 3 The pathway from tryptophan to kynurenine.

We used radiometric methods to assay both of the enzymes noted in Figure 3. Treatment of cultures with IFNγ had no effect on the activity of the second enzyme, N-formylkynurenine formamidase. The same specific activity, 80 pmol formic acid/hr per mg protein, was found in crude extracts of both treated and untreated cultures (10). In contrast, treatment with IFNγ markedly increased the activity of the first enzyme, indoleamine 2,3-dioxygenase. This enzyme could not be detected in extracts of untreated cultures. However, as shown in Figure 4, treatment with IFNγ resulted, after a brief lag, in the brisk induction of enzyme activity. The loss of tryptophan from the medium (see Fig. 2) had little effect on the induction of the enzyme since replacement of the destroyed tryptophan did not increase the amount of indoleamine 2,3-dioxygenase that was made. The observed kinetics of enzyme induction were consistent with the

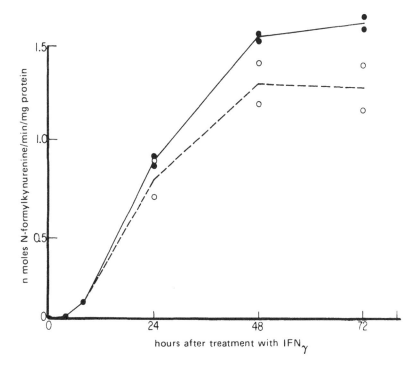

Figure 4 The induction of indoleamine 2,3-dioxygenase as a function of the time of treatment with IFNγ. Confluent monolayer cultures were treated with 32 units/ml at the same time. At various times, extracts were made for enzyme assays. Cultures incubated without supplemental tryptophan (●--●). Cultures treated with supplemental tryptophan, 20 μg/ml, at 12 hr intervals (○--○). (From Ref. 10.)

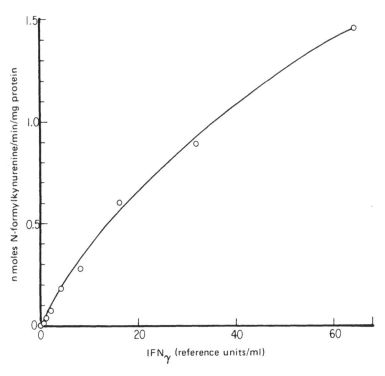

Figure 5 The effect of IFNγ concentration on the induction of indoleamine 2,3-dioxygenase in human fibroblasts. Duplicate cultures were treated with various concentrations of IFNγ and enzyme activities were measured 24 hr later. (From Ref. 10.)

previous observation (Table 2) that growth of *T. gondii* in human fibroblasts was not suppressed immediately after addition of IFNγ, but was suppressed one day later. The induction of indoleamine 2,3-dioxygenase apparently required both mRNA and protein synthesis (10) for it was blocked by low concentrations of actinomycin D (0.1 μg/ml) or of cycloheximide (10 μg/ml). As shown in Figure 5, the induction of indoleamine 2,3-dioxygenase was also a function of the amount of IFNγ supplied in the medium. The dependence of enzyme activity upon IFNγ concentration was most marked over the range of IFNγ concentrations that were effective in media with different concentrations of tryptophan (see Fig. 1).

IV. CHARACTERIZATION OF THE INDOLEAMINE 2,3-DIOXYGENASE INDUCED BY IFNγ

Two mammalian enzymes that degrade tryptophan to N-formylkynurenine have been described. The hepatic enzyme is termed tryptophan 2,3-dioxygenase be-

Table 5 Comparison of Enzymes that Degrade Tryptophan

Property	Enzyme		
	Hepatic tryptophan 2,3-dioxygenase	Intestinal indoleamine 2,3-dioxygenase	IFNγ-induced indoleamine 2,3-dioxygenase
Km, L-tryptophan (μM)	400	45	3
D-tryptophan as substrate (nmol/min/mg protein)	0	0.35	0

Source: From Ref. 10.

cause it has a narrow substrate specificity, attacking only L-tryptophan. It is inducible by high concentrations of ingested or parenterally administered tryptophan (11). The other enzyme is probably constitutive and is found in various visceral tissues, including intestinal epithelium. It is termed an indoleamine 2,3-dioxygenase because its substrates include not only L-tryptophan but also D-tryptophan and various other D- and L-indoleamines (12). The two enzymes also differ in Km for tryptophan. The hepatic enzyme has a high Km presumably to prevent it from readily exhausting the intracellular pool of tryptophan in liver cells. The intestinal indoleamine 2,3-dioxygenase has a lower Km. We found that the enzyme induced in human fibroblasts by IFNγ did not resemble either of the previously described enzymes. The human fibroblast enzyme failed to degrade D-tryptophan but did attack L-5-hydroxytryptophan, thus justifying its designation as an indoleamine 2,3-dioxygenase. The Km of the IFNγ-induced enzyme was lower than that of either of the previously described enzymes. Some properties of these three enzymes are compared in Table 5.

V. DEPLETION OF INTRACELLULAR TRYPTOPHAN

The simplest mechanism for the antitoxoplasma activity of IFNγ is starvation of the parasite for the essential amino acid tryptophan. Experiments to determine if *T. gondii* can synthesize tryptophan de novo are incomplete. We therefore turned to an indirect way to determine if the parasite was dependent upon an exogenous supply of this amino acid. Media that contained dialysed serum and reduced concentrations of tryptophan were used in infections of human fibroblasts with *T. gondii*. If tryptophan were an essential amino acid for *T. gondii*, the media with reduced tryptophan content should partially or completely starve the parasite. Data from experiments of this kind were consistent with the hypothesis that tryptophan was an essential amino acid for *T. gondii*. When the concentration of this amino acid in the medium was reduced to 0.8 μM, the growth of *T. gondii* was markedly suppressed (7).

The critical tryptophan concentration for *T. gondii* is that which prevails within the vacuole in which the parasite is growing. Since we were unable to analyze this compartment, we did the next best thing and studied the intracellular concentration of tryptophan in human fibroblasts treated with IFNγ. Since the vacuole in which *T. gondii* grows lies within the cytoplasm of its cell, the intravacuolar concentration of tryptophan is presumably most directly affected by that of the host cell cytoplasm. We briefly incubated IFNγ-treated and control cells in medium that contained [³H] tryptophan and then measured radioactive metabolites in ethanol-soluble pool extracts. Ethanol was used to extract the well-washed cells because one of the tryptophan metabolites, N-formylkynurenine, is decomposed by dilute acid. Figure 6 shows that in the control cultures, [³H] tryptophan was taken up into the soluble pool with the expected kinetics. The intracellular content of [³H] tryptophan rose rapidly in the early seconds, but then reached a plateau value at about 2 minutes. In contrast, almost no [³H] tryptophan could be detected in the ethanol-soluble pools of cultures treated with 32 units/ml IFNγ. Apparently, the radioactive amino acid was degraded as rapidly as it entered the IFNγ-treated cells. This conclusion

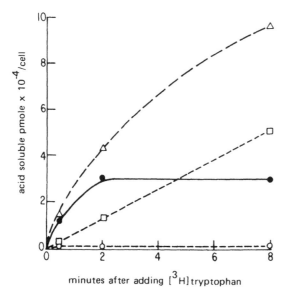

Figure 6 The fate of [³H] tryptophan in the soluble pools of control human fibroblasts and human fibroblasts previously treated for 12 hours with 32 units/ml IFNγ. Tryptophan in control cells (●——●). Tryptophan (○– –○). N-formylkynurenine (△– –△), and kynurenine (□– –□) in cells treated with IFNγ. (From Ref. 7.)

was supported by the marked accumulation of radioactive N-formylkynurenine and kynurenine within the IFNγ-treated cells (7).

VI. LACK OF EVIDENCE THAT TOXIC PRODUCTS OF TRYPTOPHAN DESTRUCTION CAN EXPLAIN THE ACTION OF IFNγ

A reasonable alternative mechanism for the antitoxoplasma activity of IFNγ is that one or more of the degradation products of tryptophan were toxic to *T. gondii* and thus suppressed its growth. We used two approaches to test this hypothesis. In the first, we assumed that the putative toxic metabolites leaked from the IFNγ-treated cells and could be found in the medium. Indeed, when the medium of IFNγ-treated cultures was transferred to fresh cultures that were then infected, inhibition of parasite growth was seen two days later (7). However, this effect was shown to result entirely from transfer of IFNγ, which is relatively stable in our medium. As shown in Table 6, when the residual IFNγ was neutralized by a specific monoclonal antibody, the medium from cultures treated with 32 units/ml IFNγ supported normal growth of *T. gondii* when used in fresh cultures. Similar results were obtained when the residual IFNγ was removed by ultrafiltration (7). Thus we were unable to detect any evidence of toxic substances in the medium of cultures treated with IFNγ. It should be noted that our experiments would not have detected such substances if they had a relatively short half-life or if they were unable to escape from the cells that produced them.

In an alternative approach, we tested the ability of the two identified metabolites of tryptophan, N-formylkynurenine and kynurenine, to block the growth of *T. gondii*. Cultures were infected with *T. gondii* in media that con-

Table 6 Absence of Toxic Metabolites in the Medium of Cultures Treated with IFNγ Provided that the Residual IFNγ was Neutralized with Antibody

IFNγ in used medium: units/ml	Anti-IFNγ treatment of used medium: anti-IFN units/ml	Percent of control growth of *T. gondii* in cultures incubated with used medium[a]
0	0	100
32	0	9
0	64	105
32	64	96

[a]Measured by the incorporation of [^{3}H]uracil.
Source: From Ref. 13.

tained various concentrations of these substances individually or together and the growth of the parasite was measured two days later. As shown in Figure 7, both N-formylkynurenine (1.6 mM) and kynurenine (3.2 mM) did suppress the growth of *T. gondii*. However, this inhibition was seen only at concentrations that far exceeded the concentration of tryptophan (0.05 mM) in the medium (7). Since the tryptophan of medium was the only source of these metabolites, their final concentration in the medium could not have exceeded the initial tryptophan concentration.

The hypothesis that N-formylkynurenine and kynurenine present in the medium of cultures treated with IFNγ act to inhibit the growth of *T. gondii* was excluded. However, our data did not eliminate the possibility that these metabo-

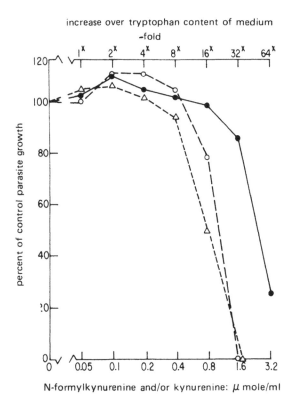

Figure 7 Effects of various concentrations of N-formylkynurenine (○– –○), kynurenine (●––●), or both together (△– –△) on the growth of *T. gondii*, measured by incorporation of [³H]uracil two days after infection. (From Ref. 7.)

lites achieved inhibitory *intracellular* concentrations in cells treated with IFNγ. Note that Figure 6 shows relatively high and still increasing intracellular concentrations of both metabolites. We assume that they accumulated in the intracellular pool for two reasons. First, activity of the interferon-induced indoleamine 2,3-dioxygenase exceeded that of the constitutive N-formylkynurenine formamidase, allowing a buildup of the first intermediate in tryptophan degradation. Second, the entry of tryptophan through its specific transporter was more rapid than the exit of tryptophan metabolites presumably by passive diffusion across the plasma membrane down the concentration gradient between the intracellular and extracellular compartments. Thus the intracellular parasites could have been exposed to concentrations of N-formylkynurenine and kynurenine that were higher than that of tryptophan in the medium. Since we have seen (Fig. 7) that higher concentrations of these substances were toxic to *T. gondii*, it was essential to analyze the intracellular pools of cultures treated with IFNγ. We determined the radioactivities of intracellular tryptophan and tryptophan metabolites during a two day IFNγ treatment of uninfected human fibroblasts incubated in medium that contained [^{3}H] tryptophan. The results are shown in Figure 8. The intracellular [^{3}H] tryptophan content of the control cultures increased only slightly over the course of the experiments, probably because cell division somewhat expanded the total volume of the soluble pool. As expected, the intracellular [^{3}H] tryptophan of the IFNγ-treated cultures fell to very low levels. This fall was not seen immediately (compare Fig. 6) because of the time required for the induction of indoleamine 2,3-dioxygenase. As the tryptophan content of the treated cells fell, the concentrations of both kynurenine and N-formylkynurenine showed a striking increase to levels that far exceeded the tryptophan content of the control cells. Then, about 12 hours after treatment with IFNγ, the intracellular concentrations of these metabolites began to fall to lower levels. The time of this fall coincided with the time at which the tryptophan content of the medium was partially exhausted (see Fig. 2) and thus less additional N-formylkynurenine could be made. The high concentrations of tryptophan metabolites within the cell were then dissipated by diffusion across the plasma membrane.

Figure 8 shows that the potentially toxic metabolites N-formylkynurenine and kynurenine reached relatively high concentration within IFNγ-treated cells. However, for several reasons, we do not think that these concentrations contributed to the antitoxoplasma activity of IFNγ. First, the intracellular N-formylkynurenine and kynurenine probably did not reach the toxic levels noted in Figure 7. Their intracellular concentrations can be estimated from the radioactivity measurements of Figure 8 because the tryptophan transport system of mammalian cells (the N system) is not driven by the sodium gradient across the plasma membrane and does not significantly concentrate this amino acid in the intracellular pool (14). Thus the intracellular tryptophan concentration of

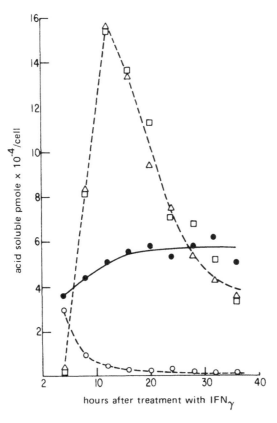

Figure 8 The fate of tryptophan in the soluble pool on control human fibroblasts and of fibroblasts treated with 32 units/ml IFNγ. The cultures were incubated with [³H] tryptophan at the time of IFNγ treatment. Tryptophan in control cells (●——●). Tryptophan (○— —○), N-formylkynurenine (□— —□), and kynurenine (△— —△) in cells treated with IFN-γ. (From Ref. 7.)

control cells should reflect that of the medium, 50 μM. We assume that the volume of the soluble pool is similar in control and in IFNγ-treated cells. Thus the ratio of total radioactivity in a tryptophan metabolite in IFNγ-treated cells and of total radioactivity in tryptophan in control cells can be used to estimate the molar concentration of the metabolite. Using the maximal ratios recorded in Figure 8, we calculated that the peak intracellular concentrations of N-formylkynurenine and of kynurenine were only 0.2 mM. This concentration is well below that required to inhibit *T. gondii* when present in the medium. There is good reason to believe that N-formylkynurenine and kynurenine enter the cell when supplied in the medium since they leak reasonably rapidly from the cell, when produced within it (Fig. 8). Our calculations suggest that although these two principal tryptophan metabolites did reach elevated concentrations in cells treated with IFNγ, these concentrations were insufficient to compromise the growth of *T. gondii*.

The second reason for thinking that the rise in intracellular N-formyl-kynurenine and kynurenine cannot account for the antitoxoplasma effect of IFNγ is that the two phenomena were not temporally coincident. The peak intracellular concentrations of tryptophan metabolites were found during the first 24 hours of IFNγ treatment. But during this interval, little or no antitoxoplasma effect was seen (Table 2). The maximal antitoxoplasma effect was actually observed when the intracellular concentrations of N-formylkynurenine and kynurenine had fallen to a lower plateau. The final argument against any toxic metabolite is that if they played a major role, increasing the tryptophan content of the medium should increase the antitoxoplasma potency of IFNγ because more toxic metabolites could be produced. However, as shown in Figure 1, increasing the concentration of tryptophan actually reduced the antitoxoplasma effect of IFNγ.

We conclude that there is no evidence for toxic tryptophan metabolites and a good deal of circumstantial evidence against their significance in the antitoxoplasma activity of IFNγ. The most reasonable explanation for the activity of IFNγ is that the intracellular parasites were starved by the depletion of an essential amino acid, tryptophan, in the cytoplasm of the host cell through the activity of a potent indoleamine 2,3-dioxygenase induced by IFNγ.

VII. LACK OF ANTITOXOPLASMA ACTIVITY OF MURINE IFNγ

All of our data obtained with human recombinant IFNγ are consistent with the hypothesis that the induction of indoleamine 2,3-dioxygenase played a central role in the antitoxoplasma activity. When we learned from Dr. Jenifer Turco (personal communication) that cultured mouse cells did not degrade tryptophan in response to treatment with murine IFNγ, the in vitro antitoxoplasma activity of murine IFN became of interest. We used murine recombinant IFNγ (Genentech, Inc.) and mouse 3T3 fibroblasts. Since no standard murine IFNγ was available, the interferon titers were based only upon parallel titrations of antiviral activity using encephalomyocarditis (EMC) virus as the challenge. The titers obtained were always within twofold of those reported by Genentech, Inc. Since *T. gondii* plaques are difficult to see in 3T3 cultures, we used only [³H] uracil incorporation to measure the growth of *T. gondii*. No inhibition of parasite growth was observed, even at the highest concentrations of murine IFNγ, 4×10^4 units/ml. In agreement with the results of Dr. Turco, no degradation of [¹⁴C] tryptophan was observed in cultured treated for two days with that concentration of murine IFNγ. Cultures of 3T3 cells treated with 4×10^4 units/ml of IFNγ showed no indoleamine 2,3-dioxygenase activity. This failure of IFNγ both to induce indoleamine 2,3-dioxygenase and to block the growth of *T. gondii* cannot be the result of an inability of the IFNγ to bind to receptors on the cell because its antiviral activity was readily detected. These results presumably reflect a striking species-specific difference in response to IFNγ. As IFNγs of other species become available, it would be of interest to determine their ability to induce indoleamine 2,3-dioxygenase in homologous fibroblasts.

VIII. IN VIVO SIGNIFICANCE

We have no evidence that induction of a tryptophan-degrading enzyme has any significance in the in vivo action of IFNγ. Murine IFNγ has recently been shown to protect mice against an otherwise fatal challenge with *T. gondii* (15). Our experiments with mouse fibroblasts showed that murine IFNγ did not induce indoleamine 2,3-dioxygenase in vivo. If, as seems likely, this result can be extrapolated to the in vivo situation, some other mechanism must be involved in the control of murine toxoplasmosis by IFNγ. IFNγ is known to be a multipotent immunological mediator (reviewed in Ref. 16). Among its many activities IFNγ functions as a macrophage-activating factor (17). Activated and nonactivated macrophages are strikingly different in their interactions with *T. gondii* in vitro. The parasite readily penetrates, grows in, and kills nonactivated macrophages. In contrast, activated macrophages phagocytose and kill *T. gondii*, predominantly by an oxidative mechanism. Activated macrophages must play a major role in suppressing infection with *T. gondii*. We do not know if degradation of tryptophan plays any role in recovery from toxoplasmosis in humans and in other species in which indoleamine 2,3-dioxygenase might be induced by IFNγ. *Toxoplasma gondii* does not infect only macrophages, but grows in most cells of the body. IFNγ has been shown to lower blood tryptophan levels in humans (18) and thus indoleamine 2,3-dioxygenase is likely to be induced in vivo as well as in vitro. If the induction of this enzyme serves to suppress the growth of *T. gondii* in cells other than macrophages, other defense mechanisms including the activation of macrophages may have greater opportunity to play their roles.

ACKNOWLEDGMENTS

Genentech, Inc., generously supplied the recombinant IFNs, which made these experiments possible. The research was supported by Grant AI-14151 from the National Institutes of Health.

REFERENCES

1. Pfefferkorn, E.R., and L.C. Pfefferkorn. 1976. *Toxoplasma gondii*: Isolation and preliminary characterization of temperature-sensitive mutants. *Exp. Parasitol. 39*:365–376.

2. Remington, J.S., and T.C. Merigan. 1968. Interferon: Protection of cells infected with an intracellular protozoan (*Toxoplasma gondii*). *Science 161*: 804–806.

3. Schmunis, G., M. Weissenbacher, E. Chowchuvech, L. Sawicki, M.A. Galin, and S. Baron. 1973. Growth of *Toxoplasma gondii* in various tissue cultures treated with In.Cn or interferon. *Proc. Soc. Exp. Biol. Med. 143*:1153–1157.

4. Ahronheim, G.A. 1979. *Toxoplasma gondii*: Human interferon studies by plaque assay. *Proc. Soc. Exp. Biol. Med. 161*:522–526.

5. Pfefferkorn, E.R. 1984. Interferon-γ blocks the growth of *Toxoplasma gondii* in human fibroblasts by inducing the host cells to degrade tryptophan. *Proc. Natl. Acad. Sci. (USA) 81*:908–912.

6. Pfefferkorn, E.R., and P.M. Guyre. 1984. Inhibition of growth of *Toxoplasma gondii* in cultured fibroblasts by human recombinant gamma interferon. *Infect. Immun. 44*:211–216.

7. Pfefferkorn, E.R., M. Eckel, and S. Rebhun. 1986. Interferon-γ suppresses the growth of *Toxoplasma gondii* in human fibroblasts through starvation for tryptophan. *Molec. Biochem. Parasitol. 20*:215–224.

8. Pfefferkorn, E.R. 1986. Interferon gamma and the growth of *Toxoplasma gondii* in fibroblasts. *Ann. Inst. Pasteur/Microbiol. 137A*:348–352.

9. Kerr, I.M., P.J. Cayley, R.H. Silverman, and M. Knight. 1982. The antiviral action of interferon. *Phil. Trans. R. Soc. Lond. B299*:59–67.

10. Pfefferkorn, E.R., S. Rebhun, and M. Eckel. 1986. Characterization of an indoleamine 2,3-dioxygenase induced by gamma-interferon in cultured human fibroblasts. *J. Interferon Res. 6*:267–279.

11. Tanaka, T., and W.G. Knox. 1959. The nature and mechanism of the tryptophan pyrrolase (peroxidase-oxidase) reaction of pseudomonas and of rat liver. *J. Biol. Chem. 234*:1162–1170.

12. Shimizu, T., S. Nomiyama, F. Hirata, and O. Hayaishi. 1978. Indoleamine 2,3-dioxygenase purification and some properties. *J. Biol. Chem. 253*: 4700–4706.

13. Pfefferkorn, E.R. 1987. The mechanism by which interferon gamma blocks the growth of *Toxoplasma gondii* in cultured fibroblasts. In *Host-Parasite Cellular and Molecular Interactions in Protozoal Infections*. Edited by K.-P. Chang and D. Snary. Springer Verlag, Berlin, Heidelberg, New York, pp. 345–354.

14. Oxender, D.L., M. Lee, P.A. Moore, and G. Cecchini. 1977. Neutral amino acid transport systems of tissue culture cells. *J. Biol. Chem. 252*:2675–2679.

15. McCabe, R.E., B.J. Luft, and J.S. Remington. 1984. Effect of murine interferon gamma on murine toxoplasmosis. *J. Infect. Dis. 150*:961–962.

16. Trinchieri, G., and B. Perussia. 1985. Immune interferon: A pleiotropic lymphokine with multiple effects. *Immunol. Today* 6:131–136.

17. Nathan, C.F., H.W. Murray, M.E. Wiebe, and B.Y. Rubin. 1983. Identification of interferon-γ as the lymphokine that activates human macrophage oxidative metabolism and antimicrobial activity. *J. Exp. Med. 158*:670–689.

18. Byrne, G.I., L.K. Lehmann, J.G. Kirschbaum, E.C. Borden, M.L. Caroll, and R.R. Brown. 1986. Induction of tryptophan degradation in vitro and in vivo: A γ-interferon-stimulated activity. *J. IFN Res. 6*:389–396.

9

Leishmania

Carol A. Nacy, Miodrag Belosevic, Monte S. Meltzer, and David L. Hoover /
Walter Reed Army Institute of Research, Washington, D.C.

I. INTRODUCTION

In the hierarchy of intracellular pathogens, the *Leishmania* surely rank at, or
near the top of the list in selectivity. In their mammalian hosts, these protozoan
parasites infect only cells of the monocyte/macrophage lineage. There they reside and replicate in phagolysosomes, the intracellular compartment of macrophages that eliminates most ingested pathogens (1,2). This extraordinary cell
preference both underscores the resistance of the parasite to classic macrophage
killing mechanisms, and explains the chronic nature of leishmanial disease: it is
exceedingly difficult to eliminate microorganisms whose stable habitat is the di-
gestive organelle of the principal effector cell of the immune system. How then
does the host resolve leishmanial infections? In certain chronic cutaneous
diseases, it does so slowly through immunologic control of macrophage antimicrobial activities that limit parasite infectivity and replication. These natural
host defenses must frequently be augmented by chemotherapy in visceral
disease. In this chapter, we will examine the immunologic regulation of certain
potent macrophage effector reactions that influence the balance between host
survival and parasite replication: those that prevent the parasite from entering its
host cell, and those that alter the intracellular environment of the macrophage so
that the parasite is unable to survive or replicate.

II. LEISHMANIA

The genus *Leishmania* is composed of several genetically and physiologically
distinct species that are introduced into the human through the bite of a sandfly

167

(Table 1). Each species enters its host in a similar manner: the sandfly tears the epidermis to obtain a blood meal, and deposits the insect stage (promastigote) onto the dermal/epidermal juncture during the process of feeding on pooled blood. It is thought that the initial cell infected by the promastigote is the Langerhans cell, the resident tissue macrophage of skin. Direct experimental evidence for this presumption, however, is still lacking. A profound alteration in phenotype and physiology of the promastigote occurs in its initial mystery host cell: the parasite loses its prominent flagellum, and therefore its motility; its physiology changes from aerobic respiration to anaerobic metabolism; and it loses its capacity to replicate outside of cells. This obligately intracellular tissue form, or amastigote, is the only form of the parasite observed in infected individuals, and it is always found within macrophages (Fig. 1).

Once the promastigote has transformed to the amastigote, the infectious cycle is initiated and the course of infection with the different leishmanial species diverges. Unknown physiologic attributes of both the host and the parasite determine whether the infection remains localized at the site of the sandfly bite or spreads systemically to lymph nodes and fiscera. Simple cutaneous disease is caused by *L. major* in the old world, *L. mexicana mexicana* in the new, and cutaneous infections by these parasites rarely have systemic complications; visceral infections are caused by *L. donovani* and *L. infantum* in the old world, *L. chagasi* in the new, and active cutaneous lesions at the sandfly bite are rarely associated with systemic infections. In between these two polar clinical syndromes is a spectrum of diseases with combinations of cutaneous and systemic manifestations. In leprosy, spectral diseases ranging from tuberculoid to lepromatous lesions are all caused by the same microorganism, *Mycobacterium leprae*. In leishmaniasis, however, each different clinical syndrome is a result of infection with a distinct species of parasite (Table 1). The host factors that may play a role in exacerbation of disease in certain of the leishmaniases are, at present, unknown.

Fundamental to establishment of intracellular infection is parasite entry into the host cell. Susceptibility or resistance to infection, then, depends on the efficiency with which infection of individual cells occurs. For *Leishmania*, which replicate only in macrophages, intracellular infection requires participation of both host cell and parasite. Infection begins with parasite attachment to macrophages. Promastigotes preferentially attach by the flagellum, with a limited number of discrete contact points along the body of the organism (3,4). Attachment is not a random phenomenon, but requires interaction of parasite membrane structures with specific receptor sites on macrophages (5): the mannose/fucose receptor and the receptor for C3bi, a complement component (6). For effective internalization, the promastigote must bind both receptors simultaneously (7). Attachment of both promastigotes and amastigotes requires effort on the part of the parasite. Attachment is inhibited by pretreatment of

Table 1 The *Leishmania*: Clinical Syndromes and Geographical Distribution

Species/subspecies	Clinical syndrome	Geographic distribution
L. donovani complex		
L. d. donovani	Visceral	India
L. d. infantum	Visceral (infantile)	Mediterranean littoral; Near/Middle East; Africa; China
L. d. chagasi	Visceral	South/Central America
L. tropica complex		
L. tropica	Cutaneous (dry, urban oriental sore)	Middle East; Mediterranean littoral
L. major	Cutaneous (moist, rural oriental sore)	Middle East; Arabia; Russia; India; Africa
L. aethiopica	Cutaneous; DCL	Ethiopia; Kenya
L. mexicana complex		
L. m. mexicana	Cutaneous; DCL rare	Mexico; Guatemala; Belize
L. m. amazonensis	Cutaneous; DCL	Brazil; Venezuela
L. braziliensis complex		
L. b. braziliensis	Cutaneous; MCL	Brazil; Peru; Ecuador; Bolivia; Venezuela; Paraguay; Colombia
L. b. panamensis	Cutaneous; MCL rare	Panama; Costa Rica; Colombia
L. b. guyanensis	Cutaneous	Guyana; Brazil; Surinam
L. b. peruviana	Cutaneous	Peru

Abbreviations: DCL = diffuse cutaneous leishmaniasis; MCL = mucocutaneous leishmaniasis.

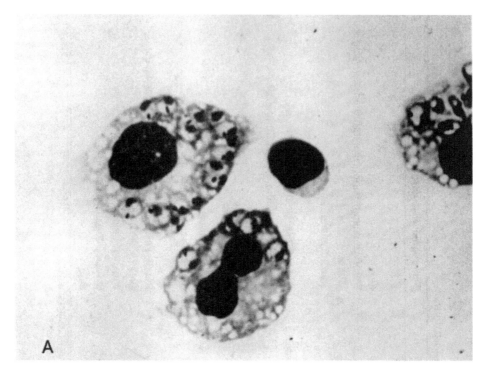

Figure 1 Resident peritoneal macrophages infected with amastigotes of *L. major* and cultured in medium for 72 hr (A) or lymphokines for 72 hr (B).

parasites with cytochalasin B (8,9); cytochalasin treatment of macrophages, in contrast, prevents internalization but not binding (9). Other inhibitors of phagocytosis also inhibit internalization of parasites. In sum, these findings suggest that leishmanial infection of macrophages is a cooperative activity initiated by the parasite, but finally accomplished by the host cell (10).

Parasites must not only bind to and enter the host cell, but must arrive at their destination alive. To do so, they must overcome the potent cytotoxicity of macrophage oxygen products. *Leishmania* promastigotes are deficient in catalase and glutathione peroxidase, which scavenge hydrogen peroxide (11). Because of this deficiency, they are readily killed by the hydrogen peroxide. They do, however, resist destruction by other potentially toxic oxygen products (11,12). Not only are promastigotes extremely sensitive to hydrogen peroxide, but they also induce its generation by mononuclear phagocytes during attachment to the cell membrane (13). As a consequence of these phenomena, at least 80% of *L. donovani* promastigotes are destroyed when incubated with monocytes or macro-

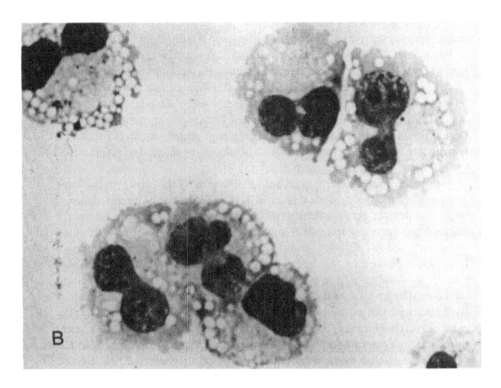

phages cultured in vitro (11,13). Given this exquisite susceptibility to oxygen-mediated destruction by macrophages from nonimmune animals, it is surprising that infection occurs at all! Recent studies on alternative pathway complement activation by promastigotes suggest that deposition of C3b on the surface of the promastigote actually protects the parasite from the toxic effects of hydrogen peroxide (14).

Despite a close family relationship, there are marked differences in susceptibility of *Leishmania* strains to oxygen-mediated damage. For example, promastigotes of *L. donovani* are threefold more sensitive to the lethal effects of hydrogen peroxide than promastigotes of *L. major*. *L. donovani* amastigotes, however, are sevenfold more resistant than promastigotes of the same strain, and survive macrophage contact to replicate intracellularly (13).

The ultimate fate of intracellular *Leishmania* depends on the physiologic state of the macrophage. In general, the events that lead to macrophage activation for leishmanicidal activity parallel those elucidated by Mackaness (15) in his pioneering work on resistance to infection with *Listeria monocytogenes*. Thus, *Leishmania* infect and replicate in resident macrophages harvested from tissues of nonimmune animals (2,16). In contrast, macrophages from immune animals previously infected with *Leishmania* destroy the parasites (17). Although macrophage cytotoxicity is clearly regulated by immune lymphocytes (18), induction of leishmanicidal activity in infected macrophages is not antigen specific: lymphocytes from *Toxoplasma*-infected animals, for example, can substitute for lymphocytes from *Leishmania*-infected animals if *Toxoplasma* antigen is present in the culture. Conversely, lymphocytes from *Leishmania*-infected animals induce macrophage cytotoxicity for *Listeria* if *Leishmania* antigen is present in the culture (19). These findings, as in the *Listeria* model, document that lymphocytes sensitive to one pathogen activate macrophages for cytotoxicity against other, unrelated pathogens.

Murine peritoneal macrophages activated in vivo by the classic macrophage activation agent, *Mycobacterium bovis* BCG (bacille Calmette-Guerin), demonstrate two distinct antimicrobial activities against amastigotes: (a) peritoneal macrophages harvested from BCG-immune animals are resistant to infection when compared to cells from untreated control animals, and (b) macrophages from BCG-immune mice that become infected clear the intracellular pathogen (20). These findings with in vivo-activated macrophages are reproduced completely in vitro by treatment of resident peritoneal macrophages from uninoculated mice with the soluble products of antigen- or mitogen-stimulated lymphocytes or lymphokines (21). If macrophages are treated with lymphokines before introduction of the parasite, the number of cells that become infected with *L. major* is reproducibly decreased; if, however, the cells are infected first, the cells are cured of their intracellular pathogens by treating them with lymphokines for 4 to 8 hr after infection (21,22). In a manner analogous to the activation of complement or blood-clotting systems, lymphokine-induced macrophage activation is remarkably complex, and occurs as a defined series of reactions. Characterization of immunoregulatory cytokines that induce macrophage antimicrobial activities, both resistance to infection and intracellular killing, has captured our attention for the last several years.

III. INDUCTION OF MACROPHAGE RESISTANCE TO INFECTION

In 1969, Miller and Twohy observed that macrophages obtained from mice inoculated with viable *M. bovis* BCG ingest fewer *L. donovani* amastigotes in vitro than macrophages from control mice (17). The implications of these studies remained unappreciated until the rediscovery in the late 1970s and early 1980s of this phenomenon in vitro with several different obligate or facultative intracellular bacteria (20,23-27). Since then, the decreased ingestion of obligate and facultative intracellular parasites by activated macrophages, operationally designated "resistance to infection," has been observed for cells exposed to immune reactions in vivo, as well as macrophages exposed to lymphokines in vitro. The immunologically regulated macrophage resistance to infection may have evolved to eliminate macrophages as host cells for replication of specialized intracellular parasites.

The characteristics that have emerged from the limited studies on this effector activity of activated macrophages include:

1. The time course of induction suggest that a synthetic or metabolic event precedes expression of the effector activity, since 4 hr minimum lymphokine treatment before introduction of the target is required to observe resistance to infection (21–23). Cells must also be exposed to lymphokine activation signals shortly after explantation; the ability of macrophages to be activated for resistance to infection decays with time in culture before addition of the activation stimuli (22).

2. Induction of the effector activity results in a long-lived phenotypic change in the activated macrophage. Intracellular and extracellular killing activities of activated macrophages are effector reactions that irreversibly decay in 24 hr (21–23,28). In contrast, activated macrophage resistance to infection persists longer than 96 hr in vitro in the absence of further stimulation (22).

3. Induction of the effector reaction is optimal with certain in vitro culture systems: macrophages maintained as a nonadherent pellet appear to express this effector reaction better than macrophages adherent to glass (29). The metabolic differences between adherent and nonadherent macrophages are legion (30,31).

4. Resident peritoneal macrophages and macrophages obtained after injection of mice with inflammatory stimuli in vivo develop equivalent resistance to infection when exposed to lymphokines for 4 to 6 hr in vitro (32). This is in marked contrast to intracellular killing activity, which is severely reduced for inflammatory macrophages infected with certain target microorganisms (33).

5. Resistance to infection is not parasite specific, but occurs with several different protozoan parasites and obligate and facultative intracellular bacteria (17,20,23–27). This effector activity, it should be noted, is principally active against microorganisms that have an absolute requirement for metabolic activity for entry into cells: resistance to infection does not occur with organisms whose infectious particles are metabolically inert, such as the *Chlamydia* (34).

Which of the immunoregulatory cytokines in our lymphokine preparations induce the macrophage effector activity that we refer to as resistance to infection? Since interferon γ (IFNγ) is a macrophage activation factor that induces a variety of effector activities, we assumed that IFNγ played a role in the induction of this effector reaction as well. As it turned out, we were right, but the induction of resistance to infection is markedly more complex, and consequently more interesting, than we had originally envisaged. We now know that resistance to infection is induced by the cooperative activites of several endogenous lymphokine signals, and that IFNγ is a mandatory but insufficient participant in this reaction.

The original studies on physicochemical characterization of the lymphokines that induce resistance to infection demonstrated a single peak of activity in the 50–55,000 M_r region following chromatography on sizing gels (21,35,36). This region of the Sephadex G-100 column also contains all the antiviral activity in lymphokines. To determine whether the IFNγ present in the activity peak was responsible for induction of resistance to infection, we tested recombinant IFNγ directly (37, Fig. 2): recombinant IFNγ is totally ineffective. The story is very complicated, however. Although IFNγ itself has no activity for induction of resistance to infection, monoclonal antibodies prepared against recombinant IFNγ completely ablate the induction of resistance to infection by lymphokines (Fig. 2). Further, adding back the ineffective IFNγ to the ineffective IFN-depleted lymphokines restores the activity of the lymphokine. Thus, it appears that

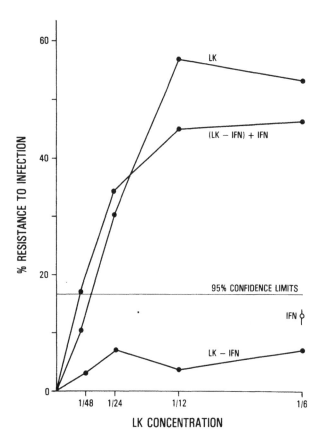

Figure 2 Dose response of lymphokines for induction of resistance to infection. Cells were treated with lymphokines, lymphokines depleted of IFNγ by immunoaffinity chromatography, or IFNγ (50 U/ml) for 6 hr and washed before introduction of the parasite. Resistance to infection was calculated as percent decrease in infected macrophages 2 hr after addition of parasites in treated cultures compared to medium-treated controls.

multiple factors in lymphokines, one of which is IFNγ, work together to regulate this macrophage effector function.

With this information in mind, we went back to the initial observation of a single activity peak in lymphokines for induction of macrophage resistance to infection, and fractionated lymphokines depleted of IFNγ by immunoaffinity chromatography on the sizing gel, polyacrylamide P-100 (38, Fig. 3). Panel a shows the single peak of activity in intact lymphokines; panel b shows the complete loss of the activity peak in lymphokines depleted of IFNγ. We then added back recombinant IFNγ to each of the inactive fractions in panel b, and found two, not one, major activity peaks for induction of resistance to infection (panel c). The second peak, at approximately 35,000 M_r, is never observed in fractionated intact lymphokines because it requires the presence of IFNγ for its activity, and IFN activity chromatographs at 50,000 M_r. Thus, there are at least two non-IFN lymphokines that are involved with IFNγ in the induction of macrophage resistance to infection: one that chromatographs with IFNγ and is approximately 50,000 M_r, and a second that is approximately 35,000 M_r. These two activities represent roughly 90% of the total activity in lymphokines for induction of resistance to infection (Fig. 3, panel d).

Clearly, characterization of the lymphokines that induce macrophage resistance to infection is a priority of our laboratory. Equally clear, however, is that the extraordinary biological activity of these lymphokines which facilitated their discovery is an impediment to isolation and purification of the factors by classic biochemical techniques. While we develop immunochemical reagents to replace the cumbersome biological assay, we examined the activity of a number of recombinant and highly purified cytokines available commercially or through collaborations (Fig. 4).

None of the cytokines alone induce macrophage resistance to infection (Fig. 4, solid bars); several of the purified lymphokines, however, are able to cooperate with IFNγ to induce this effector activity: Interleukin-2 (IL-2), granulocyte-macrophage colony-stimulating factor (GM-CSF), and the B-cell stimulatory factor-1 (BSF-1/IL-4) (Fig. 4, hatched bars). It is fascinating that lymphokines effective in the presence of IFNγ are not confined to a single category, but include cytokines with very diverse functions: a factor that regulates the differentiation of granulocytes and macrophages from bone marrow stem cells, a factor that stimulates the proliferation of antigen-stimulated T lymphocytes, and a factor that regulates the differentiation of B lymphocytes in the presence of IgM. None of these factors share significant amino acid homologies with each other, or with IFNγ; none of these factors use the same receptors as the others, or as IFNγ, in their regulatory role for other cells. And to date, none of these factors have been shown to induce a macrophage antimicrobial activity.

As the activity of each of these factors has been defined on cells other than the macrophage, we felt it important to document beyond a shadow of a doubt

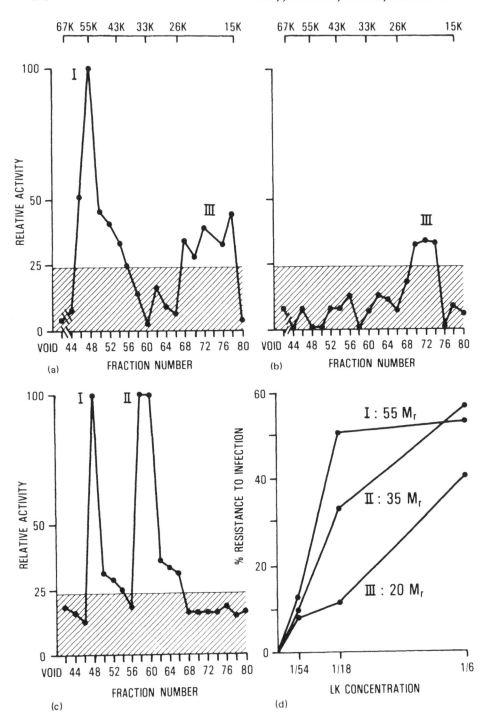

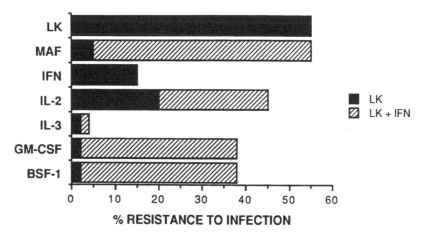

Figure 4 Cooperation of lymphokines for induction of macrophage resistance to infection. Cells were treated with the various cloned or purified lymphokines alone (solid bars) or in the presence of 50 U/ml IFNγ (hatched bars) for 6 hr, washed, and exposed to parasites. "MAF" represents lymphokines depleted of IFNγ by immunoaffinity chromatography; all cloned or purified lymphokines were used at 100 U/ml.

that the induction of resistance to infection is influenced by a direct interaction of the lymphokines on macrophages. To do this, we use three types of macrophage preparations: peritoneal cells enriched for macrophages by treatment with lymphocyte cell surface-specific antibody and complement, peritoneal cells from nude mice treated with the same antibody cocktail and complement, and bone marrow-derived macrophages cultured in colony-stimulating factor-1 (CSF-1) for 7 days (virtually 100% macrophages by several macrophage markers). Figure 5 shows the dose responses of the various effective lymphokines in the presence and absence of IFNγ with bone marrow macrophages (39). The plateau for resistance to infection is slightly different for peritoneal macrophages (range,

Figure 3 Induction of resistance to infection in macrophages treated with polyacrylamide P-100 fractions of lymphokines. (a) Whole lymphokines. Relative activity was determined by comparison of activity of each fraction to a dose response of the starting material; (b) lymphokines depleted of IFNγ by immunoaffinity chromatography; (c) lymphokines depleted of IFNγ: 50 U/ml IFNγ added to each fraction prior to treatment of macrophages; (d) Dose response of the three activity peaks for induction of resistance to infection in cells treated with 50 U/ml IFNγ: peak I represents 48% total lymphokine activity; peak II, 38%; and peak III, less than 15%.

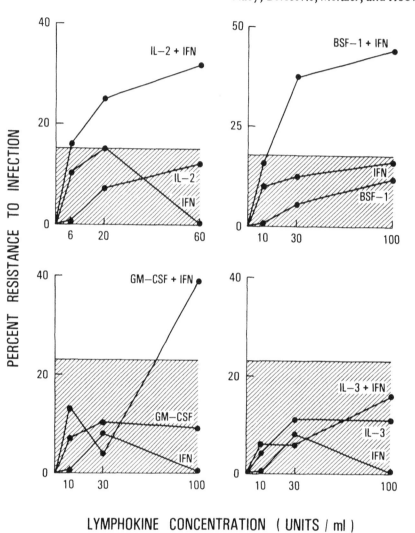

Figure 5 Dose responses of lymphokines for induction of resistance to infection in the presence or absence of IFNγ (50 U/ml) assayed on bone marrow-derived macrophages.

40–60%) and bone marrow-derived macrophages (range, 30–45%), but the dose responses for the different factors are identical. Under no circumstances have we observed resistance to infection in the absence of IFNγ, even at extraordinarily high (tens of thousands of units) doses of the other lymphokines; similarly, IFNγ remains ineffective by itself at these high doses. Results similar to those presented in Figure 5 are observed with the other two macrophage preparations. We feel confident that the participation of IL-2, GM-CSF, and BSF-1/IL-4 with IFNγ in the induction of macrophage resistance to infection represents an interaction of these factors directly with macrophages. Further evidence for the essential role that IFNγ plays in the induction of this effector reaction comes from mixing and matching experiments with the various second lymphokines: these factors only work in the presence of IFNγ, and no combination of IL-2, GM-CSF, or BSF-1/IL4 without IFN is effective.

The induction of a number of activated macrophage effector activities occurs by sequential interaction of endogenous lymphokines and accessory factors. If the sequence is changed, the reaction fails to develop (28). The absolute requirement for two lymphokines for induction of resistance to infection seemed to fit this pattern, and we attempted to identify the correct sequence by careful analysis of the kinetics of the reaction (40). Figure 6 shows the development of resistance to infection by macrophages treated with either IFNγ (panel a) or IL-2 (panel b) for 20 hr, then exposed to a short pulse of the second signal. To our surprise, the sequence of interaction of cells and lymphokines is not critical for induction of this activated macrophage effector reaction. Cells exposed to either signal for 20 hr develop resistance to infection with only a single 1–2 hr pulse with the second lymphokine (38,40). These results have been repeated with the other effective lymphokines, as well. Furthermore, expression of resistance to infection by macrophages is blocked by both treatment of cells at lower (4°C) temperature and treatment of cells with inhibitors of protein synthesis during exposure to the second signal (38). Thus, the first signal can be generated by any of the lymphokines, as can the second, as long as IFNγ is one of the lymphokines in the reaction. The sensitivity of the effector reaction to inhibitors suggests that the second signal initiates protein synthesis, but the rapidity with which this effector reaction can be observed after addition of the second signal (1 hr) suggests translation of preformed mRNA. Does the first signal simply deregulate a gene that is now transcribed into message? How do four distinct lymphokines deregulate this same gene when scores of lymphokines are not effective? What are the cellular controls that allow accumulation of message without translation? How does the second signal initiate translation and protein synthesis? All of these are intriguing questions, but questions that presently have no answers.

Several recent reports document specific receptors for IFNγ, IL-2, GM-CSF, and BSF-1/IL-4 on macrophages (41–43). We wondered if one role of the first

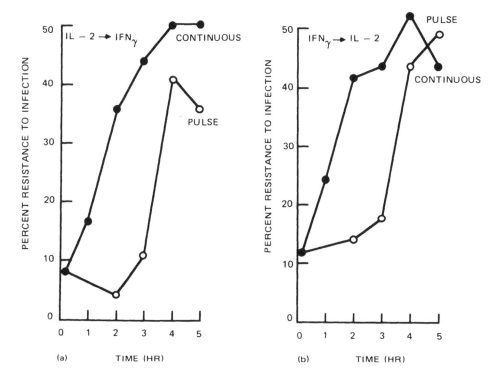

Figure 6 Time course for development of resistance to infection in macrophages exposed sequentially to IL-2 and IFNγ. Cells were exposed to the first signal for 20 hr, washed, then pulsed with the second signal for up to 6 hr (●——●) or with the second signal for 1 to 6 hr followed by medium for a total of 6 hr (○——○), washed and infected with parasites.

signal was to upregulate the receptor for the second lymphokine required for expression of resistance to infection. There is some evidence to support this hypothesis: IFNγ increases the number of a variety of receptors on several different cell types, including macrophages. However, our evidence to date suggests that IL-2 and BSF-1/IL-4 do not upregulate the IFNγ receptor. Thus, for the hypothesis to be true, IFNγ would be an obligatory first signal, and it is not. We conclude that the interaction of IFNγ and the other activating lymphokines is dependent upon unknown postreceptor event(s).

While we have much to do before we understand completely the inductive process for activated macrophage resistance to infection, it is clear that IFNγ plays an essential role in the regulation of this effector activity. We now have a number of tools available that will facilitate our studies, notably purified lymphokines and reproducibly effective pure macrophage populations, and we should be able to systematically attack the cell biology of this activated macrophage antimicrobial activity.

IV. INDUCTION OF MACROPHAGE INTRACELLULAR KILLING ACTIVITIES

Lymphokines also induce another potent antileishmanial activity in resident peritoneal macrophages: intracellular killing. This effector activity can be distinguished from resistance to infection in vitro by treating the macrophage with lymphokines after the parasite has settled into its intracellular habitat. With lymphokine treatment, the percentage of infected cells is reduced 80–100% compared to cells incubated in medium alone (Fig. 1). Much of the antileishmanial macrophage activation activity in lymphokine preparations can be attributed to IFNγ. As little as 5 IU/ml IFNγ induces maximal microbicidal activity in murine macrophages (44), and both nonglycosylated IFNγ produced by recombinant DNA technology and glycosylated forms of the molecule secreted by T-cell hybridomas are equally effective. Fifty to 60% of the activity for induction of intracellular killing in lymphokines is removed with monoclonal anti-IFNγ antibodies added to the culture fluids, or affixed to a solid support matrix to remove IFN from solution. Residual lymphokine activity is due to activation factors antigenically distinct from IFNγ (44).

The cytotoxic response of lymphokine-treated macrophages is very dramatic, but requires rather precise conditions for optimal effect. Timing, for example, is critical. Macrophage responsiveness for lymphokine-induced cytotoxicity decays rapidly: with culture for as little as 8 hr prior to lymphokine treatment, macrophage leishmanicidal activity declines 35%; with culture for 24 hr prior to treatment, cells are totally unresponsive (22). These findings suggest that as yet undefined factors present in vivo, but absent in vitro, maintain macrophages in a receptive state for the action of macrophage activation factors.

Intracellular killing is one of the many effector activities of activated macrophages that requires the sequential interaction of cells with endogenous mediators and accessory factors in a defined sequence for successful induction (28). Factors generated during an immune response "prime" the cell for killing activities, but the actual microbicidal effector function occurs after a "trigger" signal is received by the primed macrophage. Certain aspects of this priming phenomenon can be recapitulated by experiments with brief pulses of lymphokine in vitro (Fig. 7, 32). Treatment of macrophages with high concentrations of lymphokine for 4–8 hr elicits microbicidal activity at 72 hr, equivalent to that induced by treatment with lymphokine throughout the entire culture period (21). Macrophages pulsed with lymphokine for less than 4 hr, however, fail to develop microbicidal activity. During this crucial 4 hr time period of continuous

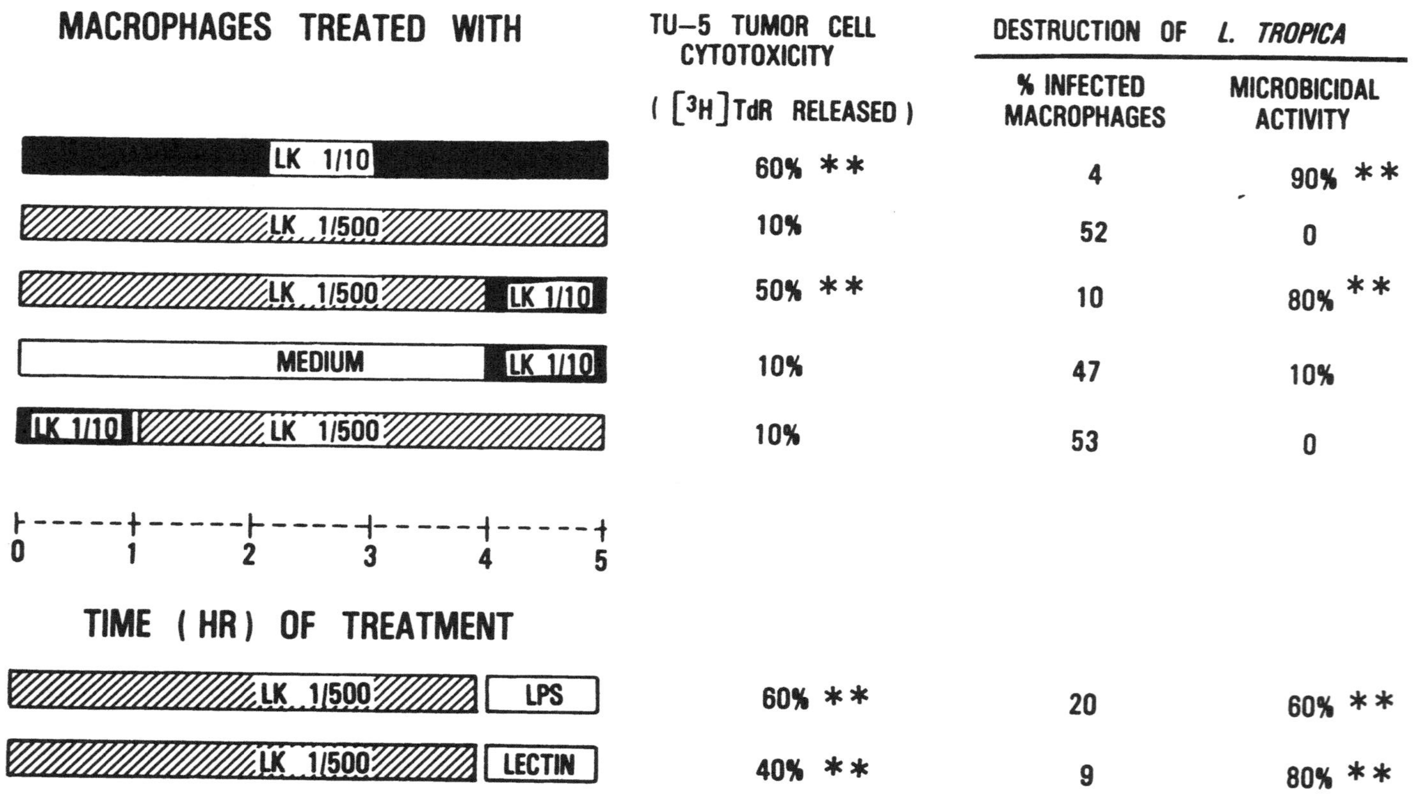

MACROPHAGES TREATED WITH

	TU–5 TUMOR CELL CYTOTOXICITY ([3H]TdR RELEASED)	DESTRUCTION OF *L. TROPICA*	
		% INFECTED MACROPHAGES	MICROBICIDAL ACTIVITY
LK 1/10	60% **	4	90% **
LK 1/500	10%	52	0
LK 1/500 LK 1/10	50% **	10	80% **
MEDIUM LK 1/10	10%	47	10%
LK 1/10 LK 1/500	10%	53	0

TIME (HR) OF TREATMENT

0 1 2 3 4 5

| LK 1/500 LPS | 60% ** | 20 | 60% ** |
| LK 1/500 LECTIN | 40% ** | 9 | 80% ** |

Figure 7 Development of activated macrophages in vitro by sequential interaction of cells and lymphokine signals: priming and triggering for tumor cytotoxicity and intracellular killing of *Leishmania* amastigotes.

exposure to low or high concentrations of lymphokine, the physiology of the resting macrophage is changed: it is "primed" for cytocidal activity, but is not yet able to kill. The primed cell is now receptive to "triggering" by a short pulse (15 min) of high-dose lymphokine or immunologically nonspecific stimuli such as lipopolysaccharide (LPS) or certain plant lectins. The requirement for several hours of priming is critical: administration of microbicidal concentrations of lymphokine for 15-60 min at the beginning of the experiment will not prime macrophages to respond to a subsequent 15 min lymphokine pulse. The order of treatment with priming and triggering agents is also distinct. Agents such as LPS, though capable of triggering the microbicidal response, do not prime the macrophage for triggering by LPS or lymphokine. These findings suggest that intervention by lymphokine may provide specificity to the microbicidal response. Further specificity is provided by distinct pathways for non-IFN lymphokines and IFNγ during the priming and triggering sequence. Non-IFN lymphokines do not prime macrophages to respond to IFNγ as a trigger, and IFNγ does not prime macrophages to respond to the non-IFN lymphokines as a trigger (36, Figs. 8 and 9).

Why redundant lymphokines for activation of macrophage antimicrobial activities? It is possible that multiple factors are required to achieve maximal cytotoxicity in vivo. Alternatively, different factors may be secreted preferentially to fine tune macrophage responses at different phases of the host response to infection. This concept, that multiple lymphokine factors participate in the regulation of macrophage leishmanicidal activity, would predict that inhibitory factors should also exist. In fact, at least one such lymphokine has been described in the leishmanicidal system. This factor, secreted by the EL-4 mouse thymoma cell line, abrogates macrophage microbicidal activity, but does not by itself enhance amastigote replication in macrophages (45). Interestingly, the suppressor lymphokine inhibits priming by both IFNγ and non-IFN lymphokines (Fig. 9). Although this molecule has so far proven difficult to characterize and purify by biochemical means, it has provided intriguing information on mechanisms of control of the antileishmania activity induced by lymphokines.

Expression of activated macrophage intracellular killing depends not only on several signals delivered to macrophages in a precise order, it also depends on the state of maturation and/or differentiation of the macrophage. Inflammatory peritoneal macrophages that have recently emigrated from the blood are both more susceptible to *L. major* infection and express less microbicidal activity in response to lymphokines than do mature, resident cells (33,46). Blood monocytes respond even more poorly than inflammatory macrophages (33). These observations may have important implications for control of *L. major* disease: macrophages that arrive early in response to inflammatory signals at the site of the sandfly bite may become infected with parasites, but be unable to kill them. Recent data from several laboratories lend support to this hypothesis. Adoptive

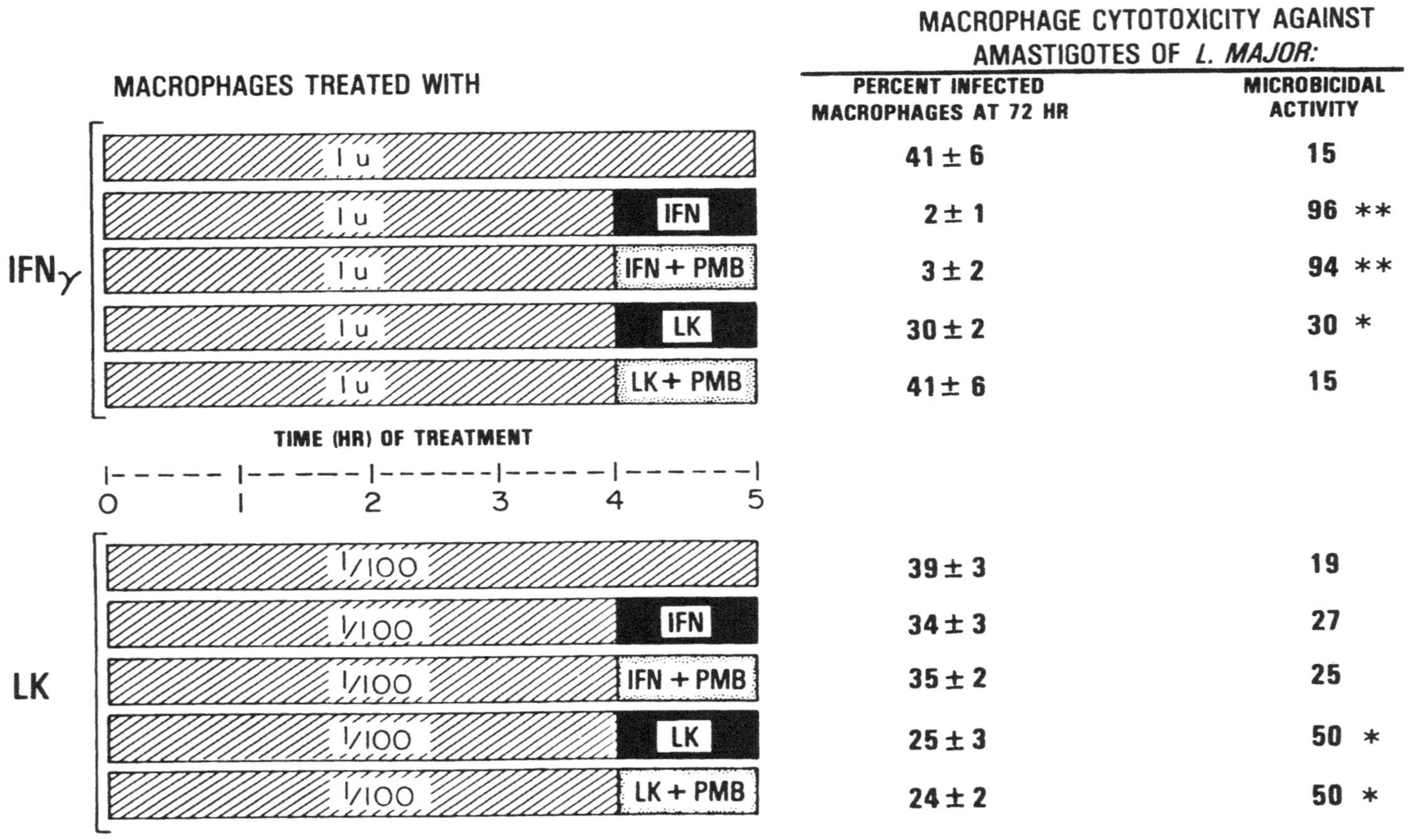

MACROPHAGE CYTOTOXICITY AGAINST AMASTIGOTES OF *L. MAJOR*:

MACROPHAGES TREATED WITH

		PERCENT INFECTED MACROPHAGES AT 72 HR	MICROBICIDAL ACTIVITY
IFNγ	1u	41 ± 6	15
	IFN 1u	2 ± 1	96 **
	IFN + PMB 1u	3 ± 2	94 **
	LK 1u	30 ± 2	30 *
	LK + PMB 1u	41 ± 6	15
LK	1/100	39 ± 3	19
	IFN 1/100	34 ± 3	27
	IFN + PMB 1/100	35 ± 2	25
	LK 1/100	25 ± 3	50 *
	LK + PMB 1/100	24 ± 2	50 *

TIME (HR) OF TREATMENT

Figure 8 Priming and triggering in vitro for intracellular killing by lymphokines depleted of IFNγ (LK) and IFNγ.

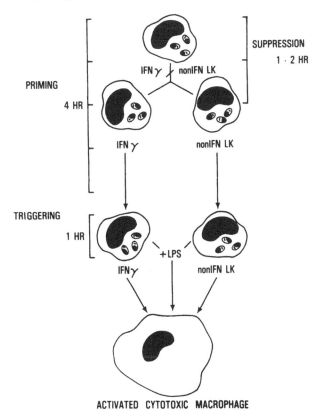

Figure 9 Sequential activation steps for induction of intracellular killing by macrophages activated by lymphokines or IFNγ.

transfer of T cells that mediate delayed type hypersensitivity to *Leishmania* causes increased lesion size in normally leishmania-resistant animals (47). These large lesions contain more macrophages that are heavily infected than cells in lesions of control animals. Chemotaxis of large numbers of immature macrophages that are incapable of leishmanicidal activity may be responsible for this paradoxically detrimental effect of sensitized lymphocytes.

One should emphasize that the above results and interpretations depend heavily on the nature of the in vitro system used to analyze interactions among *Leishmania*, macrophages, and mediators. This caveat is clearly demonstrated in findings with human monocytes. Freshly harvested human monocytes cultured as a nonadherent cell pellet, as in the preceding experiments with mouse peritoneal macrophages, respond to IFNγ and to at least one other non-IFN lymphokine to kill *L. donovani* (48). Expression of microbicidal activity requires only a brief pulse with lymphokine or IFNγ immediately before or within a few hours of infection in vitro (49). In a different assay system, adherent monocytes are cultured for 7-10 days prior to lymphokine treatment (50). During this period of culture, monocyte production of oxidative products declines dramatically (51). Production of hydrogen peroxide can, however, be boosted back to levels of freshly harvested cells by treatment with lymphokines or IFNγ. The activating agent must be administered for 3 days prior to infection to induce optimum oxidative and microbicidal effect (50). Oxidative killing may occur during parasite entry into the cell, or may occur intracellularly by oxidative and nonoxidative mechanisms. In contrast to the nonadherent cell system, microbicidal activity in the adherent cell system occurs only in response to IFNγ. Other recombinant or purified lymphokines are without effect (52,53).

V. IN VIVO EFFECTS OF IMMUNOMODULATING AGENTS IN LEISHMANIASIS

With the recognition that nonspecific immunity inhibits progression of disease caused by a number of protozoa have come efforts to enhance resistance to infection by administration of "nonspecific" immunomodulating agents. Administration of *M. bovis* BCG to BALB/c mice prior to infection with *L. donovani* reduced the number of organisms found in spleen or liver, but was not curative (54). BCG immunotherapy of cutaneous disease induced by *L. major*, however, protected BALB/c mice from the development of lethal systemic infections, and, in certain experimental protocols, prevented the development of the cutaneous lesion as well (55). Administration of glucan, a polyglucose derivative from yeast cell walls, also reduced the number of parasites in the livers and spleens of CF1 mice (56) and hamsters (57) infected with *L. donovani*. Moreover, macrophages from glucan-treated hamsters inhibited replication of *Leishmania*. The pathway by which glucan treatment induced antileishmanial activity in macrophages in vitro and presumably in vivo is unknown. Other nonspecific factors produced by specially sensitized immune cells have also been tested for therapeutic efficacy in vivo. Administration of liposome-encapsulated lymphokines, for example, reduced parasite burdens of infected mice in a murine model of new world systemic leishmaniasis (58). That IFNγ may be important in this lymphokine-induced decrease in hepatic parasites was demonstrated with antibodies to IFNγ. Antibodies administered prior to infection of mice with *L. donovani* increased the number of parasites present in liver impression smears (R.D. Schreiber, personal communication). Specific therapy with IFNγ or other purified lymphokines in experimental models of leishmaniasis has yet to be reported.

VI. SUMMARY

Macrophage activation constitutes a primary mechanism of host defense against obligately intracellular protozoan parasites such as the *Lesishmania*. The regulation of macrophage activation has been the subject of intense investigation since the early studies of Mackaness, however it is only recently that purified reagents have become available for analysis of the steps and signals required to induce antimicrobial effector activities of macrophages. Of all the characterized lymphokine macrophage-activating factors, IFNγ certainly displays the most pleotrophic effects on immune responses to invading microorganisms: it functions in the development of immune responses by regulating Ia antigen expression of antigen-presenting cells; it activates macrophages in vitro for a wide array of effector activities once immunity has become established. Does IFNγ function in vivo in these diverse capacities, or is the activity of IFNγ fine-tuned by other antagonistic molecules so that its activities are confined to a subset of possible immune functions?

The interactions of protozoan parasites with nonspecific immune mechanisms, including those regulated by IFNγ, are exceedingly complex. As a consequence, we presently understand these events at a phenomenologic rather than at a molecular level. It is likely that as our understanding deepens, we will perceive more specificity in the mechanisms that enhance or diminish antiprotozoal responses. Recognition of this specificity may provide tools for rational immunotherapy or immunoprophylaxis of disease caused by the *Leishmania*.

REFERENCES

1. Alexander, J., and K. Vickerman. 1975. Fusion of host cell secondary lysosomes with the parasitophorous vacuoles of *Leishmania mexicana* infected macrophages. *J. Protozool. 22*:502–508.
2. Chang, K.-P., and D.W. Dwyer. 1978. *Leishmania donovani*-hamster macrophage interactions in vitro: Cell entry, intracellular survival, and multiplication. *J. Exp. Med. 147*:515–530.
3. Chang, K.-P. 1979. *Leishmania donovani*: Promastigote-Macrophage surface interactions in vitro. *Exp. Parasitol. 48*:175–189.
4. Pearson, R.D., J.A. Sullivan, D. Roberts, R. Romito, and G.L. Mandell. 1983. Interaction of *Leishmania donovani* promastigotes with human phagocytes. *Infect. Immun. 40*:411–416.
5. Chang, K.-P. 1981. *Leishmania donovani*-macrophage binding mediated by surface glycoproteins/antigens: characterization in vitro by a radioisotopic assay. *Mol. Biochem. Parasitol. 4*:67–76.
6. Blackwell, J.M. 1985. Receptors and recognition mechanisms of *Leishmania* species. *Trans. Royal Soc. Trop. Med. Hyg. 79*:606–612.
7. Blackwell, J.M., R.A.B. Ezekowitz, M.B. Roberts, J.Y. Channon, R.B. Sim,

and S. Gordon. 1985. Macrophage complement and lectin-like receptors bind *Leishmania* in the absence of serum. *J. Exp. Med. 162*:324–331.

8. Aikawa, M., L.D. Hendricks, Y. Ito, and M. Jagusiak. 1982. Interactions between macrophage-like cells and *Leishmania braziliensis* in vitro. *Am. J. Pathol. 108*:50–59.

9. Wyler, D.J. 1982. In vitro parasite-monocyte interactions in human leishmaniasis: evidence of an active role of the parasite in attachment. *J. Clin. Invest. 70*:82–88.

10. Silverstein, S.C. 1977. Endocytic uptake of particles by mononuclear phagocytes and the penetration of obligate intracellular parasites. *Am. J. Trop. Med. Hyg. 26 (Suppl 6)*:161–168.

11. Murray, H.W. 1981. Susceptibility of *Leishmania* to oxygen intermediates and killing by normal macrophages. *J. Exp. Med. 153*:1302–1315.

12. Reiner, N.E., and J.W. Kazura. 1982. Oxidant-mediated damage of *Leishmania donovani* promastigotes. *Infect. Immun. 36*:1023–1027.

13. Pearson, R.D., J.L. Harcus, D. Roberts, and G.R. Donowitz. 1983. Differential survival of *Leishmania donovani* amastigotes in human monocytes. *J. Immunol. 131*:1994–1999.

14. Edelson, P.J., and D.M. Mosser. 1987. The third component of complement (C3) is responsible for the intracellular survival of *Leishmania major*. *Nature 327*:329–330.

15. Mackaness, G.B. 1962. Cellular immunity. *J. Exp. Med. 116*:381–406.

16. Nacy, C.A., and C.L. Diggs. 1981. Intracellular replication of *Leishmania tropica* in mouse peritoneal macrophages: comparison of amastigote replication in adherent and nonadherent macrophages. *Infect. Immun. 34*:310–313.

17. Miller, H.C., and D.W. Twohy. 1969. Cellular immunity to *Leishmania donovani* in macrophages in culture. *J. Parasitol. 55*:200–206.

18. Manuel, J., Y. Buchmuller, and R. Behin. 1978. Studies on the mechanism of macrophage activation: I. Destruction of intracellular *Leishmania enrietii* by macrophages activated by cocultivation with stimulated lymphocytes. *J. Exp. Med. 148*:393–407.

19. Behin, R., J. Mauel, J. Biroum-Noerjasin, and D.S. Rowe. 1975. Mechanism of protective immunity in experimental leishmaniasis of the guinea-pig: II. Selective destruction of different *Leishmania* species in activated guinea pig and mouse macrophages. *Clin. Exp. Immunol. 20*:351–358.

20. Pappas, M.G., and C.A. Nacy. 1983. Antileishmanial activities of macrophages from C3H/HeN and C3H/HeJ mice treated with *Mycobacterium bovis* strain BCG. *Cell. Immunol. 80*:217–222.

21. Nacy, C.A., M.S. Meltzer, E.J. Leonard, and D.J. Wyler. 1981. Intracellular replication and lymphokine-induced destruction of *Leishmania tropica* in C3H/HeN macrophages. *J. Immunol. 127*:2381–2386.

22. Oster, C.N., and C.A. Nacy. 1984. Macrophage activation to kill *Leishmania tropica*: Kinetics of macrophage response to lymphokines that induce microbicidal activities against *Leishmania tropica* amastigotes. *J. Immunol. 132*:1492–1500.

23. Nacy, C.A., and M.S. Meltzer. 1979. Macrophages in resistance to rickettsial infection: macrophage activation in vitro for killing *Rickettsia tsutsugamushi*. *J. Immunol. 123*:2544–2549.

24. Salvin, S.B., and S.-L. Cheng. 1971. Lymphoid cells in delayed-type hypersensitivity II. In vitro phagocytosis and cellular immunity. *Infect. Immun. 3*:548–553.

25. Hoff, R. 1975. Killing in vitro of *Trypanosoma cruzi* by macrophages from mice immunized with *T. cruzi* or BCG, and absence of cross immunity on challenge in vivo. *J. Exp. Med. 142*:299–308.

26. Horwitz, M.A., and S.C. Silverstein. 1981. Activated human monocytes inhibit the intracellular multiplication of Legionaire's disease bacteria. *J. Exp. Med. 154*:1618–1630.

27. Nacy, C.A., and M.S. Meltzer. 1982. Macrophages in resistance to rickettsial infection: strains of mice susceptible to the lethal effects of *Rickettsia akari* infection show defective microbicidal activity in vitro. *Infect. Immun. 36*: 1096–1101.

28. Ruco, L., and M.S. Meltzer. 1978. Macrophage activation for tumor cytotoxicity: development of macrophage cytotoxic activity requires the completion of a series of short-lived intermediary interactions. *J. Immunol. 121*:2035–2042.

29. Meltzer, M.S., D.L. Hoover, M.J. Gilbreath, R.D. Schreiber, and C.A. Nacy. 1986. Experimental variables for induction of activated cytotoxic macrophages. *Ann. Inst. Pasteur (Immunology) 137*:206–211.

30. Lazdins, J.K., D.K. Koech, and M.L. Karnovsky. 1980. Oxidation of glucose by mouse peritoneal macrophages: a comparison of suspensions and monolayers. *J. Cell. Physiol. 195*:191–196.

31. Pofit, J.F., and P.R. Strauss. 1977. Membrane transport by macrophages in suspension and adherent to glass. *J. Cell. Physiol. 92*:249–256.

32. Nacy, C.A., C.N. Oster, S.L. James, and M.S. Meltzer. 1984. Activation of macrophages for destruction of intracellular and extracellular parasites, *Cont. Topics Immunobiol. 13*:147–170.

33. Hoover, D.L., and C.A. Nacy. 1984. Macrophage activation to kill *Leishmania tropica*: defective intracellular destruction of amastigotes by macrophages elicited by sterile inflammatory agents. *J. Immunol. 132*:1487–1483.

34. Byrne, G.I., and C.L. Faubian. 1982. Lymphokine-mediated microbistatic mechanisms restrict *Chlamydia psittaci* growth in macrophages. *J. Immunol. 128*:469–474.

35. Nacy, C.A., E.J. Leonard, and M.S. Meltzer. 1981. Macrophages in resistance to rickettsial infections: characterization of the lymphokines that induce rickettsiacidal activity in macrophages. *J. Immunol. 126*:204–209.

36. Ralph, P., C.A. Nacy, M.S. Meltzer, N. Williams, I. Nakionz, and E.J. Leonard. 1983. Colony stimulating factors and regulation of macrophage tumoricidal and microbicidal activities. *Cell. Immun. 76*:10–21.

37. Nacy, C.A., C.E. Davis, B.A. Mock, M.J. Gilbreath, and M.S. Meltzer. 1986. Regulation of macrophage antimicrobial activities by lymphocyte products. *Clin. Immunol. Newsletter 7*:65–69.

38. Davis, C.E., M. Belosevic, M.S. Meltzer, and C.A. Nacy. 1987. Regulation of activated macrophage resistance to infection: cooperation of lymphokines for induction of macrophage resistance to infection. *J. Immunol.* (in press)

39. Belosevic, M., C.E. Davis, M.S. Meltzer, and C.A. Nacy. Characterization of the lymphokines that cooperate with interferon γ to induce activated macrophage resistance to infection with amastigotes of *Leishmania major*. (Submitted).

40. Belosevic, M., M.S. Meltzer, and C.A. Nacy. Regulation of activated macrophage antimicrobial activities: role of IL-2 in induction of macrophage resistance to infection. (Submitted).

41. Finbloom, D.S., D.L. Hoover, and L.M. Wahl. 1985. The characteristics of binding of human recombinant interferon q to its receptor on human monocytes and human monocyte-like cell lines. *J. Immunol.* 135:300–305.

42. Crawford, R.M., D.S. Finbloom, J. Ohara, W.E. Paul, and M.S. Meltzer. 1987. B cell stimulatory factor-1 (interleukin 4) activates macrophages for increased tumoricidal activity and expression of Ia antigens. *J. Immunol.* 139:135–141.

43. Grabstein, K.H., D.L. Urdal, R.J. Tushinski, D.Y. Mochizuki, V.L. Price, M.A. Cantrell, S. Gillis, and P.J. Conlon. 1986. Induction of macrophage tumoricidal activity by granulocyte-macrophage colony stimulating factor. *Science 232*:506–508.

44. Nacy, C.A., A.H. Fortier, M.S. Meltzer, N.A. Buchmeier, and R.D. Schreiber. 1985. Macrophage activation to kill *Leishmania tropica*: macrophages can be activated to kill amastigotes by both interferon gamma and non-interferon lymphokines. *J. Immunol.* 135:3505–3511.

45. Nacy, C.A. 1984. Macrophage activation to kill *Leishmania tropica*: identification of a soluble T cell factor that suppresses lymphokine-induced intracellular killing of amastigotes. *J. Immunol.* 133:448–453.

46. Fortier, A.H., D.L. Hoover, and C.A. Nacy. 1982. Intracellular replication of *Leishmania tropica* in mouse peritoneal macrophages: amastigote infection of resident cells and inflammatory exudate macrophages. *Infect. Immun. 38*:1304–1309.

47. Titus, R.G., G.C. Lima, H.D. Enger, and J.A. Louis. 1984. Exacerbation of murine leishmaniasis by adoptive transfer of parasite-specific helper T cell preparations capable of mediating *Leishmania major*-specific delayed type hypersensitivity. *J. Immunol. 133*:1594–1600.

48. Hoover, D.L., D.S. Finbloom, R.M. Crawford, C.A. Nacy, M.J. Gilbreath, and M.S. Meltzer. 1986. A lymphokine distinct from interferon gamma that activates human monocytes to kill *Leishmania donovani* in vitro. *J. Immunol. 136*:1329–1333.

49. Hoover, D.L., C.A. Nacy, and M.S. Meltzer. 1985. Human monocyte activation to kill *Leishmania donovani* amastigotes. Induction of microbicidal activity by interferon-gamma. *Cell. Immunol. 99*:500–511.

50. Murray, H.W., and D.M. Cartelli. 1983. Killing of intracellular *Leishmania donovani* by human mononuclear phagocytes. Evidence for oxygen de-

pendent and independent leishmanicidal activity. *J. Clin. Investigation 72*: 32–44.

51. Nakagawara, A., C.F. Nathan, and Z.A. Cohn. 1981. Hydrogen peroxide metabolism in human monocytes during differentiation in vitro. *J. Clin. Invest. 68*:1243.

52. Murray, H.W., B.Y. Rubin, and C.D. Rothermel. 1983. Killing of intracellular *Leishmania donovani* by lymphokine-stimulated human monocytes. Evidence that interferon γ is the activating lymphokine. *J. Clin. Invest. 72*:1506–1510.

53. Nathan, C.F., T.J. Prendergast, M.E. Wiebe, E.R. Stanley, E. Platzer, H.G. Remold, K. Welte, B.Y. Rubin, and H.W. Murray. 1984. Activation of human macrophages. Comparison of other cytokines with interferon γ. *J. Exp. Med. 160*:600–605.

54. Smrkovski, L.L., and C.L. Larson. 1977. Effect of treatment with BCG on the course of visceral leishmaniasis in BALB/c mice. *Infect. Immun. 16*: 249–257.

55. Fortier, A.H., B.A. Mock, M.S. Meltzer, and C.A. Nacy. 1987. BCG-induced protection of cutaneous and systemic *Leishmania major* infections of mice. *Infect. Immun. 55*:1707–1714.

56. Holbrook, T.W., J.A. Cook, and B.W. Parker. 1981. Immunization with *Leishmania donovani*: glucan as an adjuvant with promastigotes. *Am. J. Trop. Med. Hyg. 30*:762–768.

57. Cook, J.A., T.W. Holbrook, and W.J. Dougherty. 1982. Protective effect of glucan against visceral leishmaniasis in hamsters. *Infect. Immun. 37*:1261–1269.

58. Reed, S.G., M. Barral-Netto, and J. Inverso. 1984. Treatment of visceral leishmaniasis with lymphokine encapsulated in liposomes. *J. Immunol. 132*:3116–3119.

10

Modulatory Effects of Interferons on *Trypanosoma cruzi* Infections

Gerald Sonnenfeld / University of Louisville Schools of Medicine and Dentistry, Louisville, Kentucky

Felipe Kierszenbaum / Michigan State University, East Lansing, Michigan

I. INTRODUCTION

Trypanosoma cruzi is the etiologic agent of Chagas' disease, or American trypanosomiasis. Natural infections with this unicellular parasite are initiated by metacyclic trypomastigote forms present in the fecal fluid of the reduviid insect vectors. This fluid is deposited on the skin while the insect procures a blood meal. The flagellates contained in it may enter the body through a skin erosion or lesion, possibly including the bite site, or by contact with the conjunctival mucosa. Infections can also result from transfusion of blood from chagasic individuals (1).

T. cruzi trypomastigotes do not divide, but can invade a variety of host cells and attain cytoplasmic localization. There, the trypomastigotes transform into the multiplicative amastigote form. Amastigotes may eventually transform into new trypomastigotes that are released from bursting infected cells and these disseminate the infection to other tissues, including those of the heart and nervous system (2). While in the circulation, released trypomastigotes may be ingested by blood-sucking reduviid insects. A series of transformations in the insect gut leads to the presence of metacyclic trypomastigotes in the fecal fluid of the insects, thus completing the life cycle of the parasite.

There is no chemotherapy for Chagas' disease that has proven to be consistently effective and the few drugs currently in use have undesirable side effects (3). For these reasons, interest in potential immunological manipulations to control this parasite has remained high.

T. cruzi infection is accompanied by several immunological alterations, notably host immunosuppression detectable during the acute period of the disease (reviewed in Refs. 4,5) and purified parasites have been shown to alter lymphocyte function in vitro (6), including lymphokine production (7). Moreover, some investigators have suggested that *T. cruzi* could secrete or regulate the secretion of a bioregulatory substance that could be toxic for tumor cells, because when extracts of the trypanosome were injected into tumors of experimental animals, tumor regression was observed (8).

The notion that *T. cruzi* or *T. cruzi* infection may cause, directly or indirectly, bioregulatory effects has received additional attention in recent years. Serum levels of interferon activity have been shown to increase significantly during *T. cruzi* infection (9,10), and might be related to host defense (11) or other biological phenomena associated with the presence of the parasite. Interferons come to mind as possible mediators of *T. cruzi*-associated effects for several reasons. Since interferons can modulate immune responses (12) they may have a role in the immunological changes accompanying *T. cruzi* infection. On the other hand, interferons can have an effect on cell growth, division, and differentiation (13) that could directly influence *T. cruzi* infections. The main focus of this chapter will be an exploration of the possible interactions of the interferon system with *Trypanosoma cruzi*.

II. INTERACTIONS OF THE INTERFERON SYSTEM WITH *TRYPANOSOMA CRUZI*

One of the first reports to indicate that interferons, or interferon-like substances could be induced by *T. cruzi* came from the work of Rytel and Marsden (9), who were able to detect an antiviral activity in the sera of mice infected six days previously with the Peru strain of the parasite. This activity was never fully characterized as an interferon. Years later, other investigators confirmed and extended the finding of Rytel and Marsden. Schmuñis and co-workers showed that interferons were produced in the sera of infected mice (14). In our laboratories, we were able to show that mice infected with different *T. cruzi* isolates produced an interferon-like antiviral activity in their sera (10). The peak of this activity was found in the sera of mice 24 hr after infection with the Tulahuén isolate and 3 days postinfection with the Y isolate (10). This antiviral activity was later confirmed to be interferon α/β (11).

The reasons for the kinetic differences in production of peak interferon α/β titers could have been related to the individual biological characteristics of the various isolates, including tissue tropism. While the Peru isolate is regarded as being preferentially myotropic, the Y isolate manifests a dual preference and is generally regarded as being both myotropic and reticulotropic; the Tulahuén isolate seems to be preferentially reticulotropic (15). The amount of time required for maximal interferon α/β production could have been related to the relative number and facility of the parasites to infect phagocytic cells that are major producers of interferons. Mice of different genetic background were used in experiments with the Peru and the two other isolates, a condition which could also account for some of the noted differences.

In our work, interferon production was detectable in the sera of mice within hours or a few days after infection (10) and the activity was characterized as interferon α/β (11). More recently, however, Wietzerbin and co-workers reported that nude (nu/nu) mice and their nu/+ littermates exhibited biphasic production of interferon after *T. cruzi* infection (16). Part of this interferon was characterized as interferon γ, raising the possibility that some of the interferon γ production in this case may come from natural killer cells or nontypical T cells (17,18). Additional, uncharacterizable interferon activity was also observed (16). In any case, it appears that some strains of mice were able to produce interferon γ in response to *T. cruzi* infections.

Studies were also carried out to determine if administration of exogenous interferons could influence the outcome of experimental Chagas' disease. Mice infected with *T. cruzi* and then treated for seven days with a preparation of interferon α/β showed an alteration in the course of infection (11). There was a transient decrease in parasitemia compared to control groups, and, although 25% of both interferon α/β-treated and control infected mice died, mortality in the interferon α/β-treated mice was delayed (11).

Additional studies were carried out to determine if mice could be protected from infection with *T. cruzi* by pretreatment with interferon inducers. In one series of the studies, mice were treated with tilerone-hydrochloride, an inducer of interferon α/β, and 18 to 24 hours later, the mice were infected with *T. cruzi* trypomastigotes (19). Whereas about 50% of these animals survived for an indefinite period of time, there were no survivors in the control group (19). The increased survival time appeared to correlate with interferon production and enhanced natural killer cell activity. Furthermore, beige mice that were deficient in natural killer cell activity were not protected by tilerone-hydrochloride from trypanosome infection (19). In interpreting these data, it should be noted that tilerone-hydrochloride may have additional effects on the animals other than interferon induction.

One additional study (20) used several protocols of treating mice with the interferon inducer polyriboinosinic-polyribocytidylic acid. In this case, high doses of the inducer (50 μg) were injected repeatedly into mice that were going to be infected or were already infected with *T. cruzi*. In each case, the infection was exacerbated. However, it should be noted that repeated injections of an interferon inducer can make an animal refractory to interferon production after the first injection (21). Therefore, it is difficult to determine the role of any interferon induced in these experiments, and it is likely that other toxic effects of such high dosages of polyriboinosinic-polyribocytidylic acid contributed to the observed exacerbation.

Further evidence suggesting a role for interferon and natural killer cells in *T. cruzi* infection was provided by Hatcher and co-workers, who were able to show enhanced natural killer cell activity in a mouse model system of Chagas' disease (22). Other experiments showed that *T. cruzi* could be destroyed in vitro by cytotoxic cells identified as natural killer cells (23).

In addition, a recent report by Wietzerbin and colleagues (16) indicated that mice could be protected from *T. cruzi* infection by pretreatment with interferon γ. In this case, interferon γ and antitrypanosomal antibodies appeared to act synergistically in the protective effect (16).

As a whole, the findings discussed above strengthen the view that administration of interferons or interferon inducers could provide some protection by enhancing host resistance mechanisms against *T. cruzi*. However, it is too early to infer that interferons may have some application in the therapy of *T. cruzi* infection and further studies along these lines would be required.

III. MECHANISMS OF THE INTERACTIONS OF THE INTERFERON SYSTEM WITH *T. CRUZI*

Several groups of investigators have looked into the possible mechanisms of the effects of interferons on *T. cruzi* infections. These studies have led to some intriguing findings.

It has been established that when interferon α/β is incubated directly with the parasite, there are no detectable effects on its viability, motility, or infectivity, suggesting that these interferons lack direct toxic activity on the organism (24). Treatment of host cells with interferon did not affect the ability of the trypanosomes to infect cells and treatment of infected cells did not affect transformation of the parasite from amastigote to trypomastigote form (25). However, the presence of interferon α/β in the culture medium produced a significant reduction in the level of parasite association with rat heart myoblasts or mouse peritoneal macrophages, cells which can host *T. cruzi* (24). Pretreatment of the parasite but not of these host cells with interferon α/β reproduced the effect, indicating that interferon α/β had affected the capacity of the trypanosome to interact, at least, with myoblasts and macrophages. The active component of the preparation was interferon β, since the inhibitory effect was readily reproduced with a preparation containing only interferon β but not with one containing only interferon α (24).

Additional studies were carried out to determine the possible effects of interferon γ on in vitro infection of cultured cells with *T. cruzi*. The host cells selected for this work were the P388-D1, murine macrophage-like cell line and normal mouse resident peritoneal macrophages (26). Initially, an interferon γ preparation was used which consisted of the supernatant fluids from a T-cell hybridoma culture which also contained a trace of interleukin-2 activity (26). Treatment of cultures of the host cells with this preparation decreased both the proportion of cells engaged by parasites and the number of organisms per host cell. Since the addition of interleukin-2 to the cultures did not affect the level of parasite host-cell association, it was concluded that the active principle in the T-cell hybridoma supernatant fluids was interferon γ. This notion was confirmed by the use of pure interferon γ produced by recombinant DNA technology, which yielded similar results to those obtained with the impure interferon γ preparation (26). Moreover, interferon γ treatment enhanced the capacity of the host cell to kill *T. cruzi*. Parasite destruction appeared to be mediated by oxygen reduction metabolites. Therefore, the interferon γ-enhanced lysis of *T. cruzi* by macrophages or macrophage-like cells appears to result from the well known property of this lymphokine to activate macrophages.

Plata and his associates (27) reported that interferons and anti-*T. cruzi* antibodies synergistically protected cell cultures from infection with *T. cruzi*. This was true of both interferon α/β and interferon γ, and of infection of professional as well as nonprofessional phagocytic host cells and exposed an interesting cooperation between interferons and antibodies in vitro that may be relevant to host-resistance mechanisms.

In summary, it appears that the contributions of interferons to defensive processes in *T. cruzi* infection may be multifaceted, depending on the type of interferon and/or type of host cell involved. The immune system could also play an important role through the enhancement of interferon activities that affect directly or indirectly host-parasite interaction.

IV. THE ROLE OF INTERFERON IN NATURAL RESISTANCE TO INFECTION WITH *T. CRUZI*

The evidence described above indicates that interferon production is stimulated in mice infected with *T. cruzi*, and that exogenous interferons or interferon inducers can modify both in vivo and in vitro infections by *T. cruzi*. The concept of a role for interferons in natural resistance to a parasitic infection is difficult to reconcile with the original definition of interferon as an antiviral agent (28). However, current information supports that interferons have many additional activities, including regulation of cell growth and development and regulation of immune responses (12,13).

Since interferon α/β is produced very early in the course of *T. cruzi* infection (9,10), it seems appropriate to postulate a role for interferons in host resistance. For example, interferon β could inhibit association of the invasive trypomastigotes with host cells, providing a transient defense against infection. This transient defense could involve the participation of inflammatory cells via phagocytosis of the parasites followed by intracellular destruction (26,29-32) and/or natural killer cells whose activity is known to be enhanced by interferon (22,23). As the infection progresses, interferon α/β production is reduced to undetectable levels, and the initial interferon-mediated effects would subside. Interferon γ would likely be produced as part of the eventual immune response to the parasite and stimulate uptake and destruction of the trypanosomes by macrophages. Antibodies directed against the parasite could work in conjunction with the interferons to aid in host resistance. This hypothesis remains to be tested.

In one limited study to explore a possible role for interferon in natural resistance to *T. cruzi*, Quan and associates (33) gave anti-interferon α/β to mice and found no significant changes in host resistance to parasite infection with respect to controls. From this observation, the investigators inferred that interferon and natural killer cells probably did not play a role in natural resistance. However, this conclusion has to be viewed with caution, given the negative nature of the results. Although circulating interferon α/β was not detected in the sera of infected mice after antibody treatment, interferon α/β in local sites of infection (where it would play its most significant role) might have been unaffected. Furthermore, since the anti-interferon used should not have had an effect on interferon γ, a possible role for this type of interferon cannot be ruled out.

In several studies (14,34,35), the induction of high levels of interferon in strains of *T. cruzi*-infected mice could not be associated with better survival of those mice compared to strains in which lower levels of interferon were induced. However, in a recent study, Trischmann (35) was able to show that a recombinant inbred strain of mouse (BXH-2) that lacks the early control mechanism for *T. cruzi* infections could not produce any interferon in response to trypanosome infection. This is indirect evidence which suggests that interferon plays some role in natural resistance. The difference between resistant and susceptible strains of mice may not be the levels of interferon produced in response to *T. cruzi* infection, but, rather, the ability to utilize that interferon.

The inhibitory effects of interferon β on host cell invasion (24) observed in vitro may explain the enhanced host resistance afforded by the administration of exogenous interferon α/β (11). Similarly, the stimulatory effect of interferon γ on macrophage uptake and disposal of *T. cruzi*, also seen in vitro (26), could be a weapon in the defensive arsenal of the host. However, whether these interferon effects indeed underlie interferon-increased host defense against *T. cruzi* by the alluded mechanisms, or were designed by nature to play a role in an as yet undefined biological event secondary to *T. cruzi* infection are questions which remain to be addressed by further studies.

V. CONCLUSIONS

The studies described above clearly show that interferons are induced during the course of experimental infections with *T. cruzi* and that exogenous interferons can influence the course of infection both in cultures and in animals. Prophylactic treatment of animals with interferons affords some protection in terms of significantly reduced parasitemia, although it does not extend survival time. Therefore, therapeutic approaches using interferons in Chagas' disease is less clear. Additional investigation may clarify whether the use of more than one interferon or interferons combined with other agents will lead to a more successful treatment regimen.

In any case, a fascinating area for research to understand the basic biology of the interferon system now exists with the protozoan parasites. Where the future will lead is to be seen, but surely more exciting and perhaps unexpected data on the interferon system will be gained.

ACKNOWLEDGMENTS

The work of the authors described herein was supported by National Aeronautics and Space Administration award NCC2-213 and National Institutes of Health grants AI-14848 and AI-17041.

REFERENCES

1. Brener, Z. 1979. Present status of chemotherapy and chemoprophylaxis of human trypanosomiasis in the Western hemisphere. *Pharmacol. Therapeut.* 7:71–90.
2. Andrade, Z.A., and S.G. Andrade. 1981. Patologica. In Trypanosoma cruzi *e Doença de Chagas*. Edited by Z.A. Andrade and Z. Brener. Guanabara Koogan, Rio de Janiero, pp. 199–248.
3. Kierszenbaum, F. 1984. The chemotherapy of *Trypanosoma cruzi* infections (Chagas' disease). In *Parasitic Diseases*, Vol. II. Edited by J.M. Mansfield. Marcel Dekker, New York, pp. 133–163.
4. Kuhn, R.E. 1981. Immunology of *Trypanosoma cruzi* infections. In *Parasitic Diseases*, Vol. I. Edited by J.M. Mansfield. Marcel Dekker, New York, pp. 137–165.
5. Maleckar, J.R., and F. Kierszenbaum. 1983. Variations in cell-mediated immunity to *Trypanosoma cruzi. Ann. Trop. Med. Parasitol.* 77:247–254.
6. Maleckar, J.R., and F. Kierszenbaum. 1983. Inhibition of mitogen-induced proliferation of mouse T and B lymphocytes by bloodstream forms of *Trypanosoma cruzi. J. Immunol. 130*:908–911.
7. Beltz, L.A., and F. Kierszenbaum. 1987. Suppression of human lymphocyte responses by *Trypanosoma cruzi. Immunology 60*:309–315.
8. Coudert, J., and P. Juttin. 1950. Note sur l'action d'un lysat de Trypanosoma cruzi vis-à-vis d'un cancer greffé du vat. *Compte Rendue Soc. Biol. 144*:847–849.
9. Rytel, MW., and P.D. Marsden. 1970. Induction of an interferon-like inhibitor by *Trypanosoma cruzi. Am. J. Trop. Med. Hyg. 19*:929–931.

10. Sonnenfeld, G., and F. Kierszenbaum. 1981. Increased serum levels of an interferon-like activity during the acute period of experimental infection with different strains of *Trypanosoma cruzi*. *Am. J. Trop. Med. Hyg. 30*: 1189–1191.

11. Kierszenbaum, F., and G. Sonnenfeld. 1982. Characterization of the antiviral activity produced during *Trypanosoma cruzi* infection and protective effects of exogenous interferon against experimental Chagas' disease. *J. Parasitol. 68*:194–198.

12. Sonnenfeld, G. 1980. Modulation of immunity by interferons. *Lymphokine Rep. 1*:113–131.

13. Gresser, I. 1977. Commentary on the varied biologic effects of interferon. *Cell. Immunol. 34*:406–416.

14. Schmuñis, G.A., S. Barón, S. Gonzalez-Cappa, and M.C. Weissenbacher. 1944. El *Trypanosoma cruzi* como inductor de interferon. *Medicina (Buenos Aires) 37*:429–430.

15. Taliaferro, W.H., and T. Pizzi. 1955. Connective tissue reactions in normal and immunized mice to a reticulotropic strain of *Trypanosoma cruzi*. *J. Infect. Dis. 96*:199–226.

16. Plata, F., F. Garcia-Pons, and J. Wietzerbin. 1987. Immune resistance to *Trypanosoma cruzi*: Synergy of specific antibodies and recombinant interferon-gamma in vivo. *Ann. Inst. Pasteur/Immunol. 138*:397–415.

17. Beck, J., H. Engler, H. Brunner, and H. Kirchner. 1980. Interferon production in cocultures between mouse spleen cells and tumor cells. Possible role of mycoplasmas in interferon induction. *J. Immunol. Meth. 38*:63–71.

18. Kumar, V., J. Lust, M. Bennett, and G. Sonnenfeld. 1983. Lack of correlation between mycoplasma induced IFN-γ production in vitro and natural killer cell activity against FLD-3 cells. *Immunobiology 165*:445–458.

19. James, S.L., T.L. Kipnis, A. Sher, and R. Hoff. 1982. Enhanced resistance to acute infection with *Trypanosoma cruzi* in mice treated with an interferon inducer. *Infect. Immun. 35*:588–593.

20. Kumar, R., M. Worthington, J.G. Tilles, and W.H. Abelman. 1971. Effect of the interferon stimulator polyinosinic-polycytidylic acid on experimental *Trypanosoma cruzi* infection. *Proc. Soc. Expt'l. Biol. Med. 137*: 884–888.

21. Stringfellow, D.A. 1982. Interferon induction: Hyporesponsiveness and prostaglandins. *Texas Rep. Biol. Med. 41*:116–121.

22. Hatcher, F.M., R.E. Kuhn, M.C. Cerrone, and R.C. Burton. 1981. Increased natural killer cell activity in experimental American trypanosomiasis. *J. Immunol. 127*:1126–1130.

23. Hatcher, F.M., and R.E. Kuhn. 1982. Destruction of *Trypanosoma cruzi* by natural killer cells. *Science 218*:295–296.

24. Kierszenbaum, F., and G. Sonnenfeld. 1984. β-Interferon inhibits cell infection by *Trypanosoma cruzi*. *J. Immunol. 132*:905–908.

25. Golgher, R.R., M.S.M. Bertelli, M.L. Petrillo-Peixoto, and Z. Brener. 1980. Effect of interferon on the development of *Trypanosoma cruzi* in tissue culture "Vero" cells. *Men. Inst Oswaldo Cruz Rio de J. 75*:157–160.

26. Wirth, J.J., F. Kierszenbaum, G. Sonnenfeld, and A. Zlotnik. 1985. Enhancing effects of gamma interferon on phagocytic cell association with and killing of *Trypanosoma cruzi*. *Infect. Immun. 49*:61–66.

27. Plata, F., J. Wietzerbin, F. Garcia-Pons, E. Falcoff, and H. Eisen. 1984. Synergistic protection by specific antibodies and interferon against infection by *Trypanosoma cruzi* in vitro. *Eur. J. Immunol. 14*:930–935.

28. Isaacs, A., and J. Lindenmann. 1957. Virus Interference. I. The interferon. *Proc. Roy. Soc. Ser. B 147*:258–267.

29. Villalta, F., and F. Kierszenbaum. 1983. Role of polymorphonuclear cells in Chagas' disease. I. Uptake and mechanisms of destruction of intracellular (amastigote) forms of *Trypanosoma cruzi* by human neutrophils. *J. Immunol 131*:1504–1510.

30. Villalta, F., and F. Kierszenbaum. 1984. Role of inflammatory cells in Chagas' disease. I. Uptake and mechanisms of destruction of intracellular (amastigote) forms of *Trypanosoma cruzi* by human eosinophils. *J. Immunol. 132*:2053–2058.

31. Villalta, F., and F. Kierszenbaum. 1984. Role of inflammatory cells in Chagas' disease. II. Interactions of mouse macrophages and human monocytes with intracellular (amastigote) forms of *Trypanosoma cruzi*: uptake and mechanisms of destruction. *J. Immunol. 133*:3338–3343.

32. Lima, M.F., and F. Kierszenbaum. 1985. Lactoferrin effects on macrophage functions. I. Increased uptake and destruction of intracellular (amastigote) forms of *Trypansoma cruzi*. *J. Immunol. 134*:4176–4183.

33. Quan, P.C., B. Rager-Zinsman, M. Wittner, and H. Tanowitz. 1983. Interferon and natural killer cells in murine Chagas' disease. *J. Parasitol. 69*: 1164–1166.

34. Sonnenfeld, G., C.L. Gould, F. Kierszenbaum, A.L.W. De Gee, and J.M. Mansfield. 1986. Interferon in resistance to bacterial and protozoan infections. In *The Biology of the Interferon System, 1985*. Edited by W.E. Stewart II and H. Schellekens. Elsevier Scientific Publishers B.V., Amsterdam, pp. 291–297.

35. Trischmann, T.M. 1986. *Trypanosoma cruzi*: Early parasite proliferation and host resistance in inbred strains of mice. *Exptl. Parasitol. 62*:194–201.

11

Enhancement of Resistance to *Trypanosoma cruzi* Infection by Recombinant IFNγ

Robert E. McCabe*/Martinez Veterans Administration Medical Center, Martinez, California

Jack S. Remington / Palo Alto Medical Foundation, Palo Alto, and Stanford University Medical Center, Stanford, California

Fausto G. Araujo / Palo Alto Medical Foundation, Palo Alto, California

I. INTRODUCTION

In the acute phase of the infection of mice with *Trypanosoma cruzi*, the parasites proliferate freely within the cytoplasm of host cells. At this time there is an increase in the serum levels of interferon (1,2), and each one of the three types of this lymphokine has been shown to have some effect either on the parasite itself or on the host cells (3-6). In addition, it has been shown that recombinant interferon gamma (IFNγ) has remarkable capacity to inhibit replication of *Toxoplasma gondii* (7), leishmania (8,9), *Rickettsia prowazekii* (10), and *Chlamydia psittaci* (11) in professional and nonprofessional phagocytes.

We have demonstrated that IFNγ administered to mice infected with *Toxoplasma gondii* protected the animals against death and augmented specific antibody response (12). Moreover, macrophages from mice injected with IFNγ were activated to kill toxoplasma in vitro. Others have demonstrated that administration of IFNγ inhibits development of exoerythrocytic forms of malaria in mice, rats, and chimpanzees (13,14) and decreases counts of *Listeria monocytogenes* in spleens of mice (15). Recombinant human IFNγ has been administered to humans. In a study of cancer patients (16), intravenous recombinant interferon gamma (rIFNγ) increased blood monocyte secretion of hydrogen peroxide. In another study of AIDS patients (17), intravenous rIFNγ did not augment monocyte secretion of superoxide ion. Prospects for use of IFNγ as a therapeutic agent in humans appear promising, but IFNγ potentially may exacerbate infections due to intracellular organisms, such as mycobacteria (18,19). These data

---

Present affiliation: University of California, Davis, California

203

indicate that the effect of IFNγ needs to be investigated in vitro and in vivo for individual organisms.

In this chapter we present results of experiments to examine and to extend previous observations on the activity of recombinant interferon gamma against *T. cruzi*. Our findings revealed that time to death due to infection with *T. cruzi* was prolonged in mice injected with IFNγ. These mice had reduced levels of parasitemia and their specific antibody response was significantly augmented. In addition, administration of IFNγ to mice resulted in activation of peritoneal macrophages to inhibit the intracellular multiplication of *T. cruzi*.

II. MATERIALS AND METHODS

A. Mice and *Trypanosoma cruzi*

Swiss-Webster female mice, weighing 20–22 g each (Simonsen Laboratories, Gilroy, CA), were used in all experiments. The Y strain of *T. cruzi* (20) was used in all experiments. For in vivo experiments, blood from infected mice was collected aseptically on the seventh day of infection and the bloodform trypomastigotes (BFT) counted (21). The desired number of organisms was suspended in Hank's balanced salt solution (HBSS) containing 100 U/ml of penicillin and 100 μg/ml of streptomycin (Gibco Laboratories, Grand Island, NY). Inoculations of the experimental mice were intraperitoneal (i.p.). For in vitro experiments, BFT were isolated (22) from the blood of mice infected for seven days, and the organisms were resuspended in RPMI tissue culture medium supplemented with 10% (v/v) heat- (60°C) inactivated fetal calf serum (Gibco Laboratories) and 100 U/ml of penicillin and 100 μg/ml of streptomycin.

Amastigotes and trypomastigotes were harvested from L cell cultures and used for in vitro experiments as previously described (23).

B. Interferon and Antisera to Interferon

Recombinant murine interferon gamma was obtained from Dr. Costas Sevastopoulos (Genentech Inc., South San Francisco, CA) and was free of endotoxin as determined by the Limulus amebocyte assay. The lymphokine was titered by the method of inhibition of encephalomyocarditis virus in the L929 fibroblast cell line.

Antiserum to IFNγ was produced in rabbits and was also obtained from Genentech Inc. One milliliter of antiserum neutralized 200,000 units of antiviral activity of IFNγ as assayed in L929 cells with encephalomyocarditis virus. Rabbit anti-interferon α/β was produced in mice and was obtained from Lee Biomolecular Research Laboratories, San Diego, CA. This antiserum was titrated with anti-interferon reference reagents provided by the Antiviral Substances Program, National Institute of Allergy and Infectious Disease, National Institutes of Health, Bethesda, MD.

C. Protocol for In Vitro Experiments

Mouse peritoneal macrophage monolayers were established in eight chamber tissue culture slides (Labtek Products, Naperville, IL) as previously described (23). In some experiments, monolayers were incubated with IFNγ with or without one of the antisera to IFNγ or α/β (see below) for 18 hours. Then BFT were added to the monolayers at a parasite:macrophage ratio of 0.5:1 and the slides incubated for 24 hr in an incubator at 37°C, with 5% CO_2. Interferon gamma, with or without one of the antisera to interferon, was present during this time. Thereafter, extracellular BFT were removed from the monolayers by washing with culture medium, and incubation in the presence of interferon with or without anti-interferon sera was continued for an additional 48 hr at 37°C in 5% CO_2. Tissue culture slides were washed with phosphate-buffered saline, pH 7.2 (PBS), fixed in Bouin's solution, and stained with Giemsa stain. Five hundred cells were counted to determine the percent cells infected, and the number of intracellular organisms in 50 infected cells from at least five different areas of the slide were counted to determine the average number of amastigotes per infected cell. Each value represents the mean of the determinations for duplicate monolayers. Each experiment was repeated two to three times, except experiments that required antisera to interferons which were performed only once due to limited supply of antisera.

For experiments with antisera to interferons, IFNγ was mixed with antisera at a ratio of 1 U:1 U and the mixture incubated with the macrophage monolayers for 18 hr at 37°C in 5% CO_2 before challenge with BFT. Interferon gamma and antisera remained in contact with macrophages for the entire experiment.

D. Protocol for In Vivo Experiments

Stock IFNγ was diluted in PBS to the desired concentration immediately before intraperitoneal administration into mice. Two different protocols were used: IFNγ was administered every other day starting 48 hr before infection with *T. cruzi*; or only two doses of IFNγ were administered, one at the time of infection and the other 24 hr later.

For experiments that used antisera to interferons, IFNγ was mixed with antiserum at a ratio of 1 U:1 U and the mixture incubated for two hours at 37°C before intraperitoneal administration into mice.

E. Determination of IgG and IgM Serum Antibody Levels and Parasitemias

IgG and IgM serum antibodies specific for *T. cruzi* were determined by an enzyme-linked immunosorbent assay (ELISA) as described previously (24). Parasitemias in test and control mice were measured as described by Brener (21).

III. RESULTS

A. In Vivo Activity of IFNγ

Mice infected with 10^4 BFT and treated with 10^4 or 10^3 units of IFNγ every other day, beginning two days before infection, had prolongation of time to death when compared with control mice and mice that received either 10^2 or 10^1 units of IFNγ (Fig. 1). Statistical analysis of the differences in time to death on each day usually did not reveal significant differences. Five other similar experiments confirmed the results shown in Figure 1. In each experiment, infected mice treated with IFNγ had a slight prolongation of time to death (usually one day) when compared to control mice. None of the doses of interferon gamma, however, gave complete protection against death for any mouse; all mice eventually died of the infection, usually by day 30.

Parasitemias determined on days 6, 8, and 11 of infection (Table 1) revealed that mice treated with 10^4 units of IFNγ had significantly lower parasitemias ($p < .05$, Student's t-test) than controls on days 6 and 8, whereas mice treated

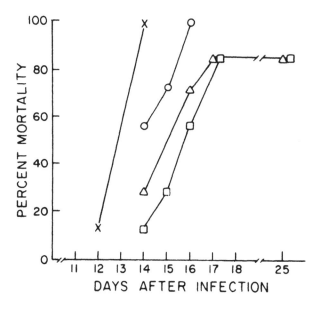

Figure 1   Effect of IFNγ on the mortality of mice due to infection with *Trypanosoma cruzi*. Mice were treated with IFNγ administered i.p. every other day starting 48 hr before infection with 10^4 BFT of the Y strain of *T. cruzi*. (x) 10 and 100 units IFNγ; (△) 10^3 units; (□) 10^4 units; (○) controls injected with PBS. There were 7 mice in each group.

Table 1 Parasitemias in Mice Injected with IFNγ and infected with *Trypanosoma cruzi*

IFNγ[a] (units)	Days after infection		
	6	8	11
None	831 ± 304[b]	10626 ± 3807	22991 ± 7136
10^4	280 ± 72[c]	3854 ± 3624[c]	14345 ± 5981
10^3	174 ± 100[c]	7628 ± 5039	10381 ± 4667[c]
10^2	667 ± 351	10046 ± 4599	38157 ± 27508
10^1	522 ± 261	9650 ± 4316	33955 ± 8056
10^4 $(0,+1)$[d]	145 ± 77[c]	4666 ± 2783[c]	29076 ± 17751
10^3 $(0,+1)$[d]	280 ± 212[c]	11689 ± 6638	36756 ± 10106

[a]Units of IFNγ administered i.p. every other day starting 48 hr before infection with 1×10^4 BFT of the Y strain of *T. cruzi.*

[b]Mean number of organisms per 5 μl of blood ± standard deviation.

[c]$p < 0.05$, Student's t-test.

[d]IFNγ administered only on day of infection and 24 hr later.

with 10^3 units of IFNγ had significantly lower parasitemia counts on days 6 and 11. Parasitemia counts in mice injected with lower doses of IFNγ were similar to controls.

IgM and IgG serum antibodies to *T. cruzi* were determined on day 14 of infection for five control mice; for five mice treated with 10^1 units of IFNγ; and for four mice treated with 10^4 units of IFNγ. No mice in the control group and only one (with a titer of 1:20) in the group that received 10^1 units of IFNγ had a positive titer for IgG. In contrast, all mice that received 10^4 units of IFNγ had titers of 1:20 or 1:40 ($p < .01$, Student's t-test). All mice had positive IgM antibody titers, but those treated with 10^4 units had somewhat higher titers ($0.05 < p < 0.10$).

Of the mice injected with two doses of IFNγ, administered at the time of infection and 24 hr later, only those that received 10^3 or 10^4 units showed a prolongation of time to death that was not statistically significant (data not shown). However, these two doses of IFNγ significantly reduced parasitemia levels (Table 1).

Antiserum to IFNγ completely neutralized the activity of IFNγ. Thus, mice injected with 5×10^3 units of interferon gamma on the day of infection and 24 hr later had significantly lower parasitemias than mice injected with the same doses of IFNγ mixed with antiserum to IFNγ (Table 2).

Table 2 Effect of Antiserum to IFNγ on the Activity of IFNγ in Inducing a
Reduction of Parasitemia

Treatment[a]	Parasitemia on day[b]	
	6	7
Controls	1714 ± 656	22158 ± 7789
IFNγ[b]	898 ± 536	8066 ± 5051[c]
IFNγ + anti-IFNγ	1971 ± 287[d]	26661 ± 7678[d]

[a]5 × 10³ units of IFNγ, (IFNγ); IFNγ mixed with antiserum to IFNγ, (IFNγ + anti-IFNγ);
   or PBS (controls) were administered on day of infection with 1 × 10⁴ BFT and 24 hr later.
[b]Mean number of organisms per 5 μl of blood ± standard deviation.
[c]$p < 0.05$ compared with controls (Student's t-test).
[d]$p < 0.05$ compared with IFN-γ.

B. In Vitro Activity of IFNγ

The activity of IFNγ on replication of *T. cruzi* in vitro was examined by incubating normal mouse peritoneal macrophages with IFNγ before, during, and after challenge of the cells with *T. cruzi*. In experiments that utilized trypomastigotes and amastigotes harvested from L cells, as little as 1 unit/ml produced significantly ($p < .05$) decreased numbers of infected cells and 10 units/ml produced both significantly decreased numbers of infected cells and decreased number of organisms per infected cell when compared with control cultures 48 hr after challenge. In macrophage cultures that had been challenged with BFT, 25 units/ml significantly inhibited the intracellular replication of amastigotes (Fig. 2). The number of infected macrophages was also significantly reduced compared with controls (Fig. 3). Antiserum to IFNγ completely abrogated the inhibitory effect of IFNγ and only partially neutralized the effect of IFNγ with respect to the number of infected macrophages. Of interest was the observation that antiserum to interferon α/β did not have any effect on the activity of interferon γ (Figs. 2 and 3). In addition, antiserum to interferon α/β or to interferon γ alone had no effect on either infection of macrophages or intracellular replication of amastigotes (Figs. 2 and 3).

Exposure of peritoneal macrophages to IFNγ for 18 hr prior to and during the challenge of the cells with BFT had no effect on internalization of the organisms by the macrophages (Table 3).

To answer the question whether administration of IFNγ induces generation of activated macrophages capable of inhibiting replication of *T. cruzi*, 5 × 10³ units of IFNγ were inoculated i.p. into mice and peritoneal cells harvested 24 hr later. Monolayers were established and challenged with BFT. After 48 hr of infection, 53% of macrophages from control mice injected with PBS were infected

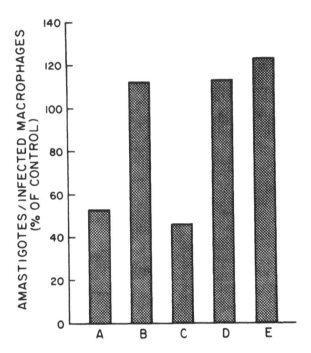

Figure 2 Effect of IFNγ and antisera to interferons on the replication of *Trypanosoma cruzi* within mouse peritoneal macrophages. Mouse peritoneal macrophage monolayers were incubated with IFNγ (A), IFNγ and antisera to IFNγ (B), IFNγ and antisera to IFNα/β (C), antisera to IFNγ (D), antisera to IFNα/β (E), and medium alone before, during, and after challenge of macrophages with BFT. Monolayers were examined 48 hr after extracellular organisms had been washed from the cultures. Results are expressed as the percent of the replication in the control monolayers. Control macrophages contained an average of 11.4 ± 0.18 (SEM) amastigotes per infected cell. Cultures treated with IFNγ and with IFNγ and antisera to IFNα/β had significantly fewer amastigotes than controls (p < .05).

with an average of 8 amastigotes per cell. In contrast, the infection rate in macrophages from mice injected with IFNγ was 36% with an average of 4 amastigotes per infected cell. Antiserum to IFNγ neutralized the ability of IFNγ to induce, in vivo, the generation of activated macrophages capable of inhibiting the intracellular replication of *T. cruzi* in vitro. Thus, control macrophages had an average of 7 amastigotes per infected cell; macrophages from IFNγ treated mice had an average of 4 amastigotes per cell; and macrophages from IFNγ-anti-IFNγ-treated mice had an average of 7 amastigotes per cell.

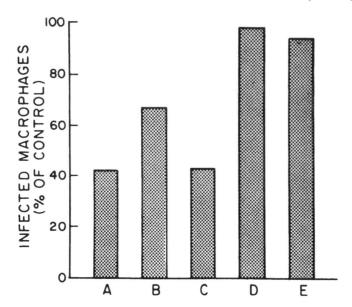

Figure 3 Effect of IFNγ on the infection of mouse peritoneal macrophages with *Trypanosoma cruzi*. Monolayers were incubated with IFNγ or antisera to interferons before, during, and after challenge with BFT. Letters indicating treatment of monolayers and expression of the results are as for Figure 2. Control cultures contained an average of 56 ± 5 (SEM) infected cells. Cultures treated with IFNγ and with IFNγ and antisera to IFNα/β had significantly fewer infected macrophages (p < .05).

Table 3 Effect of IFNγ on Internalization of *T. cruzi* by Macrophages (Mφ)

IFNγ (units)	Percent Mφ infected/amastigotes per Mφ[a]		
	hours of incubation		
	1	8	16
0	30/1.5	58/2.0	55/2.5
50	24/1.2	49/1.6	41/2.2
100	27/1.4	57/1.8	41/1.7
200	28/1.4	48/1.9	53/2.0

[a]Macrophages were harvested from the peritoneal cavities of Swiss-Webster mice and allowed to adhere to tissue culture chambers for four hours. Nonadherent cells were washed from the chambers and IFNγ added for 18 hr. At that time, monolayers were challenged with BFT. At 1, 8, and 16 hr after challenge, monolayers were fixed, stained, and examined for the number of infected macrophages and the number of amastigotes per infected macrophage. Each value is the mean of two replicate monolayers.

In the foregoing experiments, extracellular BFT were in contact with IFNγ in the culture medium before uptake by the macrophages. To examine for a direct effect of IFNγ on BFT, these organisms were incubated with IFNγ (0, 25, 100, and 200 units/ml) for 2 hr before they were used to challenge macrophages that had not been treated with IFNγ. By 48 hr of infection of the monolayers, the percent of cells infected and number of amastigotes per infected macrophage were similar in all groups of cells.

IV. DISCUSSION

A. Conclusions

Our results revealed that cloned recombinant IFNγ administered to mice infected with *T. cruzi* promoted a prolongation of the time to death and significantly decreased parasitemia levels. This effect was neutralized by antiserum to IFNγ. As previously noted (12) in infections with another intracellular parasite, *Toxoplasma gondii*, both the quantity of IFNγ administered and the time of its administration were important for the outcome of infection. Doses of 10^2 units of IFNγ and less were ineffective in murine trypanosomiasis. Two doses only of either 10^3 or 10^4 units of IFNγ administered at the time of infection and 24 hr later provided protection similar in extent to that provided by the same dosages administered every 48 hr during the course of infection. This suggests that the events that occur immediately around the time of infection are critical for determination of the fate of the host and of the invading pathogen.

Interferon gamma has been shown to suppress replication of *Toxoplasma gondii* (7), leishmania (8,9), *Rickettsia prowazekii* (10), and *Chlamydia psittaci* (11) in either professional or nonprofessional phagocytes in vitro. The mechanism suggested for this activity is that IFNγ activates macrophages (7). Recent reports have indicated that this mechanism is operative for at least *T. cruzi* (6) and *T. gondii* (25). In addition, we showed previously that as little as five units of IFNγ administered i.p. to normal mice activates peritoneal macrophages such that by 24 hr after administration the cells markedly inhibit replication and kill *T. gondii* tachyzoites in vitro (12). In this study we observed similar results with *T. cruzi*; peritoneal macrophages collected 24 hr after an i.p. injection of IFNγ markedly inhibited the intracellular replication of the organism.

The results revealed a clear inhibition of intracellular replication of *T. cruzi* in macrophages from mice injected with IFNγ. Killing of the intracellular organisms by the same macrophages was not as clear. There was a decrease in the number of infected macrophages with prolongation of incubation, which suggests that some killing may have occurred. Other investigators have demonstrated increased association of BFT with cells and also an increased killing capacity of cells treated with IFNγ (6). When BFT were exposed to macrophages in the presence of IFNγ, the percent infection of cells was reduced when compared with the percent infection of similarly plated macrophages exposed to BFT in the presence of IFNγ and antiserum to IFNγ. This observation suggests that IFNγ may have had an effect on BFT that was not sufficient to kill the organisms but prevented their interiorization. They may have associated with the macrophages, as has been demonstrated (6), but not interiorized, or they may have been interiorized and rapidly destroyed, so that they were not detected by microscopy of Giemsa-stained cell monolayers.

Of interest was the observation that both the IgM and IgG antibodies' response to T. cruzi were significantly increased in infected mice receiving injections of IFNγ. Similar observations had been reported previously with mice experimentally infected with T. gondii (12). A number of recent reports have clearly demonstrated a regulatory effect for interferon on the immune responses. Increased indirect and direct plaque-forming cellular responses, increased serum antibody response, stimulation of B-cell antibody response in vitro, and induction of immunoglobulin secretion by resting murine splenic B cells and cells from a B-cell tumor line have been reported as due to the action of IFNγ (26-29). In Chagas' disease the stimulatory effect of IFNγ on the antibody response may be of importance in the development of resistance to the infection since recent reports have demonstrated that antibody production appears to be essential both for survival during the acute infection and for the maintenance of resistance to reinfection in the chronic stage of the infection (30). The levels of IFNα and IFNβ activity are increased 24 hr after infection with T. cruzi (2). However, there are no data regarding the levels of activity of IFNγ following infection with T. cruzi. In any event, it is likely that all three types of interferon may act in concert, since it has been shown that interferon beta acts on T. cruzi trypomastigotes (3,4) and IFNα and IFNβ have a protective effect against infection with the organism (1).

Our data showed only borderline protection by IFNγ against death in this model of acute infection. Possibly this model is not optimal for demonstration of therapeutic efficacy of IFNγ. Although data are lacking for T. cruzi infection, Buchmeier and Schreiber found that endogenous IFNγ was a requirement for resolution of acute Listeria monocytogenes infection in mice (31). Havell demonstrated that transient, high levels of IFNγ are produced in serum of mice infected acutely with Listeria monocytogenes (32). In the acute infection, administration of exogenous IFNγ may provide only marginal benefit in view of the probable presence of endogenous IFNγ. Greater protection by IFNγ may be demonstrable in models of infection characterized by low levels of endogenous IFNγ. These models may include chronic infection with T. cruzi, prophylaxis of T. cruzi infection, or acute infection in hosts that have impaired ability to produce IFNγ (33).

Resistance to *T. cruzi* is due to a large number of factors related to the host and to the organism. Some host factors, such as the increased resistance of high antibody responder mice, have been defined (34). However, the roles of a large number of factors are still unclear (35,36). Because IFNγ enhances killing of intracellular parasites by phagocytic cells and increases production of specific antibodies, an interesting project for future research is determination of whether different patterns of resistance observed in inbred mice are related to differing abilities of strains to rapidly form and maintain production of IFNγ following infection with *T. cruzi*.

ACKNOWLEDGMENTS

This investigation received financial support from Grants #AI04717 and #AI18794 from the National Institutes of Health, and from the UNDP/World Bank/WHO Special Programme for Research and Training in Tropical Diseases.

Robert E. McCabe is the recipient of an Edith J. Milo Memorial Fellowship, and Fellowship Grant #AI06723 from the National Institutes of Health.

REFERENCES

1. Kierszenbaum, F., and G. Sonnenfeld. 1982. Characterization of the antiviral activity produced during *Trypanosoma cruzi* infection and protective effects of exogenous interferon against experimental Chagas' disease. *J. Parasitol. 68*:194–198.
2. Sonnenfeld, G., and F. Kierszenbaum. 1981. Increased serum levels of an interferon-like activity during the acute period of experimental infection with different strains of *Trypanosoma cruzi. Am. J. Trop. Med. Hyg. 30*: 1189–1191.
3. Golgher, R.R., M.S. Bertelli, M.L. Petrillo-Peixoto, and Z. Brener. 1980. Effect of interferon on the development of *Trypanosoma cruzi* in tissue culture vero cells. *Mem. Inst. Oswaldo Cruz (Rio de Janeiro, Brazil) 75*:157–160.
4. Kierszenbaum, F., and G. Sonnenfeld. 1984. Beta-interferon inhibits cell infection by *Trypanosoma cruzi. J. Immunol. 132*:905–908.
5. Plata, F., J. Wietzerbin, F. Garcia Pons, E. Falcoff, and H. Eisen. 1984. Synergistic protection by specific antibodies and interferon against infection by *Trypanosoma cruzi* in vitro. *Eur. J. Immunol. 14*:930–935.
6. Wirth, J.J., F. Kierszenbaum, G. Sonnenfeld, and A. Zlotnik. 1985. Enhancing effects of gamma interferon on phagocytic cell association with and killing of *Trypanosoma cruzi. Infect. Immun. 49*:61–66.
7. Nathan, C.F., H.W. Murray, M.E. Wiebe, and B.Y. Rubin. 1983. Identification of interferon gamma as the lymphokine that activates human macrophage oxidative metabolism and antimicrobial activity. *J. Exp. Med. 158*: 670–689.

8. Murray, H.W., B.Y. Rubin, and C.D. Rothermel. 1983. Killing of intracellular *Leishmania donovani* by lymphokine-stimulated human mononuclear phagocytes. Evidence that interferon gamma is the activating lymphokine. *J. Clin. Invest. 72*:1506-1510.

9. Passwell, J.H., R. Shor, and J. Shoham. 1986. The enhancing effect of interferon-beta and -gamma on the killing of *Leishmania tropica major* in human mononuclear phagocytes in vitro. *J. Immunol. 136*:3062-3066.

10. Turco, J., and H.H. Winkler. 1983. Cloned mouse interferon-gamma inhibits the growth of *Rickettsia prowazekii* in cultured mouse fibroblasts. *J. Exp. Med. 158*:2159-2164.

11. Byrne, G.I., L.K. Lehmann, and G.J. Landry. 1986. Induction of tryptophan catabolism is the mechanism for gamma-interferon-mediated inhibition of intracellular *Chlamydia psittaci* replication in T24 cells. *Infect. Immun. 53*:347-351.

12. McCabe, R.E., B.J. Luft, and J.S. Remington. 1984. Effect of murine interferon gamma on murine toxoplasmosis. *J. Infect. Dis. 150*:961-962.

13. Ferreira, A., L. Schofield, V. Enea, H. Schellekens, P. Van Der Meide, W.E. Collins, R.S. Nussenzweig, and V. Nussenzweig. 1986. Inhibition of development of exoerythrocytic forms of malaria by gamma-interferon. *Science 232*:881-884.

14. Maheshwari, R.K., C.W. Czarniecki, G.P. Dutta, S.K. Puri, B.N. Dhawan, and R.M. Friedman. 1986. Recombinant human gamma interferon inhibits simian malaria. *Infect. Immun. 53*:628-630.

15. Kiderlen, A.F., S.H.E. Kaufmann, and M-L. Lehmann-Matthes. 1984. Protection of mice against the intracellular bacterium *Listeria monocytogenes* by recombinant immune interferon. *Eur. J. Immunol. 14*:964-967.

16. Nathan, C.F., C.R. Horowitz, J. De La Harpe, S. Vadhan-Raj, S.A. Sherwin, H.F. Oettgen, and S.E. Krown. 1985. Administration of recombinant interferon gamma to cancer patients enhances monocyte secretion of hydrogen peroxide. *Proc. Natl. Acad. Sci. (USA) 82*:8686-8690.

17. Pennington, J.E., J.E. Groopman, G.J. Small, L. Laubenstein, and R. Finberg. 1986. Effect of intravenous recombinant gamma-interferon on the respiratory burst of blood monocytes from patients with AIDS. *J. Infect. Dis. 153*:609-612.

18. Douvas, G.S., D.L. Looker, A.E. Vatter, and A.J. Crowle. 1985. Gamma interferon activates human macrophages to become tumoricidal and leishmanicidal but enhances replication of macrophage-associated mycobacteria. *Infect. Immun. 50*:1-8.

19. Rook, G.A.W., J. Steele, L. Fraher, S. Barker, R. Karmali, J. O'Riordan, and J. Stanford. 1986. Vitamin D_3, gamma interferon, and control of proliferation of *Mycobacterium tuberculosis* by human monocytes. *Immunology 57*:159-163.

20. Melo, R.C., and Z. Brener. 1978. Tissue tropism of different *Trypanosoma cruzi* stains. *J. Parasitol. 64*:475-482.

21. Brener, Z. 1962. Therapeutic activity and criterion of cure in mice experimentally infected with *Trypanosoma cruzi. Rev. Inst. Med. Trop. Sao Paulo 4*:389-396.

22. Budzko, D.B., and F. Kierszenbaum. 1974. Isolation of *Trypanosoma cruzi* from blood. *J. Parasitol. 60*:1037–1038.

23. McCabe, R.E., J.S. Remington, and F.G. Araujo. 1984. Ketoconazole inhibits intracellular multiplication of *Trypanosoma cruzi* and protects mice against lethal infection with the organism. *J. Infect. Dis. 150*:594–601.

24. Araujo, F.G., and D.R. Guptill. 1984. Use of antigen preparations of the amastigote stage of *Trypanosoma cruzi* in the serology of Chagas' disease. *Am. J. Trop. Med. Hyg. 33*:362–371.

25. Wilson, C.B., and J. Westall. 1985. Activation of neonatal and adult human macrophages by alpha, beta, and gamma interferons. *Infect. Immun. 49*: 351–356.

26. Boraschi, D., S. Censini, and A. Tagliabue. 1984. Interferon-gamma reduces macrophage-suppressive activity by inhibiting prostaglandin E_2 release and inducing interleukin 1 production. *J. Immunol. 133*:764–768.

27. Leibson, H.J., M. Gefter, A. Zlotnik, P. Marrack, and J.W. Kappler. 1984. Role of gamma interferon in antibody-producing responses. *Nature 309*: 799–801.

28. Nakamura, M., T. Manser, G.D.N. Pearson, M.J. Daley, and M.L. Gefter. 1984. Effect of interferon gamma on the immune response in vivo and on gene expression in vitro. *Nature 307*:381–382.

29. Sidman, C.L., J.D. Marshall, L.D. Shultz, P.W. Gray, and H.M. Johnson. 1984. Gamma interferon is one of several direct B cell-maturing lymphokines. *Nature 309*:801–803.

30. Scott, M.T. 1981. The nature of immunity against *Trypanosoma cruzi* in mice recovered from acute infection. *Parasite Immunol. 3*:209–214.

31. Buchmeier, N.A., and R. Schreiber. 1985. Requirement of endogenous interferon gamma production for resolution of *Listeria monocytogenes* infection. *Proc. Natl. Acad. Sci. (USA) 82*:7404–7408.

32. Havell, E.A. 1986. Augmented induction of interferons during *Listeria monocytogenes* infection. *J. Infect. Dis. 153*:960–969.

33. Suzuki, F., H. Maeda, and R.B. Pollard. 1986. Suppression of interferon gamma production in mice treated with carrageenan. *Eur. J. Immunol. 16*: 375–380.

34. Kierszenbaum, F., and J.G. Howard. 1976. Mechanisms of resistance to experimental *Trypanosoma cruzi* infection. The importance of antibody and antibody-forming capacity in the Biozzi high and low responder mice. *J. Immunol. 116*:1208–1211.

35. Trischmann, T.M., H. Tanowitz, M. Wittner, and B. Bloom. 1978. *Trypanosoma cruzi*: role of the immune response in natural resistance of inbred strains of mice. *Exp. Parasitol. 45*:160–168.

36. Wrightsman, R., S. Krassner, and J. Watson. 1982. Genetic control of responses to *Trypanosoma cruzi* in mice: multiple genes influencing parasitemia and survival. *Infect. Immun. 36*:637–644.

12

Antimalarial Activity of IFNγ: Requirement for Immunity to Sporozoite Challenge

Louis Schofield and Arturo Ferreira* / New York University School of Medicine, New York, New York

I. INTRODUCTION

In this chapter, we will present recent experimental evidence concerning the antimalarial activity of interferon gamma (IFNγ). The main focus will be on the earlier stages of infection, in particular the sporozoite and exoerythrocytic (liver) stages. These studies show that recombinant IFNγ can eradicate, or substantially modify, the course of malaria infection in experimental hosts. In vitro studies have demonstrated that IFNγ acts without the participation of effector cells, by binding to surface receptors on infected cells and thereby inhibiting the intrahepatocytic development of exoerythrocytic forms (EEF). We will also present evidence demonstrating that the antimalarial activity of host IFNγ is a necessary effector mechanism for acquired immunity to sporozoite challenge.

First, however, we will review briefly some important features of parasite biology, including what is known about acquired immunity to these stages.

II. PARASITE LIFE CYCLE

The life cycle of malaria parasites within their mammalian hosts is characterized by a high degree of developmental differentiation. An uninterrupted natural infection proceeds through three morphologically distinct stages: the sporozoite, the exoerythrocytic forms (EEF), and the blood stages.

---

*Present affiliation: University of Chile, Santiago, Chile

When malaria-infected female *Anopheles* mosquitoes bite the appropriate host, the infection is initiated by the entry of sporozoites into the circulation, whence they are quickly disseminated through the body. Even in naive hosts, it appears that the majority of sporozoites are removed from circulation by the spleen (1). A smaller proportion enters the liver, to initiate the second phase of infection.

The precise details concerning the transition from sporozoite to intrahepatocytic EEF are not well understood. It is clear, however, that the process is quite rapid and that, correspondingly, the circulating sporozoite stage is shortlived. One study (2) showed the appearance of parasites within hepatocytes within two minutes of the intravenous inoculation of sporozoites. Some authors consider that the sporozoite invades hepatocytes directly (2,3) while others argue that sporozoites are taken up by Kupffer cells before entering hepatocytes (4).

Once within the hepatocyte, where it lives within a parasitophorous vacuole, probably of host plasmalemma origin, the EEF undergo a period of rapid growth and mitotic division. The duration and number of divisions varies with each species. Thus, the rodent malaria parasite *Plasmodium berghei* undergoes approximately 13 mitotic divisions over 44 hours to yield 8000 daughter forms, or EE merozoites (1,5). The human malaria parasites *P. falciparum* and *P. malariae*, on the other hand, produce 30,000 and 15,000 merozoites, in five and a half and 15 days, respectively (6).

Upon completion of EEF development, the host liver cell ruptures and the EE merozoites are released to infect circulating erythrocytes, thereby initiating the blood-stage infection. It is this stage that causes disease and can lead to the death of the host.

This highly regulated developmental sequence explains the commonly held view of stage specificity in acquired immunity to malaria. The stages differ in their cellular organization and in particular they express quite distinct surface antigens and invade different host cell types. Thus, there appear to be few antigens shared among stages which are the target of protective humoral antibodies.

III. ACQUIRED IMMUNITY TO SPOROZOITES AND EEF

When susceptible hosts are repeatedly inoculated with attenuated, X-irradiated sporozoites of a given malaria species, they acquire protective immunity to further viable sporozoite challenge. This immunity is sterile, in that no sporozoites succeed in establishing a blood-stage infection (7). However, immunized hosts remain fully susceptible to experimental challenge with blood-stage parasites, suggesting that the protective immunity is stage specific and is directed toward sporozoites or EEF. Analysis of the immune response of immunized rodents, monkeys, and humans has revealed the existence of a substantial humoral antibody titer directed against a single major parasite antigen, the circumsporozoite (CS) protein (8). Both antisporozoite sera and monoclonal antibodies raised against sporozoites from different malarial species recognize an immunodominant epitope within the CS protein that is both stage and species specific. Upon passive transfer, these monoclonal antibodies protect mice against sporozoite challenge (9), and also inhibit the invasion of hepatocytes by sporozoites in vitro. Thus, there is considerable evidence to suggest that sterile immunity to sporozoite challenge involves the neutralization of sporozoite infectivity by anti-CS protein antibodies in the serum of immunized hosts.

In addition, however, several studies indicate that cellular effector mechanisms may participate in protective immunity. Despite the neutralizing activity of monoclonal antibodies, passive transfer of immune serum to naive recipients does not confer fully protective immunity. In addition, protective immunity has been achieved in the absence of detectable antisporozoite antibodies (10). Protective immunity can also be induced by inoculation of X-irradiated sporozoites into mice whose antibody synthesis was suppressed by the administration of anti-u chain antibodies from birth (11). Furthermore, under certain conditions, adoptive transfer of immune splenic T and B cells into naive recipients, followed by further immunization, protects them against viable sporozoite challenge (12, 13). The requirement in these latter studies for further immunization of recipients following adoptive transfer of immune cells complicates interpretation of the results obtained, since the involvement of T cells could arise through a helper role in antibody production. Beyond these studies, therefore, the nature of the cellular effector mechanism is not clear. The sporozoite stage is notably short-lived, since it ends upon hepatocyte invasion soon after inoculation. It is difficult to envisage a cellular mechanism, induced by inoculation of irradiated sporozoites, that could successfully neutralize a large sporozoite challenge within this short time period. Possibly, cellular mechanisms are directed against the ensuing EEF. However, little evidence exists to indicate that the EEF of mammalian malaria species suffer a significant degree of immunological attack from either antibodies or phagocytic cell types. The parasites are intracellular and therefore not exposed to neutralizing activity by humoral antibodies. Indeed, it has been suggested that the liver acts as a "privileged site" for the development of malaria parasites, based upon the failure to detect inflammatory responses to developing EEF (14,15). Other authors have described inflammatory responses with cellular infiltration surrounding EEF, but only after the release of EE merozoites into the blood stream when the normal period of parasite development was completed (16,17).

From the studies cited above, two questions arise, namely, (a) is cellular immunity indeed required for resistance to sporozoites, and (b) if so, what is the mechanism by which cellular immunity exerts its effect? To address these questions, we have recently investigated whether lymphokines, in particular IFNγ, exert an antiparasite effect, and whether they might be involved in mediating protective immunity to sporozoite challenge.

IV. ANTIMALARIAL ACTIVITY OF INTERFERON IN VIVO

The first evidence for the antimalarial activity of interferons came from the studies of Jahiel and colleagues (18–20). These workers were able to show that inoculation of interferon inducers such as Newcastle disease virus and poly-I:poly-C protected mice against challenge with sporozoites of *P. berghei*. The greatest effect was observed when these interferon-inducing agents were administered to mice 16–24 hr after sporozoite challenge (i.e., during the period of EEF development). In contrast, no effect was observed when these agents were added during the blood-stage infection, indicating the antiparasite effect was restricted to EEF alone.

Although providing the first circumstantial evidence for the activity of lymphokines against EEF, these studies were limited in two ways: (a) by uncertainties regarding the nature of the host reaction to the inducing agents; and (b) by the inability to measure directly the effect of the agents on the course of EEF development in the liver.

In more recent studies, we undertook to examine further the antimalarial activity of lymphokines. This was made possible by the availability of recombinant proteins such as IFNγ, and also by the development of a parasite-specific, repetitive DNA probe which, via DNA extraction and hybridization, allowed for the first time the accurate measurement of parasite development in the liver of infected hosts (1).

We chose to study the activity of IFNγ by reasoning that this lymphokine, produced by activated T cells upon stimulation by antigen, would be more likely to play a role in acquired resistance to malaria sporozoites than alpha and beta interferons. In the initial experiments, female A/J mice were injected intravenously with doses of 2×10^2–2×10^5 units of murine recombinant IFNγ, 5 hours prior to challenge with 5×10^3 *P. berghei* sporozoites (21). The effect of the lymphokine treatment on the establishment of the parasite infection was determined by measuring the appearance of blood-stage parasites, utilizing the *P. berghei*-specific DNA probe. Since the assay can detect as few as 10^3 haploid parasite nuclei, it is sufficiently sensitive to detect parasites in the blood of control animals immediately after the cessation of the EE stage of development, when EE merozoites are released into the blood circulation. As shown in Figure 1, treatment of mice with recombinant murine IFNγ greatly delayed the appearance of blood-stage parasites, although all animals eventually developed parasitemia.

These observations were repeated using another host/parasite model. The daily injection of 5×10^6 units of recombinant human IFNγ into chimpanzees challenged with sporozoites of the human malaria parasite *P. vivax* resulted in a pronounced delay in the appearance of blood stage parasites (Fig. 2).

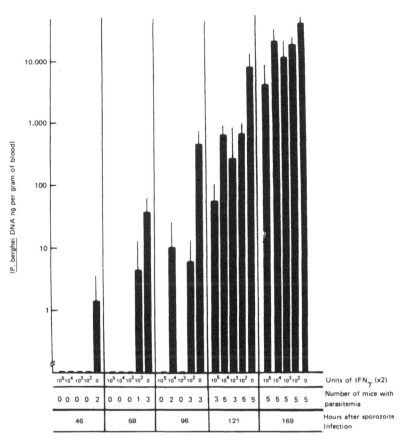

Figure 1 Parasitemia in mice treated with recombinant murine IFNγ and challenged with *P. berghei* sporozoites. Adult female A/J mice were injected with dilutions of IFNγ, or with phosphate-buffered saline, 5 hr before the intravenous injection of 5×10^3 sporozoites. The amount of parasite DNA present in the blood of these animals was measured by hybridization with ^{32}P-labelled p-263-1, a *P. berghei*-specific DNA probe. Blood smears were also prepared and examined for the presence of parasitized erythrocytes. (Reproduced from Ref. 21.)

The observed delay in patency of blood-stage parasites in IFNγ-inoculated hosts may have arisen through action of the lymphokine against sporozoites, EEF, or the blood stages themselves. Accordingly, mice were inoculated with 10^5 units of recombinant murine IFNγ 72 hr after *P. berghei* sporozoite challenge, that is, after the completion of EEF development, when all parasites had entered the blood-stage cycle. The course of parasitemia was compared with a

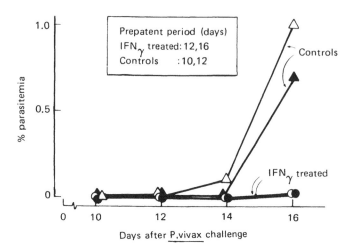

Figure 2 Inhibition of *P. vivax* infectivity in chimpanzees treated with recombinant human IFNγ. Two chimpanzees were injected intravenously with 5×10^6 U of IFNγ diluted in PBS, at ~5 hr, +2 hr, and daily on days 1 to 6; the other two received PBS alone. At time 0 the animals were challenged with 10^5 *P. vivax* sporozoites. Blood smears were made from day 8 onward, Giemsa stained, and examined for the presence of infected erythrocytes. Results are expressed as percentage parasitemia. (Reproduced from Ref. 21.)

control group receiving only phosphate-buffered saline. No statistically significant effect on the development of parasitemia was observed, indicating that under these experimental conditions, IFNγ does not effect *P. berghei* blood stages. Thus the effects of IFNγ previously observed were likely to result from activity against either sporozoites or EEF.

In order to examine this possibility further, we investigated the activity of recombinant rat IFNγ on the development of *P. berghei* EEF in the liver of Brown Norway rats. IFNγ was injected before or after challenge with 10^5 sporozoites of *P. berghei*. The subsequent growth of EEF was determined by removal of host livers 44 hours after challenge, which is the time of maximal EEF development, followed by total DNA extraction and probing with the ^{32}P-labelled, *P. berghei*-specific repetitive DNA probe. The degree of hybridization is directly proportional to the amount of parasite DNA within the host liver and is unaffected by host cell DNA (1).

As shown in Figure 3, IFNγ treatment of rats strongly inhibited the development of the *P. berghei* EEF. The degree of inhibition was related to both the dose of lymphokine and the timing of injection. Close to 100% inhibition was achieved when 1.25×10^5 units were injected 5 hours prior to sporozoite

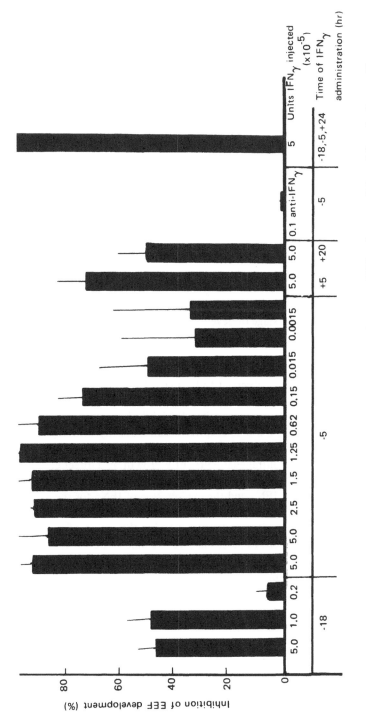

Figure 3 Inhibition of the development of *P. berghei* EEF by recombinant rat IFNγ. Groups of 4 to 6 female Brown Norway rats were injected i.v. with different doses of rat IFNγ, at different times relative to the intravenous injection of *P. berghei* sporozoites. The figure summarizes the results of several experiments; each one is compared to its own positive control. At 44 hr after challenge the rats were sacrificed and their livers removed, followed by DNA extaction and hybridization with a ^{32}P-labelled, parasite-specific DNA probe. Values expressed as means ± standard deviations, represent the percent inhibition of EEF in groups receiving IFNγ compared with control groups receiving PBS. The time of IFNγ injection is expressed in hours relative to the injection of sporozoite. In one experiment (second bar from right), the IFNγ was diluted in two neutralizing doses of polyclonal rabbit antibody to rat IFNγ. (Reproduced from Ref. 21.)

challenge. The degree of inhibition was less at lower doses, but nonetheless injection of as little as 150 units (10–20 ng of protein) was sufficient to inhibit EEF development by 30%. Thus, the dose-dependent range of inhibition of EEF by IFNγ falls over at least three logs of lymphokine concentration.

The inhibition was less marked if rat IFNγ was administered 18 hr before or several hours after sporozoite challenge, for example 73 ± 10% inhibition with 1.5×10^4 units administered 5 hr prior to challenge compared with 72 ± 10% inhibition with 5×10^5 units administered 5 hr after challenge. However, successive doses of 5×10^5 units inoculated at –18, –5, and +24 hours were sufficient to eliminate the EEF (21).

These results were substantially confirmed in an independent study (22). These authors examined the activity of recombinant human IFNγ against *P. cynomolgi* in rhesus monkeys. They found that the lymphokine exerted no effect upon the course of blood-stage parasitemia. However, when administered prior to sporozoite challenge, and during the period of EEF development, IFNγ severely depressed the appearance of blood-stage parasites, and at high doses eradicated the infection. From these observations they concluded that the effect of IFNγ was probably on the EEF stage of development.

V. ANTIMALARIAL ACTIVITY OF IFNγ IN VITRO

The in vivo reduction of EEF levels observed following IFNγ administration could arise through a number of mechanisms, including: (a) induction of splenic clearance and destruction of circulating sporozoites, (b) inhibition of sporozoite invasion into hepatocytes, (c) activation of Kupffer cells to destroy invading sporozoites, (d) destruction of intrahepatocytic EEF by activation of hepatic macrophages, and (e) inhibition of intrahepatocytic EEF growth by a direct effect upon infected cells. Each of these possibilities would have important implications for the role of endogenous IFNγ and the exploitation of IFNγ for the purposes of immunoprophylaxis.

Accordingly, we utilized an in vitro system to examine the action of recombinant IFNγ against malaria EEF. In particular, we examined whether IFNγ alone was sufficient to destroy intracellular EEF without the participation of either antibodies or immune effector cells (23).

The experimental model of choice for this purpose was that developed by Hollingdale and colleagues (24). Sporozoites of the rodent malaria parasite *P. berghei* are allowed to invade the highly differentiated human hepatoma cell line HEPG2. They transform into EEF and develop over a period of 72 hr to release merozoites capable of infecting rodent erythrocytes. For our study, we measured the growth of intracellular EEF by extraction of total DNA from cultures followed by hybridization with the parasite-specific DNA probe previously utilized in vivo (1).

Addition of recombinant human IFNγ to cultures of HEPG2 cells 6 hours prior to sporozoite invasion markedly depressed the growth of the ensuing EEF, in a concentration-dependent manner (Fig. 4). Thus, total inhibition of growth was observed with doses of the lymphokine of 10^2 U/ml or greater. The degree of inhibition decreased steadily from 10^2 to 10^{-3} U/ml, a five log range of IFNγ concentration.

The greatest inhibition was obtained when the IFNγ was added to HEPG2 cells prior to sporozoite invasion. However, a high degree of inhibition was still

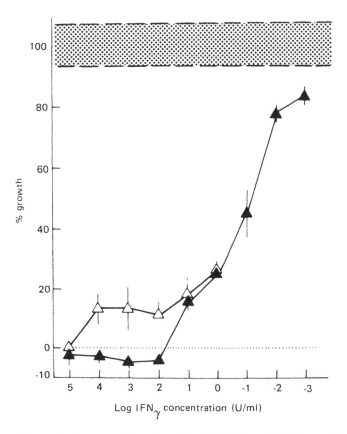

Figure 4 Concentration-dependent inhibition of EEF growth in vitro by IFNγ. The lymphokine was added at –6 hr (closed symbols) or +3 hr (open symbols). Values are means of four wells ± 1 SD expressed as percentage of growth in target cells receiving medium without lymphokine (shaded area = ± 1 SD of control). Zero growth was determined from wells receiving heat-inactivated sporozoites. All wells were harvested at +52 hr, total DNA extracted, and probed with ^{32}P-labelled parasite-specific DNA probe. (Reproduced from Ref. 23.)

obtained when the IFNγ was added to cells 3-6 hr after sporozoite interiorization had occurred (23). This demonstrates that the activity of IFNγ is directed mainly against the intrahepatocytic developing EEF.

Two further experiments were undertaken to evaluate the impact of IFNγ on EEF infectivity. Sporozoites were allowed to invade HEPG2 cells growing in 24 well plates and transform into intracellular EEF. After 6 hours, a range of IFNγ concentrations was added to the wells and maintained throughout the period of EEF development. At +70 hours, just prior to the release of infective EE merozoites, the contents of each well were inoculated into a single naive adult A/J mouse. The development of blood-stage parasitemia was monitored by the daily examination of blood films. In both experiments, EEF receiving no lymphokine uniformly gave rise to a detectable blood-stage parasitemia in four days. In the first experiment, EEF cultures exposed to from 10^3 to 10 units of IFNγ failed to infect the recipient mice. Cultures exposed to lower concentrations of lymphokine did give rise to blood-stage parasites, but with a marked delay in patency, indicating that the infecting dose was much reduced. In the second experiment, repeated as before, the capacity of IFNγ to abolish EEF infectivity appeared somewhat reduced. Total protection, implying complete inhibition of EEF, was afforded to 50% of mice receiving EEF exposed in vitro to 10^3 U/ml IFNγ, and 25% of mice receiving EEF exposed to 10^2 U/ml IFNγ. The patency of infection was still much delayed in those animals that eventually developed blood-stage parasitemia.

Taken together, these in vitro studies demonstrate that IFNγ acting alone, without the mediation of antiparasite antibodies or immune effector cells, is in some cases sufficient to eradicate totally intrahepatocytic EEF and thereby prevent the subsequent development of blood-stage infections (23).

Since IFNγ could affect intracellular parasites, we considered it likely that the antimalarial activity was mediated through the binding of the lymphokine to receptors expressed on the surface of hepatocytes and the HEPG2 cell line. To investigate the presence of such receptors, binding studies suitable for Scatchard analysis were undertaken using IFNγ labelled with ^{125}I by the Bolton-Hunter technique. The binding assays were conducted at 4°C to minimize receptor/ ligand internalization. Accordingly, HEPG2 cells express approximately 44,000 surface receptors for IFNγ each, with a Kd of 1.3×10^{-9} M (23). We also observed that upon intravenous inoculation of mice, the half life of ^{125}I IFNγ in the serum is approximately 6 minutes and 20% of the labelled lymphokine is bound by the liver, more than for any other organ (unpublished data).

As demonstrated by Pfefferkorn and colleagues (25,26), exposure of cultured fibroblasts to IFNγ leads to the expression of indoleamine deoxygenase, which in turn degrades the intracellular pool of tryptophan. This leads to the death of intracellular *Toxoplasma gondii* through tryptophan starvation. In their studies, increasing the supply of tryptophan in the culture medium reversed the antiparasite activity of IFNγ. In order to determine whether a similar mechanism was responsible for the inhibition by IFNγ of the intrahepatocytic growth of malaria EEF, we increased 8-fold the concentration of exogenous tryptophan in the culture medium of EEF exposed to IFNγ. This did not reverse the inhibitory activity of a wide range of IFNγ concentrations, indicating that under these experimental conditions tryptophan starvation does not appear to play a role in the inhibition of EEF development by IFNγ (23).

The results obtained in the in vitro studies validate and extend the previous results in vivo. In both cases, the specific activity of IFNγ against EEF is very high—as little as 150 units reduces EEF development by 30% in rats, and 10^{-3} U/ml can cause a 17% reduction of EEF development in vitro. An increase of several logs in IFNγ concentration is required to totally eradicate EEF proliferation both in vivo and in vitro. The effect of varying the time of IFNγ administration is similar in both cases, with the greatest inhibition resulting from IFNγ administration at least 6 hours prior to sporozoite inoculation. However, substantial inhibition occurs both in vivo and in vitro when IFNγ is presented a few hours after transformation of sporozoites into EEF. Although the possibility cannot be excluded that IFNγ stimulates hepatic macrophages to destroy the randomly distributed developing EEF, we consider it more likely that IFNγ acts in vivo much in the same way as it appears to do in vitro, specifically in a hormonal fashion, by binding to hepatocyte or hepatoma surface receptors and thereby inducing intracellular changes hostile to parasite development.

VI. PRODUCTION OF IFNγ IN RESPONSE TO PARASITE ANTIGENS MEDIATES RESISTANCE TO SUPERINFECTION AND IS REQUIRED FOR IMMUNITY TO SPOROZOITE CHALLENGE

The observations concerning the activity of recombinant IFNγ against malarial EEF in vivo and in vitro raise the question of whether endogenous IFNγ plays a role in the immunological control of infection. Although IFNγ may be produced in response to recognition of stage-specific antigens by sensitized T cells, it may exert its effect on a developmentally distinct phase of the parasite life cycle, by virtue of its "hormonal" action. For example, although IFNγ is very rarely detected in the serum of healthy people or patients suffering from a wide range of clinical conditions, it can indeed be detected in association with blood-stage malaria infections (27,28). It is likely that it arises from T cells sensitized to antigens of blood-stage parasites. Whatever the source of the IFNγ, we propose the hypothesis that it will prevent the establishment of subsequent natural challenge, by inhibiting the growth of intrahepatocytic EEF. This would in effect be an example of resistance to superinfection, and may account in part for the observation that immunity to malaria is of the concomitant type (premunition). If this phenomenon does occur under natural conditions it would indicate that immune states mediated through the action of lymphokines such as IFNγ need not be strictly "stage specific."

As regards sporozoite immunity, an important observation is that T cells from immunized mice respond with the production of IFNγ to stimulation by sporozoite antigens (29). Thus it is likely that in suitably immunized hosts, challenge with viable sporozoites will lead to the endogenous production of IFNγ. As an hypothesis, we propose that following sporozoite challenge, production of IFNγ by immune T cells leads to the inhibition of EEF proliferation in the liver and thereby contributes to sterile immunity. If this occurs, it may account for the postulated T-cell dependence of immunity to sporozoites.

These hypotheses may be tested experimentally. If the antimalarial activity of IFNγ is required for sterile immunity to sporozoite challenge, then destruction of IFNγ production or abrogation of IFNγ activity would lead to a reversal of sterile immunity. Abrogation of IFNγ activity in vivo may be achieved by the passive transfer of neutralizing antibodies against IFNγ. This technique has been utilized to demonstrate the requirement for endogenous IFNγ for the resolution of peritoneal infections of *Listeria* in mice (30).

Accordingly, we have recently undertaken a series of experiments designed to test whether endogenous production of IFNγ in immune hosts is a necessary component of immunity to sporozoite challenge. To do this, cohorts of mice were either immunized with multiple exposures to irradiated sporozoites, or left unimmunized (naive controls). Three weeks after the last immunization, all groups were challenged with viable sporozoites. Several minutes or several hours later, animals received a monoclonal antibody with neutralizing activity against mouse IFNγ (31), or an irrelevant control immunoglobulin. As expected, naive controls all developed blood-stage infections. Immunized animals receiving control immunoglobulin remained free of blood-stage parasites, indicating that they were solidly immune. However, immunized mice that received the monoclonal antibody directed against murine IFNγ, immediately following or several hours after the time of sporozoite challenge, developed blood stage parasitemia. This phenomenon was observed even in fully immunized animals subject to very low challenge doses of sporozoites. Thus, a small sporozoite inoculum is sufficient to trigger an IFNγ response, and neutralization of IFNγ abolishes the immune status of the host. This clearly demonstrates that in suitably immunized hosts, the endogenous activity of IFNγ, produced in response to parasite challenge, is required for the functioning of sterile immunity to sporozoites (32).

To extend these observations, we examined the impact of in vivo neutralization of IFNγ production on the course of EEF development in the livers of immune Brown Norway rats. Cohorts of rats were either immunized with successive inoculations of X-irradiated sporozoites, or left unimmunized. Several weeks after the final boost, they were challenged with viable sporozoites. The experiments contained four groups of animals; 10 immunized rats and 10 naive controls, with five animals from each group receiving control immunoglobulin and five from each group receiving DB-1, a monoclonal antibody that neutralizes rat IFNγ (31). The antibodies were administered either a few minutes or several hours following challenge with 5×10^4 *P. berghei* sporozoites. The rats were sacrificed at 44 hours and the level of *P. berghei* EEF DNA measured in the liver. Naive animals receiving either control immunoglobulin or the neutralizing monoclonal antibody to rat IFNγ (DB-1), showed a similar high degree of parasite EEF DNA in their livers. Immunized animals receiving irrelevant control Ig showed no EEF DNA at all, but in immunized animals receiving monoclonal anti-rat IFNγ, EEF proliferation was detectable at 35–43.5% of the levels in naive control rats. Thus in immunized animals subject to sporozoite challenge, endogenously produced IFNγ suppresses the development of EEF that develop from sporozoites which escape the neutralizing activity of serum antibodies (32).

We used similar experimental techniques to test the hypothesis that the production of IFNγ in response to blood-stage infection mediates resistance to superinfection by preventing the development of EEF. Brown Norway rats were inoculated with 10^6 blood-stage trophozoites of *P. vinkei*, or left uninfected. (We adapted this rodent malaria parasite from mice to rats by serial passage in juvenile Brown Norway and Sprague Dawley animals.) When approximately 0.5% of erythrocytes were infected, one group of animals received monoclonal DB-1 (with neutralizing activity to rat IFNγ), and other groups received irrelevant control Ig. Control animals without blood-stage infection also received either DB-1 or control Ig. Twenty four hours after the administration of antibodies, the animals were challenged with 5×10^4 *P. berghei* sporozoites. Forty four hours later the animals were sacrificed and their livers removed for DNA purification and hybridization with the *P. berghei*-specific DNA probe, which does not react with *P. vinkei* DNA. As expected, Brown Norway rats without *P. vinkei* blood-stage infection that received either DB-1 or control Ig, showed equally high levels of *P. berghei* EEF growth. In rats infected with *P. vinkei* blood stages and receiving control Ig, however, *P. berghei* EEF development was completely suppressed, with no detectable levels of parasite DNA in the host livers. Rats infected with *P. vinkei* and receiving DB-1, however, showed high levels of *P. berghei* EEF growth at 85% of the levels in the positive controls.

Using another technique, these observations were extended to determine whether infection with *P. berghei* blood-stage parasites resulted in IFNγ-mediated resistance to superinfection by the homologous *P. berghei* strain. Brown Norway rats infected with *P. berghei* blood stages were resistant to the further establishment of *P. berghei* EEF, as determined by sporozoite challenge followed by quantitation of EEF levels via the examination of

Giemsa/Colophonium-stained histological sections of host liver. However, a parallel group infected with *P. berghei* blood stages, but in addition receiving DB-1, showed high levels of EEF growth, comparable to the levels found in control animals without a blood-stage infection.

Thus the production of IFNγ in association with blood-stage infections of malaria parasites inhibits the development of EEF resulting from secondary sporozoite challenge, and therefore, in this experimental model, provides an effective mechanism for host resistance to superinfection with both a heterologous species and a homologous strain of the parasite (unpublished data).

VII. IMPLICATIONS OF ANTIMALARIAL ACTIVITY OF IFNγ FOR THE DEVELOPMENT OF NOVEL VACCINES

The demonstration that inoculation of experimental hosts with attenuated, X-irradiated sporozoites of a variety of malarial species confers a fully protective immunity to subsequent viable sporozoite challenge has resulted in further research aimed at understanding this phenomenon and exploiting it for the purposes of vaccination. The target of protective antibody responses is the immunodominant repeated epitope of the circumsporozoite or CS protein. Since the genes encoding these proteins have now been identified for the major human malarias, and the structure of the immunodominant epitopes elucidated, it is hoped that vaccination with recombinant proteins or synthetic peptide analogues will confer protective immunity to humans.

Recent attempts to vaccinate both rodents and humans with B-cell epitopes of sporozoite CS proteins have afforded partial, but not complete, protective immunity to recipients (13,33,34). Our studies on sporozoite-immunized hosts indicate that immunity to sporozoites is not stage specific, but requires the production of IFNγ by sensitized T cells in response to parasite antigens presented during the course of sporozoite challenge. The IFNγ then serves to inhibit the development of any subsequent EEF that arise from those sporozoites which escape the neutralizing activity of humoral antibodies to the parasite CS protein. It is likely, therefore, that the antimalarial properties of host IFNγ may usefully be incorporated into vaccination strategies. In particular, it may be possible to vaccinate humans with the appropriate parasite T- and B-cell epitopes, in conjunction with suitable adjuvants, to allow not only the production of a high level of neutralizing serum antibodies, but also the establishment of T-cell subsets that respond with the production of IFNγ to native parasite antigens presented to the host following natural sporozoite challenge.

ACKNOWLEDGMENTS

These studies were undertaken in the laboratories of Drs. Ruth and Victor Nussenzweig, to whom we are most grateful. Financial support was supplied by grants from the Agency for International Development DPE 0453-A-00-5012-00 and the National Institutes of Health P01-A117429. We also acknowledge and thank the UNDP/World Bank/WHO Special Programme for Research and Training in Tropical Diseases and the MacArthur Foundation.

REFERENCES

1. Ferreira, A., V. Enea, T. Morimoto, and V. Nussenzweig. 1982. Infectivity of *Plasmodium berghei* sporozoites measured with DNA probe. *Mol. Biochem. Parasitol. 19*:103–109.

2. Shin, S., J. Vanderberg, and J. Terzakis. 1982. Direct infection of hepatocytes by sporozoites of *Plasmodium berghei. J. Protozool. 19(3)*:448–454.

3. Shortt, H.E. 1948. The life cycle of *Plasmodium cynomolgi* in its insect and mammalian host. *Trans. R. Soc. Trop. Med. Hyg. 42*:227–230.

4. Meis, J., J. Verhave, P. Jap, and J. Th. Meuwissen. 1983. An ultrastructural study on the role of Kupffer cells in the process of infection by *Plasmodium berghei* sporozoites in the rat. *Parasitology 86*:231–242.

5. Yoeli, M., R.S. Upmanis, J. Vanderberg, and H. Most. 1966. Life cycle and patterns of development of *Plasmodium berghei* in normal and experimental rats. *Military Med. 131 (Suppl)*:900–918.

6. Bray, R.S., and P.C. Garnham. 1982. The life cycle of primate malaria parasites. *Br. Med. Bull. 38*:117–122.

7. Nussenzweig, R.S., and V. Nussenzweig. 1982. Development of sporozoite vaccines. *Phil. Trans. R. Soc. (Lond.) 289*:117–128.

8. Nussenzweig, V., and R.S. Nussenzweig. 1985. Circumsporozoite proteins of malaria parasites. *Cell 42*:401–403.

9. Potocjnack, P., N. Yoshida, R. S. Nussenzweig, and V. Nussenzweig. 1980. Monovalent fragments (Fab) of monoclonal antibodies to a sporozoite surface antigen (Pb 44) protect mice against malaria infection. *J. Exp. Med. 151*:1504–1513.

10. Spitalny, G., and R.S. Nussenzweig. 1973. *Plasmodium berghei*: relationship between protective immunity and antisporozoite (CSP) antibody in mice. *Exp. Parasitol. 33*:168–178.

11. Chen, D., R. Tigelaar, and F. Weinbaum. 1977. Immunity to sporozoite induced malaria infections in mice. I. The effect of immunization of T and B cell deficient mice. *J. Immunol. 118*:1322–1327.

12. Verhave, J., G. Strickland, H. Jaffe, and A. Ahmed. 1978. Studies on the transfer of protective immunity with lymphoid cells from mice immune to malaria sporozoites. *J. Immunol. 121*:1031–1033.

13. Egan, J., J. Weber, W. Ballou, M. Hollingdale, W. Majarian, D. Gordon, W. Malloy, S. Hoffman, R. Wirtz, I. Schneider, G. Woollett, J. Young, W. Hockmeyer. 1987. Efficacy of murine malaria sporozoite vaccines: Implications for human vaccine development. *Science 236*:435–456.

14. Garnham, P.C., and R.S. Bray. 1956. The influence of immunity on the stages (including late exoerythrocytic schizonts) of mammalian malaria parasites. *Rev. Bras. Malar. 8*:151–160.

15. Shortt, H.E., and P.C. Garnham. 1948. The pre-erythrocytic development of *Plasmodium cynomolgi* and *Plasmodium vivax*. *Trans. R. Soc. Trop. Med. Hyg. 41*:785–795.

16. Yoeli, M., and H. Most. 1965. The pre-erythrocytic development of *Plasmodium berghei*. *Nature 205*:715.

17. Jap, P., J. Meis, J. Verhave, and J. Th. Meuwissen. 1982. Degenerating exo-erythrocytic forms of *Plasmodium berghei* in rat liver: an ultrastructural and cytochemical study. *Parasitology 85*:263–269.

18. Jahiel, R., R. Nussenzweig, J. Vilcek, and J. Vanderberg. 1969. Protective effect of interferon inducers on *Plasmodium berghei* malaria. *Am. J. Trop. Med. Hyg. 18*:823–835.

19. Jahiel, R., J. Vilcek, R. Nussenzweig, and J. Vanderberg. 1968. Interferon inducers protect mice against *Plasmodium berghei* malaria. *Science 161*: 802–804.

20. Jahiel, R., J. Vilcek, R. Nussenzweig, and J. Vanderberg. 1970. Exogenous interferon protects mice against *Plasmodium berghei* malaria. *Nature (London) 227*:1350–1351.

21. Ferreira, A., L. Schofield, V. Enea, H. Schellekens, P. van der Meide, W. Collins, R. Nussenzweig, and V. Nussenzweig. 1986. Inhibition of development of exoerythrocytic forms of malaria parasites by gamma-interferon. *Science 232*:881–884.

22. Maheshwari, R., C. Czarniecki, G. Dulta, K. Pun, B. Dhawan, and R. Friedman. 1986. Recombinant human gamma interferon inhibits simian malaria. *Infect. Immun. 53*:628–630.

23. Schofield, L., A. Ferreira, R. Altszuler, V. Nussenzweig, and R.S. Nussenzweig. Interferon-gamma inhibits the intrahepatocytic development of malaria exoerythrocytic forms in vitro. *J. Immunol.* (in press).

24. Hollingdale, M., P. Leland, J. Leef, and A. Schwartz. 1983. Entry of *Plasmodium berghei* sporozoites into cultured cells, and their transformation into trophozoites. *Am. J. Trop. Med. Hyg. 32(4)*:685–690.

25. Pfefferkorn, E. 1984. Interferon-gamma blocks the growth of *Toxoplasma gondii* in human fibroblasts by inducing the host cells to degrade tryptophan. *Proc. Natl. Acad. Sci. (USA) 81*:908–912.

26. Pfefferkorn, E., M. Eckel, and S. Rebhun. 1986. Interferon-gamma suppresses the growth of *Toxoplasma gondii* in human fibroblasts through starvation for tryptophan. *Mol. Biochem. Parasito. 20*:215–224.

27. Ojo-Amaize, E., L.S. Salimonu, A. Williams, O. Akinwolere, R. Shabo, G.V. Alm, and H. Wigzell. 1981. Positive correlation between degree of parasitaemia, interferon titres, and natural killer cell activity in *Plasmodium falciparum* infected children. *J. Immunol. 127*:2296–2300.

28. Rhodes-Feulette, A., P. Druihle, M. Carnivet, M. Gentilini, and J. Peries. 1981. Presence d'interferon circulant dans les serum des malades infectes par *Plasmodium falciparum*. *C. R. Acad. Sci. Paris. 293, Serie III*:635.

29. Ojo-Amaize, E., J. Vilcek, A. Cochrane, and R. Nussenzweig. 1984. *Plasmodium berghei* sporozoites are mitogenic for murine T cells, induce interferon, and activate natural killer cells. *J. Immunol. 133*:1005–1009.

30. Buchmeier, N., and R. Schreiber. 1985. Requirement of endogenous interferon-gamma production for resolution of *Listeria monocytogenes* infection. *Proc. Natl. Acad. Sci. (USA) 82*:404-7408.

31. Van Der Meide, P., M. Dubbeld, K. Vijverberg, T. Kos, and H. Schellekens. 1986. The purification and characterization of rat gamma interferon by use of two monoclonal antibodies. *J. Gen. Virology 67*:1059-1071.

32. Schofield, L., J. Villaquiran, A. Ferreira, H. Schellekens, R. Nussenzweig, and V. Nussenzweig. 1987. γ-interferon, CD8$^+$ T cells and antibodies required for immunity to malaria sporozoites. *Nature 330*:664-666.

33. Ballou, W., S. Hoffman, J. Sherwood, M. Hollingdale, F. Neva, W. Hockmeyer, D. Gordon, I. Schneider, R. Wirtz, J. Young, G. Wasserman, P. Reeve, C. Diggs, and J. Chulay. 1987. Safety and efficacy of a recombinant DNA *Plasmodium falciparum* sporozoite vaccine. *Lancet 1*:1277-1281.

34. Herrington, D., D. Clyde, G. Losonsky, M. Cortesia, J. Murphy, J. Davis, S. Baqar, A. Felix, E. Heimer, D. Gillessen, E. Nardin, R. Nussenzweig, V. Nussenzweig, M. Hollingdale, and M. Levine. 1987. Safety and immunogenicity in man of a synthetic peptide malaria vaccine against *Plasmodium falciparum* sporozoites. *Nature 328*:257-259.

Part III

FACULTATIVE INTRACELLULAR PATHOGENS

13

Induction and Possible Functions of Interferons During Murine Listeriosis

Edward A. Havell / Trudeau Institute, Inc., Saranac Lake, New York

I. INTRODUCTION

A. A Commentary on the Clinical Use of Interferons: The Cart Before the Horse Syndrome

Interferons (IFNs) are induced proteins that render cells resistant to virus replication. Since the discovery of IFN by Isaacs and Lindenmann in 1957, studies have established that IFNs mediate many different actions in vitro, some of which could conceivably alter host resistance not only to virus, but also to other infectious agents, as well as to malignant disease. In view of these actions, considerable interest exists in determining the possible therapeutic efficacy of IFNs against infectious and malignant diseases.

The IFNs of most species are active only on cells from the same species. Thus, early clinical evaluation of the antiviral effects of IFNs were hampered by the lack of substantial quantities of human IFNs of sufficient purity. In the late 1960s, it was hoped that this problem might be overcome by using IFN-inducing agents, such as the synthetic polyribonucleotide, polyinosinic-polycytidlyic acid [poly(I):poly(C)], to induce the patient's own IFN. Unfortunately, the most effective IFN-inducing agents proved extremely toxic in vivo. During the next decade, considerable advances were made in development of methodologies for the production and purification of all three antigenically distinct human IFN classes (α,β,γ). More recently, the successful cloning and expression of various human interferon genes in bacteria enabled the production and purification of ample quantities of recombinant IFNs. These recombinant IFNs have been employed in well-controlled clinical trials.

For the most part, the results of clinical studies with IFNs have been disappointing. Many viral infections were found not to respond to IFN treatment. Of those that did, the greatest antiviral effect was observed when IFNs were used prophylactically, rather than therapeutically. The findings of in vitro studies showing that IFNs inhibited the proliferation of tumor cells and enhanced the tumoricidal mechanisms of macrophages and natural killer cells, served as the basis for the rationale behind widely heralded clinical trials that were undertaken to determine whether IFNs have antitumor actions in vivo. So far, IFN has been proven effective only in the treatment of a very rare form of cancer—hairy cell leukemia.

There may be many reasons why IFNs have not been as clinically effective as originally hoped. In the case of viral infections, it has been established that IFNs induced by virus have a very important role early in nonspecific resistance by suppressing further virus replication. It should be stressed that this action of IFN, which results in the development of a cellular state of viral resistance, does not eliminate the virus from the parasitized cell. Thus, the antiviral action of IFN serves merely to limit virus numbers until an immune response is generated, which usually terminates the infection. Symptoms of acute viral diseases are manifest just before peak titers of virus are reached in the afflicted host. Therefore, the negative results of clinical studies in which attempts were made to use IFNs to treat patients following the onset of symptoms of a given virus infection, are, in retrospect, not surprising.

Many factors may have contributed to the widespread failure of IFN in the treatment of human cancers. One major factor was the lack of evidence showing antitumor actions of IFNs in animal models. Animal studies were hindered by a lack of information about the production, characterization, and purification of animal IFNs. Furthermore, specific anti-IFN neutralizing antibodies for each IFN class were not available. These reagents would have made it possible to establish whether the administration of the different IFNs on the one hand, or the neutralization of endogenously produced IFNs by specific antibodies on the other, can alter host resistance to malignant disease. In fact, the three antigenically distinct IFN classes, first demonstrated in the human system, were not identified in mice until the early 1980s.

While an enormous amount of time and effort has been given to the study of IFNs since 1957, little is known about whether IFNs actually function in host defense against nonviral or malignant diseases. Our interest in IFNs and resistance to *Listeria monocytogenes* originated with the finding that spleen cells taken from mice at the peak of anti-*Listeria* immunity produced substantially more IFNγ than did spleen cells of uninfected mice, when stimulated in culture with polyclonal T-cell mitogens. Subsequent studies established that *Listeria*-infected mice produced all three IFN classes and acquired a greatly augmented potential for the production of these IFNs in response to stimulation with the appropriate IFN-inducing agents. Collectively, these findings suggested the possibility that IFNs function in resistance to bacterial infection. In the following pages, I have attempted to review what is known about *Listeria*-induced murine IFNs and their possible functions in nonspecific and specific anti-*Listeria* resistance mechanisms. First, however, a brief review of the properties of murine IFNs and induction of these IFNs by bacteria will be presented.

B. The Murine Interferons

Three antigenically distinct classes (α,β,γ) of human and murine IFNs are recognized (1). The IFNα and IFNβ classes retain activity following low pH (2.0) treatment, whereas IFNγ characteristically loses activity upon acidification. Based on amino acid sequence analyses, the murine IFNs appear to be analogues of the respective human IFN classes (2-7). The gene for human IFNγ is located on chromosome 12 (8) and encodes for a mature peptide of 146 amino acids (2), whereas the mature murine IFNγ peptide consists of 136 amino acids (3). At least 12 subtypes of human IFNα proteins have been reported (9), however, recombinant DNA studies indicate the existence of at least 19 human IFNα genes and pseudogenes (10). The murine IFNα gene family has been shown to consist of at least four genes (10). Recently, two distinct human IFNβ (β_1 and β_2) molecules have been demonstrated (11).

Human IFNβ_1 is produced by cells of nonlymphoid origin, whereas human IFNα is synthesized primarily by peripheral blood leukocytes in response to viral stimulation (12). In sharp contrast to the cell sources of human IFNα and IFNβ, murine cells of nonlymphoid origin produce mixtures of IFNα and IFNβ (IFNα/β) following viral stimulation (13-16). Moreover, murine macrophages have been found to synthesize IFNα/β in response to a variety of agents, many of which (e.g., endotoxin) are ineffective in eliciting IFN from other cell types. Under optimal conditions, macrophages have been shown to produce higher amounts of IFNα/β, on a per cell basis, than other murine cell types (17,18). Moreover, it appears that macrophages synthesize different proportions of IFNα and IFNβ when stimulated with different IFN-inducing agents (16).

The cellular origin of IFNγ appears to be primarily T lymphocytes (19), although natural killer cells (20) have also been reported to synthesize IFNγ. Normally, this lymphokine is secreted during the expression of T-cell-mediated immune responses to viral (21,22) or nonviral antigens (19,23). However, polyclonal T-cell mitogens, such as phytohemagglutinin, can induce nonimmune T cells to produce IFNγ (24,25). Some time ago, studies were undertaken at this Institute to determine certain physicochemical properties of murine IFNγ. It was found that the major IFNγ species produced in vitro by phytohemagglutinin (PHA)-stimulated spleen cells possessed an acidic isoelectric point (pH 5.6), and bound specifically to immobilized concanavalin A (Con A), indicating that IFNγ is a glycoprotein (26). The molecular weight of natural glycosylated IFNγ was estimated to be 38 kilodaltons (kD) under nondenaturing conditions, whereas the nonglycosylated molecule was found to be 28 kD (24). Molecular weight determinations carried out under denaturing conditions revealed the molecular weight for glycosylated IFNγ to be 20 kD, suggesting that murine IFNγ exists as a homodimer of two 20 kD subunits (27).

C. Historical Background on Interferon Induction by Bacteria

Viral induction of IFN synthesis was observed first in cell culture (28), and then in vivo (29). Since this time, a wide variety of infectious agents (viral and non-viral) (30–38), microbial components and products (30,31,39,40), and synthetic chemical compounds (41,42) have been found to be capable of inducing IFN synthesis. For a comprehensive listing of IFN-inducing agents, the reader is referred to a recent review by Monto Ho (43). A list of nonviral infectious agents capable of inducing IFN synthesis in vivo would include: protozoa (36,37), mycoplasma (34), obligate intracellular bacteria (32,33), as well as free-living gram-negative and gram-positive bacteria (30,44–49).

1. IFN Induction In Vivo

The first reports that bacteria induced IFN synthesis in vivo were made in 1964 by Youngner and Stinebring. These investigators found that following the intravenous inoculation of the gram-negative bacterium, *Brucella abortus*, high titers of IFN appeared in the serum of chickens (46) and mice (30). Subsequent studies established that animals produced IFN in response to many different bacteria including *Salmonella* (30), *Serratia* (30), *Francisella* (44), *Bordetella* (47), as well as gram-positive bacteria such as *Staphylococcus* (48), *Streptococcus* (49), and *Listeria* (44,45,50). The properties of IFNs induced by bacteria were found to be similar to those induced by viruses, in that they are stable following acid (pH 2.0) treatment, and are restricted in their antiviral actions to cells of the species from which they are derived.

In addition to finding that bacteria could elicit IFN production in vivo, Stinebring and Younger (30) also reported that bacterial endotoxin was an effective IFN-inducing agent in mice. Ho (31) simultaneously reported that this component of cell walls of gram-negative bacteria induced IFN in rabbits. Interferon was detected in the serum of animals within an hour after the intravenous infusion of endotoxin, and peak levels were reached 1–2 hours later. The appearance of endotoxin-induced IFN in the serum of mice was much more rapid than that observed for IFNs induced either by viruses or by bacteria (30). However, endotoxin-induced IFN exhibited the same properties of species specificity and stability to low pH treatment as did the serum IFNs obtained after injection of either virus or bacteria. Besides endotoxin, other bacterial components have been also reported to induce IFN in vivo (39,40).

In 1973, Salvin et al. (23) made a key discovery concerning the nature of the IFN produced during the course of a T-cell-mediated immune response. These investigators found that when mice infected with *Mycobacterium bovis* (strain BCG) were injected intravenously with either old tuberculin (OT) or living BCG cells, an IFN appeared in the circulation. This IFN, unlike other IFNs, lost all activity following acidification. This finding confirmed an earlier report made by Wheelock (24), who described the production of an acid-labile IFN in human leukocyte cultures incubated with phytohemagglutinin (PHA). Subsequent studies established that a wide variety of antigens (viral and non-viral) that evoke T-cell-mediated immune responses induced the synthesis of an acid-labile IFN, which was initially referred to as either "immune IFN" or "type II IFN" (19,21,22,50). This acid-labile IFN class is now designated as gamma IFN (IFNγ).

2. IFN Induction in Cell Cultures

The first demonstration of IFN induction by bacteria in cultured cells was made in 1964 by Hopps et al. (51). These scientists found that chick embryo cell cultures infected with the obligate intracellular parasite, *Rickettsia tsutsuga-mushi*, produced IFN. Shortly thereafter, reports appeared showing that other rickettsiae (52), and chlamydiae (53,54), which are also obligate intracellular bacteria, stimulated infected cells to produce IFN. More recently, it was shown that murine embryo fibroblasts infected with *L. monocytogenes* synthesized IFN during intracellular proliferation of this facultative intracellular bacterium (55). Evidence indicates that bacteria that induce IFN synthesis in nonphago-cytic cells of nonlymphoid origin, such as fibroblasts, must be capable of gaining entry and multiplying within these cells. Characteristically, these IFNs produced by nonphagocytic cells infected with bacteria were acid stable.

As mentioned in the preceding section, a great number of bacteria induce acid-stable IFNs in vivo. Macrophages have been shown to secrete IFN in re-sponse to stimulation with a variety of agents including viruses, double-stranded RNA, polysaccharides, and polyanions (56–58). Cultured macrophages also pro-duce IFN in response to bacteria or to endotoxin (17,18,56). Indeed, these pro-fessional phagocytes may prove to be the primary source of serum IFN produced in response to intravenous infusion of endotoxin or bacteria.

The first report of specific bacterial antigen-induced IFN synthesis in vitro was made by Green et al. (59). These investigators found that lymphocytes of individuals sensitized to different antigens, such as tetanus or diphtheria toxoids, produced IFN when incubated with the respective immunogen, whereas lympho-cytes from nonimmune donors did not. Subsequent studies demonstrated that cultures containing lymphocytes from mice immunized with a variety of bacteria, which evoke T-cell-mediated immune responses, produced IFNγ when incubated in vitro with the immunizing bacteria, or a preparation of sensitizing antigen derived from the bacteria (19,60).

Cell cultures consisting of mixtures of lymphoid and other cell types, when incubated with bacteria or endotoxin, were found to synthesize both acid-stable IFNs and IFNγ. *Listeria monocytogenes* was found to induce multiple IFN classes in cultures of human peripheral mononuclear leukocyte cultures (61). Similar IFN responses were observed in murine spleen cell cultures incubated with either *L. monocytogenes* (62) or *Corynebacterium parvum* (63). Interestingly, spleen cells from mice sensitized to one species of bacteria, when placed in culture with a different species, or with endotoxin, produced very high titers of IFNγ and other IFNs (64). The findings of Le et al. (65) could explain the mechanism(s) by which multiple IFN classes are induced by bacteria, or endotoxin, in cultures of mixed lymphoid and nonlymphoid cells. These investigators found that endotoxin-induced IFNγ synthesis in cultures of human peripheral blood leukocytes follows the induction of interleukin-1 (IL-1), which, in turn, stimulates T lymphocytes to synthesize IL-2. It has been shown that interleukin-2 stimulates T lymphocytes to synthesize IFNγ (66). Thus, induction of IFNγ by bacteria in cultures of mixed cells, containing nonimmune T lymphocytes, may occur through a similar sequence of events, with macrophages producing not only IL-1, but also the acid-stable IFN classes.

II. INTERFERON RESPONSES OF MICE DURING *Listeria monocytogenes* INFECTION

A. Effects of *Listeria* Infections on Interferon Synthesis

Soon after the discovery that bacteria could induce IFN in vivo, it was found that certain bacteria could also alter the host's responsiveness to IFN-inducing agents. Bacteria that evoke strong T-cell-mediated responses, such as *Mycobacteria*, or *Propionibacterium acnes*, greatly enhanced the potential of mice to produce IFN (67,68). Studies carried out in this laboratory (45), and elsewhere (69), have also demonstrated that mice intravenously infected with a sublethal inoculum of *L. monocytogenes*, acquire an augmented capacity to produce IFNα/β in response to an intravenous infusion of endotoxin. This enhanced potential to produce endotoxin-induced serum IFNα/β (determined 2 hr postendotoxin injection) can be detected as early as 8 hr after initiation of infection. Peak endotoxin-induced serum IFNα/β titers occur 16 hr later, when the host is producing 50 times more IFNα/β than uninfected mice. Later, the enhanced ability of the *Listeria*-infected host to synthesize endotoxin-induced IFNα/β slowly declined. Similar studies with a second IFN-inducing agent, poly(I):poly(C), which, like endotoxin, induces a peak serum IFN response 2 hr after intravenous injection, revealed that mice infected for 24 hr with *Listeria* produced only 4-8-fold more IFNα/β than did uninfected mice. Additional studies demonstrated that the enhanced ability of *Listeria*-infected mice to produce IFNα/β in response to endotoxin occurred through a T cell-independent mechanism. Moreover, mice given injections of large numbers of killed bacteria did not develop an enhanced potential to synthesize endotoxin-induced serum IFNα/β.

Listeria-infected mice also acquire the capacity to produce serum IFNγ following intravenous injection of endotoxin (45,69). Endotoxin-induced IFNγ can be detected in the serum after the levels of IFNα/β peaked. However, the highest titers of endotoxin-induced IFNγ appear in the serum of *Listeria*-infected mice 5-6 hr after endotoxin administration. IFNγ was not detected in sera of *Listeria*-infected mice following injection of poly(I):poly(C), or sera of uninfected mice after intravenous infusion of endotoxin. Mice infected with *M. bovis* (strain BCG), or mice treated with heat-killed *P. acnes*, have also been reported to produce serum IFNγ following intravenous injection of endotoxin (68,70).

Originally, our interest in the relationship between bacterial infection and the host's ability to produce IFNs stemmed from a need for large amounts of murine IFNγ for physicochemical analyses (Sect. I.B.). Before the cloning and expression of the murine IFNγ gene in bacteria (3), the primary means for the production of IFNγ involved stimulating various lymphoid cells in vitro with polyclonal T-cell mitogens. In 1981, numerous attempts were made in this laboratory to produce high-titered IFNγ preparations by stimulating cultures of spleen cells from uninfected mice with various T-cell mitogens, under different experimental conditions. The resulting IFNγ titers in the culture supernatants rarely exceeded 100 antiviral units/ml. However it was found that cultures of spleen cells established from mice infected with *L. monocytogenes* produced 20-40 times more PHA-induced IFNγ than did similarly treated spleen cells from uninfected mice (25). Further studies established that a striking temporal association existed between the augmented potential for IFNγ synthesis and the development of anti-*Listeria* immunity, in that they developed and decayed in unison. It was established, through the use of specific antibodies and complement to deplete various spleen cell populations, that T cells were required for PHA-induced IFNγ synthesis. T cells of the Ly-2⁻ phenotype were found to be the predominant source of PHA-induced IFNγ in cultures of spleen cells established from infected mice at the peak (day 6 of infection) of the anti-*Listeria* immune response. Cultures of spleen cells from *Listeria*-infected mice also produced more IFNγ in response to endotoxin, than did spleen cell cultures established from uninfected mice (64).

B. *Listeria monocytogenes*-Induced Interferons

1. IFN Induction In Vivo

Listeria monocytogenes has been shown to induce IFN in vivo. Lukas and Hruskova (44) found IFN in the serum of *Listeria*-infected chickens. Low levels of IFN were detected in the serum of mice one day after the intravenous inoculation of an immunizing *Listeria* dose (45,50). Serum IFN titers induced by *Listeria* peaked on the second day, and IFN was undetectable in the serum after the fourth day of infection. Serological analyses carried out with specific anti-IFN antibodies established that the antiviral activity in the serum of *Listeria*-infected mice was mediated by IFNα/β. At no time, during the course of an immunizing *Listeria* infection, has IFNγ been detected in the circulation (45,50, 71). However, *Listeria*-immune mice produce high levels of serum IFNγ following intravenous injection of a large *Listeria* inoculum (50).

2. IFN Induction in Cell Cultures

Cultures of spleen cells from control or *Listeria*-infected mice synthesize IFN after incubation with *Listeria*. In Table 1 are presented the *Listeria*-induced IFN titers produced by spleen cells from uninfected and 6 day *Listeria*-infected mice. The spleen cells were incubated in medium containing antibiotics, and treated with increasing numbers of viable bacteria. After 24 hr of incubation, the culture media were assayed for IFN. IFN was found in the media of all spleen cell cultures to which *Listeria* was added. Spleen cells from *Listeria*-infected mice produced at least 5 times more IFN than did similarly treated cultures of control

Table 1   *Listeria monocytogenes*-Induced IFN Synthesis in Spleen Cell Cultures

Treatment of spleen cell cultures[a]	24-hr IFN yield (units/ml) in cultures of spleen cells from	
	Uninfected mice	*Listeria*-infected mice[b]
None	<4	<4
10^7 *L. monocytogenes*	192	1024
10^5 *L. monocytogenes*	128	1024
10^3 *L. monocytogenes*	64	768
10^1 *L. monocytogenes*	<4	8

[a]2×10^7 nucleated spleen cells were incubated with the designated numbers of bacteria for 24 hours in 2 ml of RPMI 1640 medium containing 5% fetal bovine serum and antibiotics.
[b]Mice intravenously inoculated with 5×10^3 *L. monocytogenes* (serotype 1/2a) 6 days earlier.

spleen cells. Moreover, when spleen cells from *Listeria*-infected mice were incubated in antibiotic-free culture medium, they synthesized IFN, even when no bacteria were added to the cultures. This finding suggests that bacteria in the infected spleens proliferated in culture and induced IFN. It should also be mentioned that large numbers of UV-killed or heat-killed bacteria also induced IFN synthesis when added to cultures containing antibiotics and spleen cells from either *Listeria*-infected or uninfected mice. It was also found that viable or killed *L. monocytogenes* induced mixtures of IFNγ and IFNα/β in spleen cell cultures established from infected mice. Moreover, IFNγ was determined to be the predominant IFN class. These findings, showing that *L. monocytogenes* induced IFNγ synthesis in cultures of cells established from an organ which is a primary target of infecting bacteria, raise the question as to why IFNγ is not detected in the circulation of mice during an immunizing *Listeria* infection (45,50,71). One possible explanation is that following the generation and migration of sensitized T lymphocytes into infectious foci, IFNγ is produced locally in very small quantities. The levels of IFNγ produced would be dependent on both the numbers of bacteria and immune T lymphocytes within these foci. Peak *Listeria* numbers in infected organs are reached on, or shortly after, the second day of an immunizing infection, after which the number of bacteria rapidly declines (72, 73). *Listeria*-immune T lymphocytes are first detected at the time when the greatest numbers of bacteria are present in infected organs, after which the number of immune T lymphocytes increase. Thus, this inverse correlation between number of bacteria (antigenic mass) and the number of immune T lymphocytes suggests that the amount of IFNγ synthesized at any time during infection would probably be very small.

It has been shown recently that *L. monocytogenes* induces the synthesis of IFNα/β in cultures of murine embryo fibroblasts (MEF) (55). The induction of IFNα/β synthesis in MEF cells was found to be dependent on intracellular *Listeria* proliferation. Progressive intracellular proliferation ultimately resulted in the destruction of initially infected cells and the spread of infection to neighboring cells. In Table 2 are presented the IFNα/β titers produced by cultures of confluent MEF cells following initial infection with 10 or 0.5 *Listeria* per cell. The MEF cultures infected with 10 bacteria per cell synthesized the greatest amount of IFNα/β during the first 12 hr of infection. In contrast, MEF cultures infected with 0.5 bacteria per cell produced only marginal levels of IFNα/β during the first 12 hr of culture, after which the greatest rate of synthesis occurred during the next 12 hr. Following this time, the rate of synthesis declined as in the more heavily infected cultures. The decrease in IFN synthesis correlated with the progressive degeneration of MEF monolayers, as judged microscopically. Cultures of MEF cells incubated with 20 UV-killed bacteria per cell did not produce IFNα/β.

Table 2 *Listeria monocytogenes*-Induced IFNα/β Synthesis in Murine Embryo Fibroblast Cultures

Initial infecting inoculum[a]	IFN synthesis (units/ml) during progressive time periods (hr)			
	0–12	12–24	24–36	36–48
None	<4	<4	<4	<4
10 *L. monocytogenes*/fibroblast	64	32	16	4
0.5 *L. monocytogenes*/fibroblast	8	64	32	8

[a]Confluent 35 mm MEF cultures were incubated for 2 hr with bacteria in 0.5 ml of medium containing 0.25 μgs/ml of gentamicin sulfate. At the end of this time (0 hour), cultures were washed and replenished with 1 ml medium. At the end of the designated incubation period, culture media were collected, and the cultures replenished with medium, and incubated until the end of the next time interval.

III. POSSIBLE ROLES OF INTERFERONS IN HOST RESISTANCE TO *Listeria monocytogenes* INFECTIONS

A. Nonspecific and Specific Host Resistance Mechanisms to *Listeria* Infection

The foregoing studies established that during the course of a sublethal *L. monocytogenes* infection in mice, IFNα/β is present in the circulation. The results of studies showing that *L. monocytogenes* induced IFNγ synthesis in spleen cell cultures, suggest that IFNγ is probably produced in infected organs in vivo. Moreover, the infected host acquires a greatly enhanced potential for the production of the three antigenically distinct IFN classes. These findings raise questions as to the possible functions of these antiviral molecules in nonspecific and/or specific resistance to a nonviral pathogen.

The most plausible function of these *Listeria*-induced antiviral molecules would be to protect the compromised host from concurrent virus infection. However, in addition to antiviral actions, IFNs are known to mediate other actions which could alter host resistance to nonviral infections. Some actions common to all IFNs include (a) suppression of intracellular proliferation of nonviral pathogens (74–77), (b) inhibition of cell division (78–80), and (c) regulation of the generation and/or expression of immunity (81–84). In addition, there are effects which are mediated either exclusively, or preferentially by the lymphokine IFNγ. This lymphokine has been shown to (a) enhance the expression of class II histocompatibility antigens (85), (b) prime macrophages for nonspecific tumoricidal activity (86,87), (c) activate macrophage oxidative metabolism and microbicidal mechanisms (88), and (d) modulate antibody synthesis (89). Studies are beginning to yield information that may eventually explain, at the molecular level, the basis for the different biological effects of IFNγ and the other IFNs. In this connection, IFNγ has been found to bind to a cellular receptor that is distinct from the one to which other IFNs bind (90,91). Striking differences are seen in the number and kind of proteins found in cells following treatment with the different IFNs. In addition to those proteins induced in cells treated with IFNα or IFNβ, cells treated with IFNγ synthesize 12 more proteins (92). These differences between IFNγ and the other IFNs, as well as the fact that IFN is normally produced by T cells during cell-mediated immune responses to nonviral pathogens, collectively suggest that the primary function of IFNγ may be one other than rendering cells resistant to viral infection.

The response of mice to *L. monocytogenes* provides an excellent model for studying the possible roles of IFNs in nonspecific resistance and acquired specific resistance to a facultative intracellular pathogen. Expression of anti-*Listeria* resistance is dependent upon the generation of activated macrophages capable of expressing bactericidal function (72). Macrophage activation is mediated by a newly generated population of specifically sensitized T cells, first detected in the spleen on the second day of infection (93). It has been shown that peak numbers of *Listeria*-immune T cells are generated on, or around, the sixth day of infection, after which their numbers diminish as bacteria are eradicated from the organs (73,93). Studies which compared the anti-*Listeria* resistance to T-cell-intact (nu/+) mice and T-cell-deficient congenitally athymic nude (nu/nu) mice indicate that two T-cell-independent resistance mechanisms are expressed before the generation of T-cell-mediated immunity (94). Following intravenous inoculation, *Listeria* locates primarily in the liver and spleen, and during the next 12 hr a substantial proportion of the inoculum is destroyed. This initial bactericidal action is mediated by radioresistant cells, thought to be resident macrophages (94). After the first 24 hr of infection, the host develops a radiosensitive T-cell-independent mechanism that functions to limit progressive growth of bacteria that survive the initial kill. Both of these early mechanisms of resistance are expressed in T-cell-deficient nude mice. However, unlike the acute, short-lived *Listeria* infection seen in normal mice, listeriosis in nude mice is chronic, but fails to progress, presumably because of the anti-*Listeria* actions of macrophages (94–97). The macrophages that mediate this T-cell-independent resistance are believed to be freshly recruited from the circulation into infective foci. The *Listeria* numbers in the livers and spleens of nude mice after the second day of infection remain at approximately 10^5 organisms/organ for upwards to 3 weeks. Thereafter, *Listeria* proliferate and kill the T-cell-deficient host (94).

B. Anti-*Listeria* Actions of Interferons

The first suggestion that IFNs might alter the expression of anti-*Listeria* resistance came from studies which showed that mice pretreated with IFN-inducing agents were protected against a lethal *Listeria* infection. Remington and Merigan (98) reported intraperitoneal injection of different IFN-inducing agents, including poly(I):poly(C), enhanced the survival time of mice subsequently challenged with a lethal *Listeria* inoculum. Some time ago, similar studies were conducted in this laboratory to determine whether intravenous infusion of endotoxin, or poly(I):poly(C), could protect mice challenged 24 hr later with a lethal intravenous *Listeria* inoculum. It can be seen in Table 3 that untreated control mice died by the fifth day of infection, whereas most endotoxin-treated mice, and all poly(I):poly(C)-treated mice, survived. However, in sharp contrast to the enhanced anti-*Listeria* resistance exhibited by mice treated with endotoxin prior to infection, the administration of endotoxin at any time following the initiation of infection, impaired resistance. Newborg and North (99) showed that intravenous injection of endotoxin into mice with an ongoing sublethal *Listeria* infection resulted in enhanced proliferation of bacteria. These investigators presented evidence indicating that endotoxin interfered with the listericidal action of macrophages. In view of the multiple effects of IFN-inducing agents on various host resistance mechanisms (100), little could be concluded as to possible effects of IFN induced by these agents on host resistance to *Listeria*.

1. Effect of Exogenous IFNs on Listeriosis

Intravenous administration of partially purified preparations (2000 antiviral units) of either IFNα/β or IFNγ, protected mice from a lethal *Listeria* challenge given 24 hr later (101). However, these IFNs did not protect mice if given 18 hr after a lethal challenge. While these studies showed that pretreatment with partially purified preparations of IFNα/β or IFNγ enhanced anti-*Listeria* resistance, the demonstration of a causal relationship between an administered IFN and enhanced resistance to any infectious agent can be achieved only with a totally pure IFN preparation.

Table 3 Protective Actions of Interferon-Inducing Agents Against a Lethal *Listeria monocytogenes* Infection

Treatment[a] 24 hr prior to infection[b]	Number of mice surviving on day of infection						
	3	4	5	6	8	10	12
None	5/5	0/5					
PBS	5/5	1/5	0/5				
25 μg Endotoxin	5/5	5/5	5/5	5/5	3/5	3/5	3/5
25 μg Poly(I):Poly(C)	5/5	5/5	5/5	5/5	5/5	5/5	5/5

[a]Intravenous injection of agent in 0.2-cc phosphate-buffered saline (PBS).

[b]One day after treatment, mice were inoculated intravenously with 5 LD_{50}s of *L. monocytogenes*.

The successful cloning and expression of the IFN genes in *Escherichia coli* (2,3,10) has provided recombinant IFN (rIFN) preparations free of all other murine proteins. Recombinant IFNs have been used in vitro and in vivo to analyze the effects of these molecules on different cell types and host resistance to infectious and malignant diseases (86,88,102-104). Recombinant IFNγs (murine and human) were used by Kiderlen et al. (103) to determine whether they could enhance host resistance to *Listeria* infection. Graded concentrations of murine rIFNγ (rMuIFNγ) were given either intravenously, or in the left hind footpad of mice. One day later, the pretreated mice were injected in the same sites with a suspension of *Listeria* in the same concentration of rMuIFNγ used in the pretreatment procedure. Mice treated with rMuIFNγ in the left hind footpad were also simultaneously infected in the untreated hind right footpad. Two days later, *Listeria* numbers were determined in the infected footpads, or in the spleens of mice infected intravenously. It was found that the numbers of bacteria in the spleens of mice pretreated intravenously with rMuIFNγ, or in the MuIFNγ-treated footpads, were lower than the corresponding controls. The rMuIFNγ used in this study also activated the tumoricidal mechanisms of cultured macrophages. This in vitro observation led to the conclusion that the enhanced anti-*Listeria* resistance observed in rMuIFNγ-treated mice was due to rMuIFNγ effecting enhanced macrophage bactericidal activity. However, there is a possibility that the enhanced resistance exhibited by mice pretreated with rMuIFNγ may have been due in part to contaminants present in the rMuIFNγ preparation. This possibility is raised because these investigators observed similar anti-*Listeria* effects regardless of whether mice were treated with rMuIFNγ or with recombinant human γ IFN (rHuIFNγ). But Murray et al. (102) clearly showed that whereas rMuIFNγ activates murine peritoneal macrophages, as judged by increased oxidative metabolism, rHuIFNγ does not. Therefore, the finding that mice pretreated with either rMuIFNγ, or rHuIFNγ preparations exhibited enhanced resistance to *Listeria* infection raises the possibility that this effect was not mediated by the recombinant proteins. One plausible explanation for the enhanced anti-*Listeria* resistance observed in vivo, is that these rIFNγ preparations became contaminated with endotoxin during extraction of the rIFNγs from the bacteria (*E. coli*) in which the respective IFN genes were cloned and expressed. If so, the endotoxin could have had the same anti-*Listeria* effects in vivo as did the endotoxin preparation used in the studies presented in Table 3.

2. Effect of Anti-MuIFNγ Antibody Treatment on Listeriosis

Considerable interest has centered on the role of IFNγ in macrophage activation. In vitro studies have shown that this lymphokine causes macrophage activation, as judged by different criteria. It was found, for instance, that rIFNγ preparations were effective in enhancing the cytolytic actions of cultured macrophages for tumor cells (86). Through the use of a monoclonal anti-MuIFNγ neutralizing antibody (MAb), IFNγ was identified as the factor in crude murine lymphokine preparations responsible for activating the tumoricidal mechanisms of peritoneal macrophages (87). In addition to activating the tumoricidal mechanisms of macrophages, it has been demonstrated that IFNγ also enhanced the oxidative metabolism of these cells (102). Macrophages exposed to IFNγ have also been reported to acquire an enhanced ability to kill microorganisms (102), including *L. monocytogenes* (105).

In view of the studies showing that IFNγ activates macrophages in culture, and the fact that activated macrophages generated during the expression of anti-*Listeria* immunity effect the elimination of bacteria from the host (72,73), it is possible that *Listeria*-induced IFNγ causes macrophage activation in vivo. Thus, if IFNγ functions in host resistance to *Listeria* infection, then infected mice in which IFNγ is neutralized by administering an excess of specific anti-IFNγ antibody should theoretically exhibit an increased level of infection, and therefore, a normal immunizing infection could become lethal, or chronic, as in athymic nude mice (94–96).

A number of hybridomas have been generated which secrete anti-MuIFNγ monoclonal neutralizing antibodies (87,105–107). Buchmeier and Schreiber (71) treated mice with an anti-MuIFNγ MAb (H22.1) 1 day after initiation of a sublethal intraperitoneal *Listeria* infection. These investigators reported increased *Listeria* proliferation in the spleens and peritoneal cavities of H22.1-treated mice. Indeed, this is a very exciting finding because it indicates a role for IFNγ in anti-*Listeria* resistance. However, there is reason to express a degree of caution in interpreting these experimental findings. First, the *Listeria* numbers in the peritoneal cavities of mice on the fourth day of infection differed by 10,000-fold between identically treated groups in separate experiments. Such a discrepancy, unfortunately, serves to detract from the results presented for the other experimental groups in this important study. Secondly, 10^7 bacteria were found in the spleens of the H22.1-treated mice on the sixth day of infection. However, only 33% of the H22.1 MAb-treated animals were reported to have died in this study. It is the general finding that mice die soon after the numbers of a virulent strain of *L. monocytogenes* exceed 10^6 per spleen (72,73). A series of studies of similar design are currently underway in this laboratory. The rat anti-MuIFNγ MAb (R4-6A2) (87) is being used to treat mice intravenously infected with an immunizing *Listeria* inoculum. So far, these studies indicate that the R4-6A2 MAb has little or no effect on the course of *Listeria* infection. However, these findings do not contradict the results of experiments in which the H22.1 MAb was used, because these two MAbs may interact with different functional domains on the IFNγ molecule. Indeed, the available evidence indicates that there are at least two functional domains on MuIFNγ molecules, one for antiviral activity, and the other for macrophage activation (105). Each functional domain is associated with a distinct epitope. Therefore, the negative findings obtained in vivo with the R4-6A2 MAb could conceivably be due to the inability of this MAb to block the functional site which elicits macrophage activation. Therefore, it is of utmost importance that these anti-IFNγ neutralizing antibody studies be repeated, preferably with a polyvalent monospecific anti-MuIFNγ antibody, so that a role for IFNγ in host resistance to *Listeria* can be conclusively established.

3. Inhibition of Intracellular *Listeria* Proliferation by IFNs

Interferons can elicit a cellular state of resistance to the intracellular multiplication of obligate and facultative intracellular bacteria. It has been shown that IFN-treated cells failed to support the intracellular proliferation of *Shigella* (75), *Chlamydia* (76), and *Rickettsia* (74,77). Histological examination of livers obtained from mice 1-2 days following the intravenous injection of *Listeria* revealed large numbers of *Listeria* within both phagocytic cells and cells normally considered to be nonphagocytic (e.g., hepatocytes) (108). Indeed, ongoing studies elsewhere in this Institute indicate that a large proportion of the bacterial load is contained in nonprofessional phagocytes. The presence of bacteria in nonprofessional phagocytes that probably lack the mechanisms required to destroy internalized bacteria could constitute an important aspect in the pathogenesis of listeriosis and other infections, if intracellular bacteria proliferate and spread to neighboring cells. Studies carried out in vitro have shown that *L. monocytogenes* proliferates intracellularly in fibroblasts, and spreads to contiguous uninfected cells (55). Obviously, if such an infection process occurs in vivo, there must be a mechanism by which the intracellular proliferation of this bacterium is controlled in these cells. Otherwise, bacteria would multiply and spread an overwhelming infection to other nonprofessional phagocytes before specific immunity could be generated.

Studies have been initiated in this laboratory to determine whether IFNα/β, or IFNγ, can induce a state of anti-*Listeria* resistance in murine embryo fibroblast (MEF) cells. Cultures of confluent MEF cells in 35 mm plastic dishes were incubated for 24 hr with increasing concentrations of IFNα/β, or IFNγ in medium containing the antibiotic gentamicin sulfate. The concentration of gentamicin sulfate used in these studies is listericidal for extracellular bacteria, but not for bacteria within MEF (55). Following the 24 hr IFN treatment procedure, the MEF were infected for 2 hr with 0.5 ml of a suspension of bacteria at a ratio of 10 bacteria per MEF. Two hours later, the medium was aspirated, the monolayers washed 3 times, and then overlayed with 2 ml of medium containing 1% agarose and gentamicin sulfate. Microscopic examination of the monolayers at this time revealed that less than 1% of the MEF were associated with bacteria. The mean number of bacteria per infected MEF was 1. Following solidification of the overlay, cultures were inverted and incubated. Two days later, *Listeria* plaques were enumerated. In Table 4 are the results of this study. It can be seen that in the IFNγ-treated MEF cultures, the number of *Listeria* plaques was reduced in a dose-dependent manner, and as little as 0.5 unit of IFNγ antiviral activity caused a decrease in plaque number. Based on relative antiviral activities, IFNα/β proved considerably less effective than IFNγ in suppressing *Listeria* plaque numbers. In other studies, it has been shown that the IFNγ-induced state of anti-*Listeria* resistance develops over a period of 18 hr, and anti-MuIFNγ MAb inhibits this action of IFN.

The *Listeria* plaques in the IFNα/β-treated MEF cultures were smaller than those in untreated MEF cultures. No difference in plaque size was observed between untreated and IFNγ-treated MEF cultures. A decrease in plaque size indicates that intracellular *Listeria* proliferation was suppressed, whereas a reduction in plaque number suggests that an early event in the infection process was altered, such as the initial bacterial interaction with the MEF cell, or the internalization process. It has been determined that the numbers of *Listeria* associated with IFNγ-treated or untreated cells did not differ immediately after the initial infection procedure, indicating that IFNγ treatment may interfere with the *Listeria* internalization process. Indeed, there is a precedent for such an IFNγ-mediated inhibitory effect on bacterial uptake, for it has been shown that the

Table 4 Inhibition of *Listeria monocytogenes* Plaque Formation in Murine Embryo Fibroblasts Treated with Interferons

Cell treatment 24 hr prior to infection (units/ml)[a]	Mean plaque number[b]	% Plaque reduction
None	22	—
IFNγ 500	2	91
50	3.6	8.4
5	8	6.4
0.5	15.2	31
0.05	22.3	0
IFNα/β 500	12	46
50	15	32
5	23	0
0.5	22	0
0.05	23	0

[a]Details given in text.
[b]Plaque numbers determined 48 hr after initial infection procedure.

uptake of *Shigella* spp. into cells was greatly inhibited by IFNγ pretreatment (109).

In view of the inhibitory effects of IFNγ and IFNα/β on the internalization and intracellular proliferation of *Listeria*, it is possible that these antiviral molecules produced during listeriosis may prove important in inhibiting bacterial proliferation in nonprofessional phagocytic cells, such as liver hepatocytes. A nonspecific defense mechanism of this type could serve to restrict bacterial growth until an effective T-cell-mediated immune response can be mounted by the host to resolve the infection.

IV. CONCLUSION

A normal tendency is to postulate an in vivo function for an IFN based on effects observed in cell cultures. Unfortunately, as evidence from other in vitro studies confirms the original observation, the postulation as to the in vivo function of an IFN becomes almost dogma. What has become obvious to this author is that it is immensely difficult in just establishing whether an IFN functions in a resistance mechanism in vivo, let alone elucidating the actual role of the IFN in vivo. The first step in determining whether an IFN has a role in a host resistance mechanism to an infectious or malignant disease is to determine whether the IFN is produced during the course of the disease. The second step, is to determine whether a causal relationship exists between the IFN and the expression of resistance. In theory, this would require that it be shown that administering pure IFN increases resistance, or that blocking endogenously produced IFN with specific antibodies decreases resistance. However, such approaches have technical difficulties. As mentioned earlier in this chapter, endotoxin either enhances or suppresses anti-*Listeria* resistance, depending on the time of infection that it is administered. In view of the ubiquity of endotoxin, preparations of IFN (and for that matter antibody) should be assayed for endotoxin prior to use. If endotoxin is found, appropriate control preparations consisting of equivalent endotoxin concentrations must be tested for an effect on resistance. If resistance to *Listeria* infection is suppressed following the administration of a specific anti-IFN neutralizing antibody, the possibility that immune complexes may interfere with the expression of resistance must be considered. Virgin et al. (110) have shown that immune complexes formed in vivo can suppress resistance to *Listeria* infection. Thus, until it is shown that this does not happen, one could conclude erroneously, through the use of specific neutralizing anti-IFN antibody, that endogenously produced IFN performs an essential role in anti-*Listeria* resistance of the host.

In view of the foregoing, I believe that functions for the various IFNs in resistance to infectious or malignant diseases have yet to be established. Moreover, in view of the multiple actions ascribed to IFNs, the elucidation of the roles of each IFN class in various aspects of host defense will offer a challenge for years to come.

ACKNOWLEDGMENTS

The work presented from this laboratory was supported in part by Public Health Service grants AI-23544 and RR-05705 from the National Institutes of Health, and contract N00014-83-C-0407 from the Office of Naval Research.

REFERENCES

1. Stewart, W.E., J.E. Blalock, D.C. Burke, C. Chany, J.K. Dunnick, E. Falcoff, R.M. Friedman, G.J. Galasso, W.K. Joklik, J.T. Vilcek, J.S. Youngner, and K.C. Zoon. 1980. Interferon nomenclature. *Nature 286*:110.
2. Gray, P.W., D. Leung, D. Pennica, E. Yelverton, R. Najarian, C.C. Simonsen, R. Derynck, P.J. Sherwood, D.M. Wallace, S.L. Berger, A.D. Berger, A.D. Levinson, and D.V. Goeddel. 1982. Expression of human immune interferon cDNA in *E. coli* and monkey cells. *Nature 295*:503–508.
3. Gray, P.W., and D.V. Goeddel. 1983. Cloning and expression of murine IFNγ cDNA. *Proc. Natl. Acad. Sci. (USA) 80*:5842–5846.
4. Rinderknecht, E., B.H. O'Conner, and H. Rodriguez. 1984. Natural human interferon. Complete amino acid sequence and determination of sites of glycosylation. *J. Biol. Chem. 11*:6790–6797.
5. Knight, E., M.W. Hunkapiller, B.D. Korant, R.W.F. Hardy, and L.E. Hood. 1980. Human fibroblast interferon: amino acid sequence analysis and amino terminal amino acid sequence. *Science 207*:525–526.
6. Zoon, K.C., M.E. Smith, P.J. Bridgen, C.B. Anfinsen, M.W. Hunkapiller, and L.E. Hood. 1980. Amino terminal sequence of the major component of human lymphoblastoid interferon. *Science 207*:527.
7. Taira, H., R.J. Broeze, B.M. Jayaram, P. Lengel, M.W. Hunkapiller, and L.E. Hood. 1980. Mouse interferons: amino terminal amino acid sequences of various species. *Science 207*:528–529.
8. Naylor, S.L., A.Y. Sakaguchi, T.B. Shows, M.L. Law, D.V. Goeddel, and P.W. Gray. 1983. Human immune interferon is located on chromosome 12. *J. Exp. Med. 57*:1020–1027.
9. Berg, K. 1984. Identification, production and characterization of murine monoclonal antibody (LO-22) recognizing 12 native species of human alpha interferon. *J. IFN Res. 4*:481–491.
10. Shaw, G.D., W. Boll, H. Taira, N. Mantei, P. Lengel, and C. Weissman. 1983. Structure and expression of cloned murine IFN genes. *Nucleic Acids Res. 11*:555–571.
11. Sehgal, P.B., L.T. May, I. Tamm, and J.T. Vilcek. 1987. Human β2 interferon and B-cell differentiation factor BSF-2 are identical. *Science 235*: 731–732.

12. Havell, E.A., B. Berman, C.A. Ogburn, K. Berg, K. Paucker, and J. Vilcek. 1975. Two antigenically distinct species of human interferon. *Proc. Natl. Acad. Sci. (USA) 72*:2185–2187.

13. Yamamoto, Y. 1981. Antigenicity of mouse interferons: two distinct molecular species common to interferons of various sources. *Virology 111*: 312–319.

14. Yamamoto, Y., and Y. Kawade. 1980. Antigenicity of mouse interferons: distinct antigenicity of the two L cell interferon species. *Virology 103*:80–88.

15. Kawade, Y., M. Aguet, and M.G. Tovey. 1982. Antigenic correlations between components of C243 and L cell interferons. *Antiviral Res. 2*:155–159.

16. Havell, E.A., and G.L. Spitalny. 1983. Endotoxin-induced interferon synthesis in macrophage cultures. *J. Reticuloendothel. Soc. 33*:191–198.

17. Havell, E.A., and G.L. Spitalny. 1980. The induction and characterization of interferon from pure cultures of murine macrophages. *Ann. NY Acad. Sci. 350*:413–421.

18. Havell, E.A., and G.L. Spitalny. 1982. Antigenic properties of murine interferons from cells of nonlymphoid origin. *Tex. Rep. Biol. Med. 156*:112–127.

19. Sonnenfeld, G., A. Mandel, and T. Merigan. 1979. In vitro production and cellular origin of murine type II interferon. *Immunology 36*:883–890.

20. Handa, K., R. Suzuki, H. Matsui, Y. Shimizu, and K. Kumagai. 1983. Natural killer (NK) cells as a responder to interleukin 2 (IL2). Il-2 induced interferonγ production. *J. Immunol. 130*:988–992.

21. Valle, M.J., A.M. Bobrove, S. Strober, and T.C. Merigan. 1975. Immune specific production of interferon by human T cells in combined macrophage-lymphocyte cultures in response to herpes simplex antigen. *J. Immunol. 114*:435–446.

22. Valle, M.J., G.W. Jordan, S. Haahr, and T.C. Merigan. 1975. Characteristics of immune interferon produced by human lymphocyte cultures compared to other human interferons. *J. Immunol. 115*:230–244.

23. Salvin, S.B., J.S. Youngner, and W.H. Lederer. 1973. Migration inhibitory factor and interferon in the circulation of mice with delayed hypersensitivity. *Infect. Immun. 7*:68–74.

24. Wheelock, E.F. 1965. Interferon-like virus-inhibitor induced in human leukocytes by phytohemagglutinin. *Science 149*:310–311.

25. Havell, E.A., G.L. Spitalny, and P.J. Patel. 1982. Enhanced production of murine interferon γ by T cells generated in response to bacterial infection. *J. Exp. Med. 156*:112–127.

26. Havell, E.A., and G.L. Spitalny. 1984. The glycoprotein nature of murine gamma interferon. *Arch. Virol. 80*:195–207.

27. Havell, E.A., and G.L. Spitalny. 1984. Two molecular weight species of murine gamma interferon. *Virology 129*:508–513.

28. Isaacs, A., and J. Lindenmann. 1957. Virus interference. I. The interferon. *Proc. R. Soc. Biol. Sci. 147*:258-267.

29. Isaacs, A., J. Lindenmann, and R.C. Valentine. 1957. Virus interference II. Some properties of interferon. *Proc. R. Soc. Biol. Med. 147*:268-273.

30. Stinebring, W.R., and J.S. Youngner. 1964. Patterns of interferon appearance in mice injected with bacteria and bacterial endotoxin. *Nature 204*:712-715.

31. Ho, M. 1964. Interferon-like viral inhibitor in rabbits after intravenous administration of endotoxin. *Science 146*:1472-1474.

32. Merigan, T.C., and L. Hanna. 1966. Characteristics of interferon induced *in vitro* and *in vivo* by a TRIC agent. *Proc. Soc. Exp. Biol. Med. 122*:421-424.

33. Kazar, J. 1966. Interferon-like inhibitor in mouse sera induced by rickettsiae. *Acta Virol. 10*:277.

34. Rinaldo, C.R., B. Cole, J.C. Overall, and L.A. Glasgow. 1974. Induction of interferon in mice by mycoplasmas. *Infect. Immun. 10*:1296-1302.

35. Kleinschmidt, W.J., C. Cline, and E.B. Murphy. 1964. Interferon production by statolon. *Proc. Natl. Acad. Sci. (USA) 52*:741-744.

36. Rytel, M., and T.C. Jones. 1966. Induction of interferon in mice infected with *Toxoplasma gondii. Proc. Soc. Exp. Biol. Med. 123*:859-8862.

37. Huang, K.Y., W.W. Schultz, and F.B. Gordon. 1968. Interferon induced by *Plasmodium berghei. Science 162*:123-125.

38. Stinebring, W.R., and P.M. Absher. 1970. Production of interferon following an immune response. *Ann. NY Acad. Sci. 173*:714-725.

39. Kato, N., I. Nakashima, and F. Ohta. 1972. Interferon production in mice by the capsular polysaccharide of *Klebsiella pneumoniae. Infect. Immun. 12*:1-9.

40. Grossberg, S.E., G. Burleson, P. Morahan, and P. Jameson. 1972. A bacterial protein inducing antiviral resistance and high titers of interferon. *Prog. Immunobiol. Stand. 5*:274-278.

41. Field, A.K., A.A. Tytell, G.P. Lampson, and M.R. Hilleman. 1967. Inducers of interferon and host resistance. II. Multistranded synthetic polynucleotide complexes. *Proc. Natl. Acad. Sci. (USA) 58*:1004-1009.

42. Field, A.K., A.A. Tytell, G.P. Lampson, M.M. Nemes, and M.R. Hilleman. 1970. Double-stranded polynucleotides as interferon inducers. *J. Gen. Physiol. 56*:905-909.

43. Ho, M. 1984. Induction and inducers of interferon. In *Interferon*, volume 1, General and Applied Aspects. Edited by A. Billiau. Elsevier Science Publishers B.V., Amsterdam, pp. 79-144.

44. Lukas, B., and J. Hruskova. 1967. A virus inhibitor circulating in the blood of chickens induced by *Francisella tularensis* and *Listeria monocytogenes. Folio Microbiol. 12*:157-160.

45. Havell, E.A. 1986. Augmented induction of interferons during *Listeria monocytogenes* infection. *J. Infect. Dis. 153*:960-969.

46. Youngner, J.S., and W.R. Stinebring. 1964. Interferon production in chickens infected with *Brucella abortus. Science 144*:1022-1023.

47. Borecky, L., and V. Lackovic. 1967. The cellular background of interferon production in vivo. Comparison of interferon induction by Newcastle disease virus and *Bordetella pertussis. Acta. Virol. 11*:150–158.

48. Degre, M., and H. Dahl. 1974. Interferon production and prevention of viral infections in mice by components of a mixed bacterial vaccine. *Acta Pathol. Microbiol. Scand. 82*:904–910.

49. Baron, S., V. Howie, M. Langford, E.M. Macdonald, G.J. Stanton, J. Reitmeyer, and D.A. Wiegent. 1982. Induction of interferon by bacteria, protozoa, and viruses: defensive role. *Tex. Rep. Biol. Med. 156*:150–157.

50. Nakane, A., and T. Minagawa. 1984. The significance of alpha/beta interferons and gamma interferon produced in mice infected with *Listeria monocytogenes. Cell. Immunol. 88*:29–40.

51. Hopps, H.E., S. Kohno, and J.E. Smadel. 1964. Production of interferon in tissue cultures infected with *Rickettsia tsutsugamushi. Bact. Proc. 64*:115.

52. Kohno, S., M. Kohase, H. Sakata, Y. Shimizo, M. Hikata, and A. Shishido. 1970. Production of interferon in primary chick embryonic cells infected with *Rickettsia mooseri. J. Immunol. 105*:1553–1558.

53. Merigan, T.C., and L. Hanna. 1966. Characteristics of interferon induced *in vitro* and *in vivo* by a TRIC agent. *Proc. Soc. Exp. Biol. Med. 122*:421–424.

54. Kozikowska, E.H., and N. Hahon. 1970. Interferon induction by psittacosis agent in guinea pig leukocyte cultures. *Infect. Immun. 2*:731–734.

55. Havell, E.A. 1986. Synthesis and secretion of interferon by murine fibroblasts in response to intracellular *Listeria monocytogenes. Infect. Immun. 54*:787–792.

56. Borecky, L., V. Lackovic, N. Fuchsberger, and V. Hajnicka. 1974. Stimulation of interferon production in macrophages. In *Activation of Macrophages.* Edited by W.H. Wagner and H. Hahn. American Elsevier Publishing Co., New York, pp. 111–122.

57. Schultz, R.M., J.D. Papamatheakis, and M.A. Chirigos. 1977. Interferon: an inducer of macrophage activation by polyanions. *Science 197*:674–676.

58. Smith, T.J., A.S. Lubinecki, J.A. Armstrong, and S.B. Russ. 1973. Interferon induced by endotoxin and Newcastle disease virus in rabbit macrophage and kidney cell cultures. *Proc. Soc. Exp. Biol. Med. 142*:481–486.

59. Green, J.A., S.R. Cooperband, and S. Kibrick. 1969. Immune specific induction of interferon production in cultures of human blood lymphocytes. *Science 164*:1415–1417.

60. Kaufmann, S.H.E., H. Hahn, R. Burger, and H. Kirchner. 1983. Interferon-γ production by *Listeria monocytogenes*-specific T cells active in cellular antibacterial immunity. *Eur. J. Immunol. 13*:265–168.

61. Nakane, A., and T. Minagawa. 1981. Alternative induction of IFNα and IFNγ by *Listeria monocytogenes* in human peripheral blood mononuclear leukocyte cultures. *J. Immunol. 126*:2139–2142.

62. Nakane, A., and T. Minagawa. 1983. Alternative induction of alpha/beta interferons and gamma interferon by *Listeria monocytogenes* in mouse spleen cell cultures. *Cell. Immunol. 75*:283–291.

63. Evans, S.R., and H.M. Johnson. 1981. The induction of at least two distinct types of interferon in mouse spleen cell cultures. *Cell. Immunol. 64*:64072.

64. Saito, M., T. Yamaguchi, T. Ebina, M. Koi, E. Aonuma, H. Usami, and N. Ishida. 1983. In vitro production of immune interferon (IFNγ) by murine spleen cells when different sensitizing antigens are used in vivo and in vitro. *Cell. Immunol. 78*:379–386.

65. Le, J., D. Hendriksen-DeStefano, and J. Vilcek. 1986. Bacterial lipopoly-saccharide-induced interferon-γ production: roles of interleukin 1 and interleukin 2. *J. Immunol. 136*:4525–4530.

66. Yamamoto, J.K., W.L. Farrar, and H.M. Johnson. 1982. Interleukin 2 regulation in mitogen induction of immune interferon (IFNγ) in spleen cells and thymocytes. *Cell. Immunol. 66*:333–341.

67. Youngner, J.S., and W.R. Stinebring. 1965. Interferon appearance stimulated by endotoxin, bacteria, or viruses in mice pretreated with *Escherichia coli*, endotoxin, or infected with *Mycobacterium tuberculosis*. *Nature 208*: 456–458.

68. Okamura, H., K. Kawaguchi, K. Shoji, and Y. Kawade. 1982. High-level induction of gamma interferon with various mitogens in mice pretreated with *Propionibacterium acnes*. *Infect. Immun. 38*:440–443.

69. Nakane, A., and T. Minagawa. 1985. Sequential production of alpha and beta interferons and gamma interferon in the circulation of *Listeria monocytogenes*-infected mice after stimulation with bacterial lipopolysaccharide. *Microbiol. Immunol. 29*:659–669.

70. Wada, M., H. Okamura, K. Nagata, T. Shimoyama, and Y. Kawade. 1985. Cellular mechanisms in *in vivo* production of gamma interferon induced by lipopolysaccharide in mice infected with *Mycobacterium bovis* BCG. *J. IFN Res. 5*:431–443.

71. Buchmeier, N., and R.D. Schreiber. 1985. Requirement of endogenous interferon-γ production for resolution of *Listeria monocytogenes* infection. *Proc. Natl. Acad. Sci. (USA) 82*:7404–7408.

72. Mackaness, G.B. 1962. Cellular resistance to infection. *J. Exp. Med. 116*: 381–406.

73. North, R.J. 1974. Cell mediated immunity and the response to infection. In *Mechanisms of Cell-Mediated Immunity*. Edited by R.T. Mc Cluskey and S. Cohen. John Wiley and Sons, New York, pp. 185–219.

74. Kazar, J., P. Krautwurst, and F. Gordon. 1971. Effect of interferon and interferon inducers on infections with a nonviral intracellular microorganism *Rickettsia akari*. *Infect. Immun. 3*:819–823.

75. Gober, L., A. Friedman-Kien, E.A. Havell, and J. Vilcek. 1972. Suppression of the intracellular growth of *Shigella flexneri* in cell cultures by interferon preparations and polyinosinic-polycytidylic acid. *Infect. Immun. 5*:370–376.

76. Rothermel, C., G.I. Byrne, and E.A. Havell. 1983. Effect of interferon on the growth of *Chlamydia trachomatis* in mouse fibroblasts (L cells). *Infect. Immun. 39*:362–370.

77. Turco, J., and H.H. Winkler. 1983. Comparison of the properties of anti-rickettsial activity and interferon in mouse lymphokines. *Infect. Immun. 42*:27–32.

78. Paucker, K., K. Cantell, and W. Henle. 1962. Quantitative studies on viral interference in suspended L cells. II. Effect of interfering viruses and interferon on the growth rate of cells. *Virology 17*:324–334.

79. Knight, E. 1976. Antiviral and cell growth inhibitory activites reside in the same glycoprotein of human fibroblast interferon. *Nature 262*:302–303.

80. Fleischmann, W.R. 1982. Potentiation of the direct anticellular activity of mouse interferons: mutual synergism and interferon concentration dependence. *Cancer Res. 42*:869–875.

81. Sonnenfeld, G., A. Mandel, and T.C. Merigan. 1977. The immunosuppressive effect of type II mouse interferon preparations on antibody production. *Cell. Immunol. 34*:193–206.

82. Sonnenfeld, G., A. Mandel, and T.C. Merigan. 1978. Time and dosage of immunoenhancement by murine type II interferon preparations. *Cell. Immunol. 40*:285–293.

83. DeMaeyer-Guignard, J., A. Cachard, and E. DeMaeyer. 1975. Delayed-type hypersensitivity to sheep red cells: inhibition of senstization by interferon. *Science 190*:574–578.

84. DeMaeyer, E., J. DeMaeyer-Guignard, and M. Vandeputte. 1975. Inhibition by interferon of delayed-type hypersensitivity in the mouse. *Proc. Natl. Acad. Sci. (USA) 72*:1753–1758.

85. Collins, T., A.J. Korman, C.T. Wake, J.M. Ross, W. Fiers, K.A. Ault, M.A. Gimbrone, J.L. Strominger, and J.S. Prober. 1984. Immune interferon activates multiple class II major histocompatibility complex genes and the associated invariant chain gene in human endothelial and dermal fibroblasts. *Proc. Natl. Acad. Sci. (USA) 81*:4917–4921.

86. Pace, J.L., S.W. Russell, B.A. Torres, H.M. Johnson, and P.G. Gray. 1983. Recombinant mouseγ interferon induces the priming step in macrophage activation for tumor cell killing. *J. Immunol. 130*:2011–2014.

87. Spitalny, G.L., and E.A. Havell. 1984. Monoclonal antibody to MuIFNγ inhibits lymphokine-induced antiviral and macrophage tumoricidal activities. *J. Exp. Med. 159*:1560–1565.

88. Nathan, C.F., H.W. Murray, M.E. Wiebe, and B.Y. Rubin. 1983. Identification of interferon-γ as the lymphokine that activates human macrophage oxidative metabolism and antimicrobial activity. *J. Exp. Med. 158*:670–676.

89. Brunswick, M., and P. Lake. 1985. Obligatory role of gamma interferon in T cell-replacing factor-dependent, antigen-specific B cell responses. *J. Exp. Med. 161*:953–971.

90. Aguet, M., F. Belardelli, B. Branchard, R. Marcucci, and I. Gresser. 1982. Mouse γ interferon and cholera toxin do not compete for common receptor site of α/β interferon. *Virology 117*:541–544.

91. Raziuddin, A., F.H. Sarkar, R. Dutkowski, L. Shulman, F.H. Ruddle, and S.L. Gupta. 1984. Receptors for human α and β interferon but not for γ

interferon are specified by human chromosome 21. *Proc. Natl. Acad. Sci. (USA) 81*:5504–5508.

92. Weil, J., C.J. Epstein, L.B. Epstein, J. Sedmak, J.L. Saban, and S.E. Grossberg. 1983. A unique set of polypeptides is induced by γ interferon in addition to those induced with α or β interferons. *Nature 301*:437–439.

93. North, R.J. 1973. The cellular mediators of anti-*Listeria* immunity as an enlarged population of short-lived replicating T cells; kinetics of their production. *J. Exp. Med. 138*:342–355.

94. Newborg, M., and R.J. North. 1980. On the mechanism of T cell-independent anti-*Listeria* resistance in nude mice. *J. Immunol. 124*:571–580.

95. Zinkernagel, R., and R. Blanden. 1975. Macrophage activation in mice lacking thymus-derived (T) cells. *Experientia 31*:591–593.

96. Emerling, P., J. Finger, and J. Bockemul. 1975. *Listeria monocytogenes* infection in nude mice. *Infect. Immun. 12*:437–439.

97. Heymer, B., H. Hof, P. Emerling, and H. Finger. 1976. Morphology and time course of experimental listeriosis in nude mice. *Infect. Immun. 14*:832–835.

98. Remington, J.S., and T.C. Merigan. 1970. Synthetic polyanions protect mice against intracellular bacterial infection. *Nature 226*:361–363.

99. Newborg, M.F., and R.J. North. 1979. Suppressive effect of bacterial endotoxin on the expression of cell-mediated anti-*Listeria* immunity. *Infect. Immun. 24*:667–672.

100. Morrison, D.C., and J.L. Ryan. 1979. Bacterial endotoxins and host immune responses. In *Advances in Immunology*, Volume 28. Academic Press, Inc., New York, pp. 294–450.

101. Havell, E.A. 1986. Interferons as components of the host response to non-viral infection. In *Mechanisms of Host Resistance to Infectious Agents, Tumors, and Allografts*. Edited by R.M. Steinman and R.J. North. The Rockefeller University Press, New York, pp. 334–355.

102. Murray, H.W., G.L. Spitalny, and C.E. Nathan. 1985. Activation of mouse peritoneal macrophages *in vitro* and *in vivo* by interferon-γ. *J. Immunol. 134*:1619–1622.

103. Kiderlen, A.F., S.H.E. Kaufmann, and M-L. Lohmann-Matthes. 1984. Protection of mice against the intracellular bacterium *Listeria monocytogenes* by recombinant immune interferon. *Eur. J. Immunol. 14*:964–967.

104. Broukaert, P., G.G. Leroux-Roels, Y. Guisez, J. Tavernier, and W. Fiers. 1985. In vivo anti-tumor activity of recombinant human and murine TNF, alone and in combination with murine IFN-γ, on a syngeneic murine melanoma. *Int. J. Cancer 38*:763–769.

105. Schreiber, R.D., L.J. Hicks, A. Celada, N.A. Buchmeier, and P.W. Gray. 1985. Monoclonal antibodies to murine γ-interferon which differentially modulate macrophage activation and antiviral activity. *J. Immunol. 134*:1609–1618.

106. Prat, M., B. Gribaudo, P.M. Comoglio, G. Cavallo, and S. Landolfo. 1984.

Monoclonal antibodies against murine γ interferon. *Proc. Natl. Acad. Sci. (USA) 81*:4515-4519.

107. Russell, J.K., M.P. Hayes, J.M. Carter, B.A. Torres, B.N. Dunn, S.W. Russell, and H.M. Johnson. 1986. Epitope and functional specificity of monoclonal antibodies to mouse interferon-γ: the synthetic peptide approach. *J. Immunol. 136*:3324-3328.

108. North, R.J. 1970. The relative importance of blood monocytes and fixed macrophages to the expression of cell-mediated immunity to infection. *J. Exp. Med. 132*:521-534.

109. Niesel, D.W., C.B. Hess, Y.J. Cho, K.D. Klimpel, and G.R. Klimpel. 1986. Natural and recombinant interferons inhibit epithelial cell invasion by *Shigella* spp. *Infect. Immun. 52*:828-833.

110. Virgin, H.W., G.F. Wittenberg, G.J. Bancroft, and E.R. Unanue. 1985. Suppression of immune response to *Listeria monocytogenes*: mechanism(s) of immune complex suppression. *Infect. Immun. 50*:343-353.

14

Role of Interferon in Immunity to Mycobacteria

Peter W. Andrew / University of Leicester, Leicester, England

Douglas B. Lowrie* / Hammersmith Hospital, London, England

I. INTRODUCTION

The genus *Mycobacterium* currently contains 53 recognized species (1). Although many of these species are nonpathogenic saprophytes, the genus contains a number of pathogens of man and animals. The two most important human pathogens are *Mycobacterium tuberculosis* and *Mycobacterium leprae*, respectively the etiological agents of human tuberculosis and leprosy. Other mycobacterial species, for example *M. intracellulare*, can cause disease which is clinically undistinguishable from tuberculosis in humans. Other species cause disease in other animals. For example, *M. bovis* and *M. paratuberculosis* are serious causes of disease in cattle (2).

Tuberculosis and leprosy are ancient diseases of man. Archaeological evidence shows that man has suffered from tuberculosis since Egyptian or even neolithic times (3-5). Leprosy may not be as ancient as tuberculosis. It is probably only a relatively recent arrival (6).

These diseases have been deeply ingrained into the psyche of man. This is not surprising in view of the fatalities and disfigurement associated with them and is reflected in the extensive references to these diseases in popular and scientific literature (3). Tuberculosis became epidemic following the growth of urbanization that accompanied the industrial revolution (7). At its peak, in the middle of the nineteenth century, tuberculosis was responsible for around 20% of all deaths in England (3). Leprosy, while not having the high mortality of tuberculosis, has perhaps engendered more fear than tuberculosis because of its

*Present affiliation: National Institute for Medical Research, London, England

263

ability to cripple and disfigure. Reference to the bible highlights the social impact of this disease.

Although these diseases have declined in the socially advantaged countries in recent times, tuberculosis and leprosy still rank as two of the world's most common diseases. Recent estimates put the number of cases of tuberculosis at around 10,000,000 worldwide, with 3,000,000 deaths per year (8,9). It is probable that these numbers will increase in the coming years (9). There are, perhaps, 12,000,000 cases of leprosy (10). Even in the United States and Western Europe, tuberculosis remains a public health problem with more than 30,000 new cases of pulmonary tuberculosis in the United States in 1980 resulting in almost 3000 deaths (11). In the United Kingdom, tuberculosis kills more than 850 people per year, more than all other notifiable infectious diseases combined (12). In addition, with the ever increasing numbers of individuals with suppressed immunity, infections with more unusual mycobacteria are becoming more common (13-16).

Although *M. leprae* was the first etiological agent of infectious diseases to be described (17), it was the pioneering work of Koch (18) on *M. tuberculosis* that provided the stimulus for an explosion of work toward understanding the pathology and immunology of chronic infectious diseases.

II. IMMUNITY TO MYCOBACTERIAL INFECTION

Specific antibodies are produced during mycobacterial infection (19,20), but current thinking supports the early conclusions (21,22) that antibodies are not essential for effective immunity to mycobacterial infection (20). This conclusion is based mainly on the observation that immunity cannot be transferred by serum (22,23). However, there is still debate over the possibility that humoral immunity could have some subordinate role in modulating cell-mediated immunity (24,25).

Acquired immunity can be transferred between experimental animals by sensitized lymphocytes (26-28), so that over the last several decades the cell-mediated immune system has been considered central to immunity to mycobacterial infection. There have been many reviews of the evidence supporting the preeminence of cell-mediated immunity involving T cells and activated macrophages in determination of resistance to mycobacterial infection (29-33).

There is a spectrum of disease and immunological responsiveness associated with leprosy. At one end there is well-developed cell-mediated immunity in tuberculoid disease. There is progressive loss of cell-mediated immunity in moving across the spectrum, until in lepromatous disease, cell-mediated immunity is absent. Tuberculoid leprosy is a relatively benign disease characterized by the presence of few lesions containing few bacilli and many lymphocytes. In lepromatous disease there are many lesions packed with bacilli and little evidence of mononuclear cell infiltrates (34,35). A similar but less clear spectrum also can be seen in tuberculosis (36). It is this immunological spectrum that determines the clinical manifestations that result from infection.

However, there is an irony. The whole concept of cell-mediated immunity grew out of the work of Lurie (37,38) on mycobacterial infections in the rabbit, and then need for interaction between lymphocytes and macrophages for effective immunity was established during studies on antimycobacterial immunity (26,39–43). Yet, in recent times, advances in understanding the details of antimycobacterial immunity has tended to lag behind understanding of immunity to other microorganisms. Two reasons of practicality can be proposed. First, mycobacteria are not the ideal organisms for studies of macrophage-parasite interaction. They grow only slowly in culture, taking weeks to form colonies on solid media. They grow as clumps in liquid culture which causes problems in counting bacteria. With *M. leprae* there are further problems. It has never been grown in vitro (1,2). This fact in itself introduces obvious complications in determining antimicrobial effects.

The second problem is more fundamental. Attempts to demonstrate that macrophages kill mycobacteria in vitro have met with limited success. Several groups had been able to activate human or mouse mononuclear phagocytes to slow the growth of mycobacteria by using either sensitized lymphocytes or supernatants from cultures of these lymphocytes (44–49). Others had shown an inhibition of replication of *M. microti* and *M. lepraemurium* in human and mouse cells (50). While Alexander and Smith (51) reported that mixed lymphocyte cultures could produce a transient phase of slow killing of *M. lepraemurium* in mouse bone marrow-derived macrophages, it was not until the work of Walker and Lowrie (52) that it was shown conclusively that macrophages could be activated by lymphocytes to kill mycobacteria. Unfortunately, attempts to exploit this observation and determine the active ingredient within a lymphokine soup have only been partially successful. Although there have been reports of other ways of activating macrophages to kill mycobacteria (53), it is still true that most recent reports describe, at best, only a state of 100% stasis without kill (54).

Therefore, when we come to evaluate the evidence for a role of interferon in antimycobacterial immunity we find it is often, in large measure, circumstantial and inconclusive. This chapter may therefore appear more negative than others in this book because the evidence leads us to the conclusion that, at present, the best that can be said of the role of interferon in antimycobacterial immunity, especially in man, is "case not proven."

The first question we will ask in looking for evidence is, could interferon production be part of the normal immune response to infection with mycobacteria?

III. INTERFERON PRODUCTION IN RESPONSE TO MYCOBACTERIA

The first direct evidence that interferon (IFN) production was associated with an immune response was provided by Glasgow (55). It gradually became clear that not only did mycobacteria, as cell wall fractions in oil, or as an infection, sensitize mice to release extra interferon on subsequent challenge with classical interferon inducers such as bacterial endotoxin or Newcastle disease virus (56,57), but also that mycobacterial antigen itself could induce both interferon α/β or interferon γ production in specifically sensitized animals. Mycobacteria were effective eliciting agents in various forms, as whole, live bacteria (58) or as old tuberculin (59), purified protein derivative (PPD) (60-62), or as cell wall fractions (63-65).

It is clear from several studies that it is T lymphocytes with helper/inducer surface phenotype that induce macrophages to show increased antimicrobial activity (32,50,66,67) including antimycobacterial activity (68-71). Therefore, it is significant that lymphocytes isolated from sensitized animals secrete IFNγ in response to mycobacterial antigen (72-74). However, there has been debate as to which subset of T lymphocytes produce IFNγ. Lymphocytes with helper/inducer or with suppressor/cytotoxic phenotype have both been reported as producers of IFNγ (75-81). Indeed, recent experiments have suggested that cells of suppressor/cytotoxic phenotype, as well as those of helper/inducer phenotype, can passively transfer resistance between mice (82).

This debate may be resolved by the recent advances in techniques for maintenance of T-cell lines and clones. There are now T-cell clones specifically reactive to purified *M. tuberculosis* (83) or *M. leprae* antigens (84,85) or a mixed antigen preparation from *M. bovis* BCG or *M. tuberculosis* (86). These clones produced IFNγ in response to antigenic stimulation (86,87). To date, all of the clones have been of a helper/inducer phenotype except in the work of Mustafa and colleagues (86) where one clone out of 121 raised was not of this phenotype. Whether clones of suppressor/cytotoxic phenotype will be equally active in interferon production awaits solution of the problem of raising more clones of this phenotype.

Having established that mycobacteria are potent inducers of interferon production we can now ask if the production of interferon correlates with resistance to mycobacterial infections.

IV. CORRELATION OF INTERFERON PRODUCTION WITH RESISTANCE TO INFECTION

As long ago as 1947, in a study of 18 inbred strains of mice, it was shown that mouse strains differ in their ability to halt the progression of mycobacterial disease (88). It also seems that production of interferon is influenced directly by genotype. Neta and Salvin (89) in a survey of 11 strains of mice found that they could be split into two groups on the basis of levels of IFNγ found in the serum of mice that had been sensitized with *M. bovis* BCG cell walls in oil and challenged with old tuberculin. The difference between the groups may be due to differences in efficacy of sensitization. Pretreatment with Freund's complete adjuvent converted low-responder, C3H/He mice to high responders. They found that high responders were of H-2^d haplotype (C57Bl, Balb/c and DBA/2) while low responders were of H-2^k haplotype (C3H/He, AKR and CBA/Ca). Two groups (90,91) have shown that C3H/He mice are low responders due to failure of antigen presentation. Do these data correlate with resistance of the mouse strains to mycobacterial disease? The current state of evidence does not allow this question to be answered satisfactorily. One problem centers around what we mean by resistance.

In considering resistance to mycobacterial disease two separate aspects must be considered; the extent of bacterial multiplication and the extent of tissue damage. Tissue damage in mycobacterial diseases is caused almost entirely by the reactions of cell-mediated immune response known as delayed-type hypersensitivity (DTH). The immune response that results in inhibition and destruction of the mycobacteria is frequently, but not inevitably, accompanied by DTH immunopathology. DTH does not necessarily confer immunity (25,92). Therefore, interferon may contribute to either antibacterial action or to DTH or to both. We must also bear in mind that the antibacterial response to mycobacterial infection may consist of two parts, an innate resistance and an acquired resistance.

Substantial differences have been found between inbred mouse strains in studies where resistance to mycobacterial disease has been measured in terms of changes in bacterial numbers in the tissues. A notable example is where genetic analysis of resistance to a low dose of *M. bovis* BCG (93,94), *M. lepraemurium* (95,96), or *M. intracellulare* (97) indicated a single resistance gene or group of genes which was not linked to the H-2 locus. This resistance is innate, evidently occurring early, before the expression of acquired immunity, although not before sensitization was initiated (98). However, the pattern of resistance contrasted with the pattern of interferon production found by Neta and Salvin (89). Thus, there is no evidence that innate resistance is dependent on IFNγ production. Indeed, there appears to be an inverse relationship between this innate resistance and interferon production. For example, Neta and Salvin (89) found that AKR and C3H/He mice were low interferon producers in response to mycobacterial antigen, yet they carry the *Bcg^r* gene (93) and also Balb/c and C57BL mice carry the *Bcg^s* but were high interferon responders to mycobacterial antigen. Maybe mice with the *Bcg^r* gene rely on innate immunity to control mycobacterial infection, and therefore normally do not need to respond to mycobacterial antigen by producing IFNγ.

Brett and Butler (74) reported a direct correlation between the T-cell-dependent ability of mice to control the course of *M. lepraemurium* infection and the ability of lymph node cells from these mice to produce interferon in response to heat-killed *M. lepraemurium*. This looks like good evidence in favor of a role for interferon in cell-mediated immunity, but again the picture is not clear. Although Alexander and Curtis (99) agree with the definition of C57BL mice and Balb/c mice as being resistant and susceptible, respectively, to *M. lepraemurium* infection, several workers have shown no difference in the susceptibility of these mice, whatever the dose of *M. lepraemurium* (95,96, 100-102). However, one can take the view that it is not necessary to have an absolute definition of resistance to accept the evidence of Brett and Butler (74), the behavior of the animals during the experiments in question being more important. Like the findings from estimates of bacterial numbers, assessments of DTH do not always indicate a correlation with interferon production. For example, the mouse strain variations in granuloma formation in response to *M. bovis* BCG cell walls reported by Yamamoto and Kakinuma (103) is similar to the IFNγ production spectrum of Neta and Salvin (89). However, there were differences, and these must carry greater weight here. AKR mice, for example, were high granuloma formers but low producers of IFNγ. Likewise, despite the substantial differences between Balb/c and C57Bl mice in interferon release from lymph node cells (74), there was little difference in foot-pad swelling or in vitro lymphoproliferative responses (74,99).

Unfortunately, in looking for these correlations there are difficulties. Although several laboratories have shown a genetic influence on interferon production in response to mycobacterial antigen, there is variation in the reported data. For example, whereas Neta and Salvin (89) and Adelman and co-workers (104) report Balb/c mice as high interferon producers, others, using similar techniques, have found them to be low producers (62,72,105). There are further problems. Not only can low responders be converted to high responders by altering the method of immunizing (89), but some mice, for example CBA/Ca, can be classified as low responders on the basis of serum levels of interferon after challenge with PPD and as high responders when interferon production by their isolated spleen cells is measured (72). Furthermore, the class of interferon produced can vary with time after challenge (62). Such variations indicate that care must be exercised in making correlations between data from different laboratories. Experiments of the type done by Brett and Butler (74) provide their own internal control. Overall though, it appears that interferon is not involved in innate resistance to mycobacterial infection. What these experiments tell us about its role in acquired immunity or DTH is an open question. The small number of experiments that have been done suggest it may have a role in reducing bacterial numbers, but not in DTH, in the mouse.

In man, it is not so easy to do experiments on the modulation of interferon production, but as an alternative we can try to evaluate if mycobacterial disease is associated with abnormalities in interferon production.

V. INTERFERON PRODUCTION DURING MYCOBACTERIAL DISEASE IN HUMANS

There have been two recent publications reporting a deficiency in IFNγ production during active pulmonary tuberculosis (106,107). However, again the situation is not straightforward. Onwubalili and colleagues (106) found no difference between peripheral blood mononuclear cells (PBMC) of patients and controls in the mean production of IFNα/β induced by Newcastle disease virus or IFNγ induced by streptococcal enterotoxin A or PPD. However, they identified a subgroup of 9 patients, out of 25, who had lower titers of IFNγ in response to PPD. The significance of this is hard to ascertain. There was no difference in the severity of disease between high and low responders.

In the study by Vilcek and colleagues (107), PBMC from 9 of 14 patients produced a low titer of IFNγ in response to at least one of the three mitogens tested and also had defective IFNγ production after PPD. Again the authors attempted to correlate clinical severity of disease with the degree of defect in interferon production, without success. In addition, in the context of discussing a role for interferon, perhaps the finding of normal interferon production in the presence of severe, advanced disease is more instructive? These findings mitigate against interferon having either a beneficial or harmful role in the development of the disease in humans.

In considering these data we must also bear in mind the concept of local immunity. The sequestration of antigen-specific cells within sites in the tissues during tuberculosis has been proposed (108,109). Such a phenomenon could deplete the peripheral blood of cells reactive to mycobacterial antigen. Conversely, suppressor cell action, which may impair interferon production, has been reported to be greater in peripheral blood than in tuberculous lesions (110). In a study of 18 patients with tuberculous pleurisy, Shimokata and colleagues (73) found that peripheral blood T lymphocytes from only four patients produced interferon in response to PPD and then only at a low titer. In contrast, with T lymphocytes isolated from pleural effusions, 17 patients produced a high titer of interferon.

Defective interferon production has been found associated with *Mycobacterium avium intracellulare* and *M. leprae* infections. Murray and Roberts (111) reported that on average, PBMC from patients with disseminated *M. avium intracellulare* infection, in conjunction with acquired immune deficiency syndrome, were far less responsive to mycobacterial antigen in terms of IFNγ production than PPD-positive controls. However, yet again some patients (2 of 15) had normal interferon production in association with severe disseminated disease.

As mentioned earlier, leprosy is a disease typified by a clinical spectrum that reflects the immunological capacity of the patient. Tuberculoid patients have a well-developed cell-mediated response while lepromatous patients do not. The defect does not appear to be in macrophage effector function, three groups having reported that macrophages, from either end of the spectrum, are equally competent (112-115). Therefore, such a spectrum offers an opportunity for investigating the mechanism of control of macrophage effector function. Indeed, Nogueira and co-workers (116) investigated production of IFNγ, in response to *M. leprae* antigen, by PBMC from leprosy patients. It is of interest that PBMC from 17 of 18 lepromatous patients failed to respond to *M. leprae* with production of IFNγ, whereas PBMC from most tuberculoid patients had a large response.

At first sight some of these observations look encouraging. However, there are two points, apart from those already discussed, that must be borne in mind when considering such data. There is the consideration of production of multiple lymphokines by individual T cells. There is much evidence from T-cell clones for this phenomenon (86,87). Therefore, there is likely to be decreased production of other lymphokines in parallel with a defect in production of IFNγ. The other lymphokines may be more critical. The second problem is more difficult. One cannot easily separate cause and effect when considering these data. There is no doubt that low production of IFNγ can be a consequence of infection. There are several suggestions of induction of suppressor cells during mycobacterial infections (117-122). The development of antigenic tolerance and anergy are well-recognized features of heavily infected animals (123,124). Anergy often quickly appears during progressive tuberculosis (31,117). There is some correlation between degree of infection and anergy in tuberculosis and immune responsiveness returns after effective chemotherapy (125). There is evidence that chemotherapy can restore interferon responses in previously low interferon producers (106).

What is missing here is evidence that low interferon production enhances disease, or evidence that a preexisting, genetic defect enhances susceptibility. An alternative approach is to look at the effect of administration of interferon to infected patients. First though, before dealing with the small amount of information that is available from studies with humans, we will look at the effect of treatment of other animals.

VI. INTERFERON TREATMENT OF MYCOBACTERIAL INFECTIONS

If interferon has any potential role in protective immunity to mycobacterial infection we could expect some protection from infection and remission of disease after administration of IFN. Such an approach has proved effective in protecting mice against *Toxoplasma gondii* (126) and *Listeria monocytogenes* (127) infections. There have been several recent reports of attempts to evaluate interferon as a therapeutic agent in the treatment of mycobacterial disease, either alone or in conjunction with conventional chemotherapy.

Spleen cells taken from mice with a chronic infection of *M. intracellulare* fail to produce interleukin-2 (IL-2) or IFNγ in response to *M. intracellulare* antigen (128). Treatment of these chronically infected mice with seven daily doses of IFNγ significantly reduced numbers of viable bacilli in the lungs and spleens. These results certainly indicate that IFNγ has the potential to act to strengthen a protective immune response that has decreased secondarily to advancing infection, though it leaves open the question of whether or not it is the main physiological agent. We will deal with this point later in the chapter. Other attempts to enhance antimycobacterial immunity with exogeneous interferon have had more limited success.

Khor and colleagues (129) found a very small reduction in numbers of viable *M. tuberculosis* in spleens and in lungs of mice was achieved by administration of IFNγ. This reduction was greatest when interferon was given prophylactically, one day before infection. Interferon γ had no effect on the course of an established infection of *M. tuberculosis*. Banerjee and colleagues (130) also reported that prophylactic interferon had some protective effect. In this case, when the number of *M. bovis* BCG recovered from spleens of infected mice was measured. Again, interferon had no effect on an established infection. However, this study was taken further by looking at the effect of interferon on the course of infection in athymic mice. Such mice are severely depleted of functional T lymphocytes (131–133), hence any effect of interferon would be occurring in the virtual absence of other T-cell products. Interferon γ gave no protection against *M. bovis* BCG to such mice whether given before or after infection. Indeed, interferon increased the numbers of bacteria recovered from infected animals. This can be taken as evidence against IFNγ being the only macrophage-activating factor having a role in antimycobacterial immunity and suggests that it may act only in conjunction with other T-cell products that are absent in athymic mice.

Nomaguchi and co-workers (134) found no evidence in support of interferon having a role in protection of mice against *M. lepraemurium* infection. In their alternative approach they found that treatments that increase serum levels of IFNα/β or IFNγ had no effect on the survival of infected mice.

The most adventurous attempt to look at the effect of interferon in vivo was the administration of IFNγ to lepromatous leprosy patients (135). Three daily doses of IFNγ were injected directly into a single, cutaneous lesion. After six days a biopsy was taken from the injection site and from a control site. In the six subjects tested, histological changes toward tuberculoid-type features were seen. In four of the six there was evidence for some decrease in the numbers of acid-fast bacilli found in the tissue. These are encouraging results, but again we must play the devil's advocate. As the authors themselves note, the method for determining acid-fast bacilli within the biopsy tissue was not unambiquous. The problems associated with the counting of stained mycobacteria in tissue that can lead to apparant, though false, disappearance of bacteria have been described by Lefford (133). He pointed out that care must be taken in evaluating single time point data. It is such data that Nathan and his colleagues (135) report.

It has been thought for a long time that it is the macrophage that is responsible ultimately for the destruction of pathogenic mycobacteria, possibly by mechanisms involving hydrogen peroxide. There is now a substantial amount of evidence to support this hypothesis (53,136-138). Therefore, crucial experiments assess whether interferon modulates the antimycobacterial mechanisms of the macrophage.

VII. IN VITRO EFFECTS OF INTERFERON ON THE ANTI-MYCOBACTERIAL ACTIVITY OF THE MACROPHAGE

Interferon enhances the respiratory burst of macrophages (139,140) including those collected from animals infected with mycobacteria (115,135,141). If we accept the presumption for an involvement of peroxide in killing of mycobacteria by macrophages we might consider this effect of interferon as indirect evidence for a role in immunity. However, while the case for peroxide involvement in killing of *M. tuberculosis* and *M. microti* is substantial, in no instance is it conclusive (53). We can speculate that IFNγ also may enhance oxygen-independent antimycobacterial systems. In murine peritoneal macrophages treated with IFNγ, 68% of phagosomes containing *M. leprae* fused with lysosomes, compared with 27% in untreated macrophages (142). However, we still do not know to what extent lysosomal contents are effective antimycobacterial agents.

What of the direct evidence for an effect of IFNγ on the intracellular survival of mycobacteria within macrophages? As discussed previously, most reports of macrophage activation by lymphocytes or their products have been of inhibition of replication (44,46,47,49), although there are some reports of killing of mycobacteria (52,143). Recent studies of the effect of IFNγ also have tended to reveal growth inhibition rather than killing. These reports bring us to the conclusion that there is reasonable evidence that IFNγ is effective in inducing antimycobacterial systems in mouse macrophages, but for human macrophages there is no evidence from in vitro data that interferon is effective.

Khor and co-workers (143) looked at the intracellular survival of *M. microti* in mouse peritoneal macrophages pretreated with IFNγ before infection. They found survival was unaffected by interferon in the three days following infection, but that an effect of interferon was seen by a killing of mycobacteria during the first 15 minutes of infection. The effect had subsided by 30 minutes. Interestingly, they found that if interferon was given after phagocytosis was complete then only a slow onset of bacteriostasis was induced. These are potentially very important observations. The ability to selectively induce either cidal or stasis systems should be an enormous aid toward understanding the details of macrophage antimycobacterial mechanisms. However, it must be said that as yet we have no evidence that early antimycobacterial killing occurs in other systems we are exploring, specifically, guinea pig alveolar macrophages with *M. tuberculosis* (53) or human monocyte-derived macrophages and *M. phlei* (A.K. Robertson and P.W. Andrew, unpublished data).

Rook and his colleagues in a series of publications reported their investigations in which they used changes in mycobacterial ability to incorporate radiolabelled uracil as a guide to changes in numbers of intracellular bacteria. When radiolabel incorporation was impaired this was interpreted as due to bacterial growth inhibition or stasis (144). They found that mouse peritoneal macrophages could be activated by crude lymphokine culture supernatants or by I-A matched PPD-reactive T-cell lines to cause stasis of *M. tuberculosis* and *M. intracellulare*. Evidence that this was a true stasis rather than a balance of multiplication and death came from a parallel decrease in the effectiveness of an antibiotic that is probably dependent on bacterial growth for its action (54). They then compared the effects of IFNγ on the ability of human and murine mononuclear phagocytes to inhibit proliferation of *M. tuberculosis*.

Interferon γ treatment caused normal mouse peritoneal macrophages to markedly inhibit the multiplication of *M. tuberculosis* (145,146). Uptake of uracil was reduced to 10–20% of the uptake by the bacteria from nonactivated macrophages. With human cells the picture is more disappointing. The effect of IFNγ on human macrophages obtained by bronchioalveolar lavage was, at best, only that of a weak and variable agent for activating antimycobacterial systems (146). The mean result was no effect of interferon, although in two of seven experiments a significant decrease in uracil uptake compared with uptake by bacteria from nontreated cells was seen. Whether or not this result was achieved because the two individuals were smokers whose lungs were therefore environments which primed macrophages to be responsive to interferon is an open question.

The picture is the same with other human mononuclear phagocytes. If cultured human monocytes were given IFNγ there was a small inhibition of growth of *M. tuberculosis* (147). However, even the largest effects seen here were significantly less than those seen with mouse peritoneal macrophages. If cells were pretreated with IFNγ before infection with *M. tuberculosis*, there was again, only a small inhibition of growth of the bacteria. In contrast to the observations of the murine system of Khor and her co-workers (143), antimycobacterial mechanisms that may be effective at the moment of phagocytosis were not induced in human mononuclear phagocytes by interferon.

With freshly explanted monocytes a very variable effect of IFNγ was seen (147). Significant inhibition of *M. tuberculosis* was seen with cells from some individuals, though again less than with murine cells. However, with cells from other individuals IFNγ treatment resulted in a significant increase in growth of *M. tuberculosis*. Although this may seem a surprising observation, it has been seen by others. It was reported by Douvas and colleagues (148) that they could activate human monocyte-derived macrophages with IFNγ to kill *Leishmania donovani* promastigotes or mouse tumor cells. In contrast, IFNγ pretreatment of human macrophages enhanced the replication of *M. tuberculosis* within these cells. We have found the same effect in our own laboratories with other species of mycobacteria. Interferon γ treatment of human monocyte-derived macrophages enhanced the intracellular survival of *M. microti* (L. Walker and D.B. Lowrie, unpublished data). We found also that *M. phlei* had enhanced replication in IFNγ-pretreated human macrophages even though, like others (149), we can activate these cells with IFNγ to inhibit the replication of *Toxoplasma gondii* (A.K. Robertson and P.W. Andrew, unpublished data). In fact, it appears that the nonactivated macrophage is a more favorable culture medium for growth of mycobacteria than RPMI 1640 culture medium with 10–20% human serum. Both *M. tuberculosis* and *M. phlei* multiply faster in nonactivated macrophages than in the culture medium (145, A.K. Robertson and P.W. Andrew, unpublished data). Thus, in humans, IFNγ may not be beneficial, but may be detrimental. Rather than activating macrophages to be antimycobacterial, IFNγ can activate macrophages to be a more favorable environment for mycobacterial growth.

Even if stasis rather than kill is the usual result of interferon, this may not be advantageous. As discussed by Lowrie (137) and by Altes and colleagues (54), induction of stasis may render the bacteria less vulnerable to antibiotics and to killing systems activated by other agents. Indeed, it has been shown that induction by lymphokines of macrophage systems causing mycobacterial stasis can protect intracellular mycobacteria from the effects of antibiotics (54). Although Khor and colleagues (143) showed that IFNγ had no effect on the efficacy of rifampicin, an antibiotic effective against bacteria that are only growing intermittently, there was some evidence that IFNγ may influence the action of isoniazid, an antibiotic which may be more dependent on a continuously dividing bacterial population for effectiveness. However, this influence on isoniazid was only seen in vitro, the action of the antibiotic was uneffected in vivo (129). Overall then, at this stage the evidence neither confirms nor precludes a role for interferon in immunity to mycobacterial infections. What of other possible mediators of this immunity?

VIII. ARE THERE ALTERNATIVES TO INTERFERON?

The evidence in support of interferon discussed in this chapter is consistant with the hypothesis that substances other than interferon may have a role in activating antimycobacterial immunity, either independently of or in concert with interferon. Indeed, some of the evidence presented is directly supportive of this hypothesis, such as the observation that IFNγ by itself could not reconstitute the antimycobacterial immunity of athymic mice (130). There is other evidence that macrophage-activating factors other than interferon may function in antimycobacterial immunity.

We have shown that IFNγ is not the only macrophage-activating factor released by human T-cell clones specifically responsive to purified antigen from *M. tuberculosis* (87) while Milon and co-workers (150) have shown that supernatants from the murine T-cell line, EL4, depleted of IFNγ could still activate mouse peritoneal macrophages to inhibit the replication of *M. bovis* BCG. The hypothesis of multiple macrophage-activating factors is not confined to mycobacterial systems. The conclusion that there are multiple macrophage-activating factors produced by lymphocytes also has been drawn from studies of other infections (151–156).

We have also found that we can activate mouse peritoneal macrophages to kill *M. microti*, without the addition of interferon or any other lymphokine, by maintenance of macrophages in a serum-free medium (157). Rook and co-workers (147) found that human monocytes could be activated by metabolites of vitamin D_3 to inhibit the growth of *M. tuberculosis*, 1,25-dihydroxy vitamin D_3 being most effective. This metabolite did not cause monocytes to become more receptive to IFNγ, rather it had an additive effect with each having a similar influence. These observations are consistent with reports suggesting a role for vitamin D in the pathogenesis of tuberculosis (158,159) and may reawaken interest in this vitamin for treatment of tuberculosis (158).

IX. SUMMARY

In summary, while we might anticipate from work on other infections that interferon would have a central role in immunity to tuberculosis and leprosy, appraisal of the available evidence suggests that arrival at such a conclusion is premature. The evidence is not only inconclusive in support of a role for interferon, it is not inconsistent, with other factors being of greater importance. What these other factors are remains a mystery.

REFERENCES

1. Wayne, L.G., and G.P. Kubica. 1986. Mycobacterium. In *Bergey's Manual of Systematic Bacteriology*. Edited by P.H.A. Sneath, N.S. Mair, and M.E. Sharpe. Williams & Wilkins, Baltimore, vol. 2, pp. 1436–1457.

2. Grange, J.M. 1983. The mycobacteria. In *Topley & Wilsons Principles of Bacteriology, Virology and Immunity*, 7th ed. Edited by M.T. Parker. Edward Arnold, London, pp. 60–93.

3. Dubos, R., and J. Dubos. 1952. *The White Plague – Tuberculosis, Man and Society*. Little, Brown & Co., Boston.

4. Morse, D., D.R. Brothwell, and P.J. Ucko. 1964. Tuberculosis in ancient Egypt. *Am. Rev. Resp. Dis. 90*:524–541.

5. Zimmerman, M.R. 1979. Pulmonary and osseous tuberculosis in an Egyptian mummy. *Bull. N.Y. Acad. Med. 55*:604–608.

6. Godal, T. 1984. Leprosy. In *Bacterial Vaccines*. Edited by R. Germanier. Academic Press, London, pp. 419–430.

7. Youmans, G.P. 1979. Epidemiology of tuberculosis. In *Tuberculosis*. Edited by G.P. Youmans. W.B. Saunders and Co., Philadelphia, pp. 356–369.

8. World Health Organization. 1982. Tuberculosis control: report of a joint IUAT/WHO study group. *W.H.O. Tech. Rep.* Series 671.

9. Stylbo, K. 1983. Tuberculosis and its control: lessons to be learned from past experience, and implications for leprosy control programmes. *Ethiop. Med. J. 21*:101–122.

10. Lincoln, E.M., and L.A. Gilbert. 1972. Diseases in children due to mycobacteria other than *Mycobacterium tuberculosis. Am. Rev. Resp. Dis. 105*:603–716.

11. Collins, F.M. 1984. Tuberculosis. In *Bacterial Vaccines*. Edited by R. Germanier. Academic Press, London, pp. 373–418.

12. Office of Population Censuses and Surveys. 1984. *Communicable Disease Statistics*. HMSO, London.

13. Lichtenstein, I.J.H., and R.R. MacGregor. 1977. Mycobacterial infections in renal transplant recipients. Report of five cases and review of the tuberculosis. *Clin. Exp. Immun. 27*:230–237.

14. Cohen, R.J., M.K. Samoszuk, D. Busch, and M. Lagios. 1983. Occult infections with M. intracellulare in bone-marrow biopsy specimens from patients with AIDS. *N. Engl. J. Med. 308*:1475–1476.

15. Croxson, T.S., D. Ebantes, and D. Mildran. 1983. Atypical mycobacteria and Kaposi's sarcoma in the same biopsy specimens. *N. Engl. J. Med. 308*: 1476.

16. Wong, B., F.F. Edwards, T.E. Kiehn, E. Whinbey, H. Donnely, E.M. Bernard, J.W.M. Gold, and D. Armstrong. 1985. Continuous high grade *Mycobacterium avium-intracellulare* bacteremia in patients with acquired immune deficiency syndrome. *Am. J. Med. 78*:35–40.

17. Hansen, G.H.A. 1874. Indberltung til det Norske medicinske Selskab i Christiania om en med understottelse af selskabet foretagen reise for

anstille undersogelser angaende spedalskhedens arsager, tildels ud forte sammen med forstander Hartwig. *Norsk. Mag. Laegevidensk 4*:1–88.

18. Koch, R. 1982. Die aetiologie der tuberkulose. *Berl. Klin. Wochenschr. 19*:221–230.

19. Middlebrook, G., and R.J. Dubos. 1948. Specific serum agglutination of erythrocytes sensitized with extracts of tubercle bacilli. *J. Exp. Med. 88*: 521–528.

20. Youmans, G.P. 1979. Nature of the specific acquired immune response in tuberculosis. In *Tuberculosis*. Edited by G.P. Youmans. W.B. Saunders and Co., Philadelphia, pp. 285–301.

21. Rich, A.R. 1955. *The Pathogenesis of Tuberculosis*, 2nd ed. Blackwell Scientific Publications, Oxford.

22. Raffel, S. 1955. The mechanism involved in acquired immunity to tuberculosis. In *CIBA Foundation Symposium on Experimental Tuberculosis, Bacillus and Host With an Addendum on Leprosy*. Edited by G.E.W. Wolstenholme, M.P. Cameron, and C.M. O'Connor. J.A. Churchill, London, pp. 261–282.

23. Reggiardo, Z., and G. Middlebrook. 1974. Failure of passive serum transfer of immunity against aerogenic tuberculosis in rabbits. *Proc. Soc. Exp. Biol. Med. 145*:173–175.

24. Barksdale, L., and K-S. Kim. 1977. Mycobacterium. *Bact. Rev. 41*:217–372.

25. Lefford, M.J. 1981. Immunology of *Mycobacterium tuberculosis*. In *Immunology of Human Infection*, Part 1. Edited by A.J. Nahmias and R.J. O'Reilly. Plenum Press, New York and London, pp. 345–368.

26. Lefford, M.J. 1975. Transfer of adoptive immunity to tuberculosis in mice. *Infect. Immun. 11*:1175–1181.

27. North, R.J. 1973. Importance of thymus derived lymphocytes in cell mediated immunity to infection. *Cell. Immun. 7*:166–176.

28. North, R.J. 1974. T cell dependence of macrophage activation and mobilization during infection with *Mycobacterium tuberculosis. Infect. Immun. 10*:66–71.

29. Turk, J.L. 1969. Cell mediated immunological process in leprosy. *Bull. W.H.O. 41*:779–792.

30. Collins, F.M. 1982. The immunology of tuberculosis. *Am. Rev. Resp. Dis. 25*:42–49.

31. Chaparas, S.D. 1982. The immunology of mycobacterial infections. *CRC Crit. Rev. Microbiol. 9*:139–197.

32. Collins, F.M., and S.G. Campbell. 1982. Immunity to intracellular bacteria. *Vet. Immun. Immunopath. 3*:5–66.

33. Lagrange, P.H., B. Hurtrel, M. Brandely, and P.M. Thickstun. 1983. Immunological mechanisms controlling mycobacterial infections. *Bull. Eur. Physiopath Resp. 19*:163–172.

34. Ridley, D.S. 1974. Histological classification and the immunological spectrum of leprosy. *Bull. W.H.O. 51*:451–465.

35. Bullock, W.E. 1978. Leprosy: a model of immunological perturbation in chronic infection. *J. Inf. Dis. 137*:341–354.

36. Lenzini, L., P. Rottoli, and L. Rottoli. 1983. The spectrum of human literature. *Rev. Infect. Dis. 5*:216–226.

37. Lurie, M.B. 1928. The fate of human and bovine tubercule bacilli in various organs of the rabbit. *J. Exp. Med. 48*:155–182.

38. Lurie, M.B. 1942. Studies on the mechanism of immunity in tuberculosis. The fate of tubercule bacilli ingested by mononuclear phagocytes derived from normal and immunized animals. *J. Exp. Med. 75*:247–267.

39. Suter, E. 1953. Multiplication of tubercle bacilli within mononuclear phagocytes in tissue cultures derived from normal animals and animals vaccinated with BCG. *J. Exp. Med. 97*:235–247.

40. Suter, E. 1956. Interaction between phagocytes and pathogenic microorganisms. *Bact. Rev. 20*:94–132.

41. Mackaness, G.B. 1954. The growth of tubercle bacilli in monocytes from normal and vaccinated rabbits. *Am. Rev. Tuber. 69*:495–504.

42. Mackaness, G.B. 1968. The immunology of antituberculosis immunity. *Am. Rev. Resp. Dis. 97*:337–344.

43. Lefford, M.J. 1975. Delayed hypersensitivity and immunity in tuberculosis. *Am. Rev. Resp. Dis. 111*:243–246.

44. Patterson, R.J., and G.P. Youmans. 1970. Demonstration in tissue culture of lymphocyte mediated immunity to tuberculosis. *Infect. Immun. 1*:600–603.

45. Klun, C.L., and G.P. Youmans. 1973. The effect of lymphocyte supernatant fluids on the intracellular growth of virulent tubercle bacilli. *J. Ret. Soc. 13*:263–274.

46. Klun, C.L., and G.P. Youmans. 1973. The induction by *Listeria monocytogenes* and plant mitogens of lymphocyte supernatant fluids which inhibit the growth of *Mycobacterium tuberculosis* with macrophages in vitro. *J. Ret. Soc. 13*:275–285.

47. Turcotte, R., Y. DesOrmeaux, and A.F. Borduas. 1976. Partial characterization of a factor extracted from sensitized lymphocytes that inhibits the growth of *Mycobacterium tuberculosis* within macrophages in vitro. *Infect. Immun. 14*:337–344.

48. Crowle, A.J., and M. May. 1981. Preliminary demonstration of human tuberculo-immunity in vitro. *Infect. Immun. 31*:453–464.

49. Zlotnik, A., and A.J. Crowle. 1982. Lymphokine induced mycobacteriostatic activity in mouse pleural macrophages. *Infect. Immun. 37*:786–793.

50. Godal, T., R.J.W. Rees, and J.O. Lamvik. 1971. Lymphocyte-mediated modification of blood-derived macrophage function; Inhibition of growth of intracellular mycobacteria with lymphokines. *Clin. Exper. Immun. 8*: 625–637.

51. Alexander, J., and C.C. Smith. 1978. Growth of *Mycobacterium lepraemurium* in nonstimulated and stimulated mouse peritoneal derived and bone marrow derived macrophages in vitro. *Infect. Immun. 22*:631–636.

52. Walker, L., and D.B. Lowrie. 1981. Killing of *Mycobacterium microti* by immunologically activated macrophages. *Nature 293*:69–70.

53. Lowrie, D.B., P.S. Jackett, and P.W. Andrew. 1985. Activation of macrophages for antimycobacterial activity. *Infect. Immun. 11*:195–203.

54. Altes, C., J. Steele, J.L. Stanford, and G.A.W. Rook. 1985. The effect of lymphokines on the ability of macrophages to protect mycobacteria from a bactericidal antibiotic. *Tubercle 66*:261–266.

55. Glasgow, L.A. 1966. Leukocytes and interferon in the host response to viral infections. Enhanced response of leukocytes from immune animals. *J. Bact. 91*:2185–2191.

56. Imanishi, J. 1981. Enhanced production of interferon in mice infected with *Mycobacterium bovis* BCG. *Biken J. 24*:123–126.

57. Kato, N., I. Nakashima, M. Ohta, S. Naito, and T. Kojima. 1979. Interferon and cytotoxic factor released in the blood of mice infected with *Mycobacterium bovis* BCG. Enhanced production of interferon and appearance of cytotoxin stimulated by capsular polysaccharide of *Klebsiella pneumoniae* or bacterial lipopolysaccharide. *Microbiol. Immun. 23*:383–394.

58. Salvin, S.B., J.S. Youngner, and W.J. Leder. 1973. Migration inhibitory factor and interferon in the circulation of mice with delayed hypersensitivity. *Infect. Immun. 7*:68–75.

59. Salvin, S.B., E. Ribi, D.L. Granger, and J.S. Younger. 1975. Migration inhibitory factor and type II interferon in the circulation of mice sensitized with mycobacterial components. *J. Immun. 114*:354–359.

60. Stineberg, W.R., and P.M. Absher. 1970. Production of interferon following an immune response. *Ann. NY Acad. Sci. 173*:714–718.

61. Epstein, L.B., M.J. Cline, and T.C. Merigan. 1971. PPD stimulated interferon: in vitro macrophage lymphocyte interaction in the production of a mediator of cellular immunity. *Cell Immun. 2*:602–613.

62. Shibukawa, N., H. Hashimoto, A. Shibusawa, and Y. Kojima. 1977. Induction of interferon by stimulation with tuberculin in mice previously infected with *Mycobacterium tuberculosis* strain BCG. *Kitasato Arch. Exp. Med. 50*:23–29.

63. Nagano, Y., K. Mizunoe, N. Maehara, and Y. Kumazawa. 1971. Induction of virus inhibiting factor or interferon by cell fractions of *Mycobacterium tuberculosis. Jpn. J. Micro. 15*:542–544.

64. Nakane, A., and T. Minagawa. 1982. Induction of alpha and beta interferons during the hyporeactive state of gamma interferon by *Mycobacterium bovis* BCG cell wall fraction in *Mycobacterium bovis* BCG-sensitized mice. *Infect. Immun. 36*:966–970.

65. Winters, W.D., and S.C. Harris. 1982. Interferon induction in healthy and tumor-bearing dogs by cell walls of *Mycobacterium bovis* strain BCG. *Am. J. Vet. Res. 43*:1232–1237.

66. Hahn, H., and S.H.E. Kaufmann. 1981. The role of cell mediated immunity in bacterial infections. *Rev. Infect. Dis. 3*:1221–1250.

67. Kaufmann, S.H.E., and V. Brinkman. 1984. Attempts to characterize the T cell population and lymphokine involved in the activation of macrophage oxygen metabolism in murine listeriosis. *Cell Immun. 88*:545–550.

68. Rook, G.A.W., B.R. Champion, J. Steele, A.M. Varey, and J.L. Stanford. 1985. I-A restricted activation by T cell lines of antituberculosis activity in murine macrophages. *Clin. Exp. Immun. 59*:414–420.

69. Bloom, B.R., and T. Godal. 1983. Selective primary health care: strategies for control of disease in the developing world. V. Leprosy. *Rev. Infect. Dis. 5*:765–780.

70. Lowe, C., S.J. Brett, and R.J.W. Rees. 1985. Adoptive cell transfer of resistance to *Mycobacterium leprae* infections in mice. *Clin. Exp. Immun. 61*:336–342.

71. Kaufmann, S.H.E., and I. Flesch. 1986. Function and antigen recognition pattern of L3+4+ T cell clones from *Mycobacterium tuberculosis* immune mice. *Infect. Immun. 54*:291–296.

72. Huygen, K., and K. Palfliet. 1983. Strain variation in interferon γ production of BCG sensitised mice challenged with PPD. *Cell. Immun. 80*:329–334.

73. Shimokata, K., H. Kawachi, L.F. Kishmoto, F. Maeda, and Y. Ito. 1982. Local cellular immunity in tuberculous pleurisy. *Am. Rev. Resp. Dis. 126*:822–824.

74. Brett, S.J., and R. Butler. 1986. Resistance to *Mycobacterium lepraemurium* is correlated with the capacity to generate macrophage activating factor(s) in response to mycobacterial antigens in vitro. *Immun. 59*:339–345.

75. Sonnenfeld, G.A., D. Mandel, and T.C. Merigan. 1979. In vitro production and cellular origin of murine Type II interferon. *Immunology 36*:883–890.

76. Chang, T-W, D. Testa, P.C. Kung, L. Perry, H.J. Dreskin, and G. Goldstein. 1982. Cellular origin and interactions involved in γ interferon production induced by OKT3 monoclonal antibody. *J. Immun. 128*:585–589.

77. Matsuyama, M., K. Sugamura, Y. Kawade, and Y. Hinuma. 1982. Production of immune interferon by human cytotoxic T cell clones. *J. Immun. 129*:450–451.

78. Kasahara, T., J.J. Hooks, S.F. Dougherty, and J.J. Oppenheim. 1983. Interleukin 2-mediated immune interferon (IFNγ) production by human T cells and T cell subsets. *J. Immun. 130*:1784–1789.

79. Palacios, R., O. Martinez-Maza, M. DeLey. 1983. Production of human immune interferon (HuIFN-γ) studied at the single cell level. Origin, evidence for spontaneous secretion and effect of cyclosporin A. *Eur. J. Immun. 13*:221–225.

80. Shimokato, K., H. Kishimoto, E. Takagi, and H. Tsunekawa. 1986. Determination of the T cell subset producing γ interferon in tuberculous pleural effusion. *Microbiol. Immun. 30*:353–361.

81. Torres, B.A., J.K., and H.M. Johnson. 1982. Cellular regulation of gamma interferon production: Lyt phenotype of the supressor cell. *Infect. Immun. 35*:770–776.

82. Orme, I.M. 1987. The kinetics of emergence and loss of mediator T lymphocytes acquired in response to infection with *Mycobacterium tuberculosis. J. Immun. 138*:293-298.

83. Matthews, R., A. Scoging, and A.D.M. Rees. 1985. Mycobacterial antigen-specific human T-cell clones secreting macrophage activating factors. *Immunology 54*:17-23.

84. Mustafa, A.S., H.K. Gill, A. Nerland, W.J. Britton, V. Melira, B.R. Bloom, R.A. Young, and T. Godal. 1986. Human T cell clones recognize a major *M. leprae* protein antigen expressed in *E. coli. Nature (Lond.) 319*:63-66.

85. Ottenhoff, T.H.M., P.R. Klatser, J. Ivanyi, D.G. Elferink, M.Y.L. de Wit, and R.R.P. de Vries. 1986. *Mycobacterium leprae* specific protein antigens defined by cloned human helper T cells. *Nature 319*:66-68.

86. Mustafa, A.S., G. Kvalheim, M. Degre, and T. Godal. 1986. *Mycobacterium bovis* BCG induced human T cell clones from BCG vaccinated healthy subjects. Antigen specificity and lymphokine production. *Infect. Immun. 53*:491-497.

87. Andrew, P.W., A.D.W. Rees, A. Scoging, N. Dobson, R. Matthews, J.T. Whittall, A.R.M. Coates, and D.B. Lowrie. 1984. Secretion of a macrophage activating factor distinct from interferon by human T cell clones. *Eur. J. Immun. 14*:962-964.

88. Pierce, C., R.J. Dubos, and G. Middlebrook. 1947. Infection of mice with mammalian tubercle bacilli grown in tween-albumin liquid media. *J. Exp. Med. 86*:159-173.

89. Neta, R., and S.B. Salvin. 1980. In vivo release of lymphokines in different strains of mice. *Cell. Immun. 51*:173-178.

90. Kakinuma, M., K. Onoe, R. Yasumiza, and K. Yamamoto. 1983. Strain differences in lung granuloma formation in response to a BCG cell wall vaccine in mice, failure of antigen presentation by low responder macrophages. *Infect. Immun. 50*:423-431.

91. Nakamura, R.M., T. Tokunaga, and S. Yamamoto. 1980. Difference in antigen presenting ability of macrophages between high and low responder mice in delayed-type hypersensitivity to *Mycobacterium bovis* BCG. *Infect. Immun. 27*:268-170.

92. Lovik, M., and O. Closs. 1982. Repeated delayed-type hypersensitivity reactions against *Mycobacterium lepraemurium* antigens at the infection site do not effect bacillary multiplication in C3H mice. *Infect. Immun. 36*:768-774.

93. Gros, P., E. Skamene, and A. Forget. 1981. Genetic control of natural resistance to *Mycobacterium bovis* (BCG) in mice. *J. Immun. 127*:2417-2421.

94. Gros, P., E. Skamene, A. Forget, and B. Taylor. 1983. Host response to infection with *Mycobacterium bovis* (BCG) in mice: genetic study of natural resistance. *Adv. Exp. Med. Biol. 162*:183-188.

95. Brown, I.N., A.A. Glynn, and J. Plant. 1982. Inbred mouse strain resistance to *Mycobacterium lepraemurium* follows the Ity/hsh pattern. *Immun. 47*:149-156.

96. Skamene, E., P. Gros, A. Forget, P.J. Patel, and N. Nesbitt. 1984. Regulation of resistance to leprosy by chromosome 1 locus in the mouse. *Immunogenetics 19*:117-124.

97. Goto, Y., R.M. Nakamura, H. Takahashi, and T. Tokunaga. 1984. Genetic control of resistance to *Mycobacterium intracellulare* infection in mice. *Infect. Immun. 46*:135-140.

98. Orme, I.M., and F.M. Collins. 1984. Demonstration of acquired resistance in Bcgr inbred mouse strains infected with a low dose of BCG montreal. *Clin. Exp. Immun. 56*:81-88.

99. Alexander, J., and J. Curtis. 1979. Development of delayed hypersensitivity responses in *Mycobacterium lepraemurium* infections in resistant and susceptible strains of mice. *Immunology 36*:563-567.

100. Lefford, M.J., P.J. Patel, L.W. Poulter, and G.B. Mackness. 1977. Induction of cell mediated immunity to *Mycobacterium lepraemurium* in susceptible mice. *Infect. Immun. 18*:654-659.

101. Lagrange, P.H., and B. Hurtrel. 1979. The influence of BCG vaccination on murine leprosy in C57Bl/6 and C3H mice. *Ann. Immun. (Inst. Past.). 130C*:687-709.

102. Hoffenbach, A., P.H. Lagrange, and M-A, Bach. 1985. Strain variation of lymphokine production and specific antibodies in mice infected with *Mycobacterium lepraemurium. Cell. Immun. 91*:1-11.

103. Yamamoto, K., and M. Kakinumi. 1978. Genetic control of granuloma response to oil-associated BCG cell wall vaccine in mice. *Microbiol. Immun. 22*:335-348.

104. Adelman, N., S. Cohen, and T. Yoshida. 1978. Strain variation in murine MIF production. *J. Immun. 121*:209-212.

105. Virelizier, J-L. 1982. Murine genotype influences the in vitro production of γ (immune) interferon. *Eur. J. Immun. 12*:988-990.

106. Onwubalili, J.K., G.M. Scott, and J.A. Robinson. 1985. Deficient immune interferon production in tuberculosis. *Clin. Exp. Immun. 59*:405-413.

107. Vilcek, J., A. Khon, D. Henriksen-Destefano, A. Zemtsov, D.M. Davidson, M. Davidson, and A.E. Friedman-Kien. 1986. Defective γ interferon production in peripheral blood leukocytes of patients with acute tuberculosis. *J. Clin. Immun. 6*:146-151.

108. Schlossman, S.F., H.A. Levin, R.E. Rocklin, and J.R. David. 1971. The compartmentalization of antigen-reactive lymphocytes in desensitized guinea pigs. *J. Exp. Med. 134*:741-750.

109. Rook, G.A.W., J.M. Carswell, and J.L. Stanford. 1976. Preliminary evidence for the trapping of antigen specific lymphocytes in the lymphoid tissue of anergic tuberculosis. *Clin. Exp. Immun. 26*:129-132.

110. Ellner, J.J. 1978. Pleural fluid and peripheral blood lymphocyte function in tuberculosis. *Ann. Intern. Med. 89*:932-933.

111. Murray, H.W., and R.B. Roberts. 1986. T cell responses to mycobacterial antigen in AIDS patients with MAI infection. *Clin. Res. 34*:6771.

112. Drutz, D.J., M.J. Cline, and L. Levy. 1974. Leukocyte antimicrobial function in patients with leprosy. *J. Clin. Invest. 53*:380-386.

113. Horwitz, M.A., W.R. Levis, and Z.A. Cohn. 1984. Defective production of monocyte activating cytakines in lepromatous leprosy. *J. Exp. Med. 159*: 666-678.

114. Sharp, AK, and D.K. Banerjee. 1984. Macrophage activity in *Mycobacterium leprae* infection. *Acta Leprol. 11*:259-266.

115. Sharp, A.K., and D.K. Banerjee. 1985. Hydrogen peroxide and superoxide production by peripheral blood monocytes in leprosy. *Clin. Exp. Immun. 60*:203-206.

116. Noqueira, N., G. Kaplan, E. Levy, E.N. Samo, P. Kushner, A. Granelli-Piperno, L. Viera, V. Colomer Gould, W. Levis, R. Steinman, Y.K. Yip, and S.A. Cohn. 1983. Defective γ interferon production in leprosy. Reversal with antigen and interleukin 2. *J. Exp. Med. 158*:2165-2170.

117. Kantor, F.S. 1975. Infection, anergy and cell mediated immunity. *N. Engl. J. Med. 292*:629-634.

118. Ellner, J.J. 1978. Suppressor adherent cells in human tuberculosis. *J. Immun. 121*:2573-2579.

119. Katz, P., R.A. Goldstein, and A.S. Fauci. 1979. Immunoregulation in infection caused by *Mycobacterium tuberculosis*: the presence of suppressor monocytes and the alteration of subpopulation of T lymphocytes. *J. Infect. Dis. 140*:12-21.

120. Watson, S.R., and F.M. Collins. 1980. Development of suppressor T cells in mice heavily infected with mycobacteria. *Immunology 39*:367-373.

121. Mustafa, A.S., and T. Godal. 1985. BCG induced suppressor T cells. Optimal conditions for in vitro induction and mode of action. *Clin. Exp. Immun. 62*:474-481.

122. Mustafa, A.S., and T. Godal. 1983. In vitro induction of human supressor T cells by mycobacterial antigens. BCG activated OKT4+ cells mediate supression of antigen induced T cell proliferation. *Clin. Exp. Immun. 52*: 29-37.

123. Gerson, R.K., and K. Kondo. 1971. Infectious immunological tolerance. *Immunology 21*:903-914.

124. Bullock, W.E. 1976. Anergy and infection. *Adv. Intern. Med. 21*:149-173.

125. Bhatnager, R., A.N. Malaviya, S. Narayanan, P. Rajgopalan, R. Kumar, and O.P. Bharadwaj. 1977. Spectrum of immune response abnormalities in different clinical forms of tuberculosis. *Am. Rev. Resp. Dis. 115*: 207-212.

126. McCabe, R.E., B.J. Luft, and J.S. Remington. 1984. Effect of murine interferon gamma on murine toxoplasmosis. *J. Infect. Dis. 150*:961-967.

127. Kiderlen, A.F., S.H.E. Kaufmann, and M.L. Lohmann-Matthes. 1984. Protection of mice against the intracellular bacterium *Listeria monocytogenes* by recombinant immune interferon. *Eur. J. Immun. 14*:964-967.

128. Edwards, C.K., H.B. Hedegaard, A. Zlotnik, P.R. Gangadharam, R.B. Johnston, and M.J. Pabst. 1986. Chronic infection due to *Mycobacterium intracellulare* in mice associated with macrophage release of prostaglandin E2 and reversed by injection of indomethacin, muramyl dipeptide or interferon γ. *J. Immun. 136*:1820-1827.

129. Khor, M., D.B. Lowrie, A.R.M. Coates, and D.A. Mitchison. 1986. Recombinant interferon-gamma and chemotherapy with isoniazid and rifampicin in experimental murine tuberculosis. *Br. J. Exp. Pathol. 67*: 587–596.

130. Banerjee, D.K., A.K. Sharp, and D.B. Lowrie. 1986. The effect of gamma-interferon during Mycobacterium bovis (BCG) infection in athymic and euthymic mice. *Micro Pathol. 1*:221–224.

131. Ranges, G.E., G. Goldstein, E.A. Boyse, M.P. Schield. 1982. T cell development in normal and thymopentin treated nude mice. *J. Exp. Med. 156*:1057–1064.

132. Lancaster, R.D., G.R.F. Hilson, A.C. McDougall, and M.J. Colston. 1983. *Mycobacterium leprae* infection in nude mice-bacteriological and histological responses to primary infection and large inocula. *Infect. Immun. 39*:865–872.

133. Lefford, M.J. 1985. *Mycobacterium lepraemurium* infection of nude athymic (nu/nu) mice. *Infect. Immun. 49*:190–196.

134. Nomaguchi, H., K. Kohsaka, Y. Mujata, T. Morn, T. Ito, K. Nagata, H. Okamura, and K. Shoji. 1985. Induction of γ interferon with various mitogens in mice pretreated with *Mycobacterium lepraemurium*. *Int. J. Lep. 53*:706–707.

135. Nathan, C.F., G. Kaplan, W.R. Levis, A. Nusrat, M.D. Witman, S.A. Sherwin, C.K. Job, C.R. Horowitz, R.M. Steinman, and Z.A. Cohn. 1986. Local and systemic effects of intradermal recombinant interferon γ in patients with lepromatous leprosy. *N. Engl. J. Med. 315*:6–15.

136. Lowrie, D.B. 1983. How do macrophages kill tubercle bacilli? *J. Med. Micro. 16*:1–12.

137. Lowrie, D.B. 1983. The macrophage and mycobacterial infections. *Trans. Roy. Soc. Trop. Med. Hyg. 77*:646–655.

138. Lowrie, D.B. 1983. Mononuclear phagocyte-mycobacterium interaction. In *Biology of the Mycobacteria*, Vol. 2. Edited by J. Stanford, and C. Ratledge. Academic Press, London, pp. 235–278.

139. Nathan, C.F., H.W. Murray, M.E. Wiebe, and B.Y. Rubin. 1983. Identification of interferon γ as the lymphokine that activates human macrophage oxidative metabolism and antimicrobial activity. *J. Exp. Med. 158*:670–689.

140. Murray, H.W., G.L. Spialny, and C.F. Nathan. 1985. Activation of mouse peritoneal macrophages in vitro and in vivo by interferon γ. *J. Immun. 134*:1619–1622.

141. Kaplan, G., C.F. Nathan, R. Gandhi, M.A. Horwitz, W.R. Levis, and Z.A. Cohn. 1986. Effect of recombinant interferon gamma on hydrogen peroxide releasing capacity of monocyte derived macrophages from patients with lepromatous leprosy. *J. Immun. 137*:983–987.

142. Sibley, L.D., S.G. Franzblau, and J.L. Krahenbuhl. 1987. Intracellular fate of *Mycobacterium leprae* in normal and activated mouse macrophages. *Infect. Immun. 55*:680–685.

143. Khor, M., D.B. Lowrie, and D.A. Mitchison. 1986. Effects of recombinant interferon and chemotherapy with isoniazid and rifampicin on infections of mouse peritoneal macrophages with *Listeria monocytogenes* and *Mycobacterium microti* in vitro. *Br. J. Exp. Pathol. 67*:707–717.

144. Rook, G.A.W., and S.R. Rainbow. 1981. An isotope incorporation assay for the antimycobacterial effects of human monocytes. *Ann. Immun. (Inst. Past.) 132D*:281–289.

145. Rook, G.A.W., J. Steele, M. Ainsworth, and B.R. Champion. 1986. Activation of macrophages to inhibit proliferation of *Mycobacterium tuberculosis*—comparison of the effects of recombinant γ interferon on human monocytes and murine peritoneal macrophages. *Immunology 59*:333–338.

146. Steele, J., K.C. Flint, A.L. Pozniak, B. Hudspith, M. Mcl-Johnson, and G.A.W. Rook. 1986. Inhibition of virulent *Mycobacterium tuberculosis* by murine peritoneal macrophages and human alveolar lavage cells: the effects of lymphokines and recombinant-γ interferon. *Tubercle 67*:289–294.

147. Rook, G.A.W., J. Steele, L. Fraher, S. Barker, R. Karmali, J. O'Riordan, and J. Stanford. 1986. Vitamin D$_3$, gamma interferon and control of proliferation of *Mycobacterium tuberculosis* by human monocytes. *Immunology 57*:159–163.

148. Douvas, G.S., D.L. Looker, A.E. Vatter, and A.J. Crowle. 1985. Gamma interferon activates human macrophages to become tumoricidal and leishmanicidal but enhances replication of macrophage associated mycobacteria. *Infect. Immun. 50*:1–8.

149. Murray, H.W., B.Y. Rubin, S.M. Carriero, A.M. Harris, and E.A. Jaffee. 1985. Human mononuclear phagocyte antiprotozoal mechanisms: oxygen-dependent vs oxygen-independent activity against intracellular *Toxoplasma gondii. J. Immun. 134*:1982–1988.

150. Milon, G., M. Lebastard, and G. Marchal. 1985. T dependent production and activation of mononuclear phagocytes during murine BCG infection. *Immun. Lett. 11*:189–194.

151. Meltzer, M.S., W.R. Benjamin, and J.J. Farrar. 1982. Macrophage activation for tumour cytotoxicity: induction of macrophage tumouricidal activity by lymphokines from EL4, a continuous T cell line. *J. Immun. 129*:2802–2807.

152. Gemsa, D., K.M. Debatin, W. Kramer, C. Kubellea, W. Deimann, V. Kees, and P.H. Krammer. 1982. Macrophage activating factors from different T cell clones induce distinct macrophage functions. *J. Immun. 131*:833–844.

153. Nacy, C.A., S.L. James, W.R. Benjamin, J.J. Farrar, W.T. Hockmeyer, and M.S. Meltzer. 1983. Activation of macrophages for microbicidal and tumoricidal effector functions by soluble factors from EL4, a continuous T cell line. *Infect. Immun. 40*:820–824.

154. Miyamoto, D., N. Nakamura, Y. Ishii, Y. Kobayashi, and T. Osawa. 1987.

Establishment of a human T-cell hybridoma that produces human macrophage activating factor for superoxide production and translation of messanger RNA of the factor in *Xenopus laevis* oocyte. *Mol. Immun. 24*: 239–245.

155. Hoover, D.L., D.S. Finbloom, R.M. Crawford, C.A. Nacy, M. Gilbreath, and M.S. Meltzer. 1986. A lymphokine distinct from interferon γ that activates human monocytes to kill *Leishmania donovani* in vitro. *J. Immun. 136*:1329–1333.

156. Lee, J.C., L. Rebar, P. Young, F.W. Ruscetti, N. Hanna, and G. Poste. 1986. Identification and characterisation of a human T cell line-derived lymphokine with MAF-like activity distinct from interferon γ. *J. Immun. 136*:1322–1328.

157. Walker, L., P.W. Andrew, and D.B. Lowrie. 1984. Mycobactericidal activation of mouse peritoneal macrophages by maintenance in a serum free medium. *Eur. J. Clin. Invest. 14*:34.

158. Davies, P.D.O. 1985. A possible link between vitamin D deficiency and improved host defence to *Mycobacterium tuberculosis. Tubercle 66*: 301–306.

159. Grange, J.M., P.D.O. Davies, R.C. Brown, J.S. Woodhead, and T. Kardjito. 1985. A study of vitamin D levels in Indonesian patients with untreated pulmonary tuberculosis. *Tubercle 66*:187–191.

15

Legionella pneumophila

Marcus A. Horwitz / Center for the Health Sciences, UCLA School of Medicine, Los Angeles, California

I. INTRODUCTION

Legionella pneumophila is a gram-negative bacterial pathogen and the causative agent of Legionnaires' disease (1). The organism is a facultative intracellular parasite that multiplies in human monocytes and alveolar macrophages (2,3). Under tissue culture conditions, and presumably in vivo, multiplication is exclusively intracellular (2).

The importance of interferon gamma (IFNγ) to the immunobiology of *L. pneumophila* derives from its capacity to activate the mononuclear phagocyte, the effector arm of cell-mediated immunity. In view of this, in this chapter I shall first discuss the role of cell-mediated immunity in host defense against Legionnaires' disease and the influence of monocyte activation on intracellular *L. pneumophila*. I shall then review our studies on the influence of IFNγ on *L. pneumophila*—mononuclear phagocyte interaction. Finally, I shall summarize studies on the combined effects of IFNγ and antibiotics on intracellular *L. pneumophila*, studies which have important implications for the therapeutic use of IFNγ.

II. ROLE OF CELL-MEDIATED IMMUNITY IN LEGIONNAIRES' DISEASE

Patients with Legionnaires' disease respond to the infection with both humoral and cell-mediated immune responses. Humoral immunity appears to play a minor role in host defense. This conclusion is based on an in vitro analysis that revealed that antibody fails to serve any of three important functions for the host (4,5). First, antibody fails to promote complement killing of *L. pneumophila* (4). Second, antibody fails to promote effective killing of *L. pneumophila* by phagocytes. In the presence of complement, antibody promotes phagocytosis of *L. pneumophila* by human polymorphonuclear leukocytes, monocytes, and alveolar macrophages, but such phagocytes kill only a modest proportion of intracellular bacteria. Third, antibody fails to inhibit intracellular multiplication. Antibody- and complement-coated *L. pneumophila* multiply intracellularly in human monocytes and at the same rate as *L. pneumophila* that enter monocytes in the absence of specific antibody.

In contrast, cell-mediated immunity appears to play a major role in host defense against *L. pneumophila* (6,7). Patients with Legionnaires' disease expand a pool of lymphocytes that recognize *L. pneumophila* antigens. When incubated with *L. pneumophila* antigens, these lymphocytes proliferate and produce monocyte-activating cytokines, that is cytokines with the capacity to activate normal monocytes such that they inhibit *L. pneumophila* intracellular multiplication (7). Cytokines from either antigen- or mitogen-induced supernates of mononuclear cell cultures so activate human monocytes (6,7). Mitogen-induced cytokines also activate human alveolar macrophages such that these mononuclear phagocytes inhibit *L. pneumophila* intracellular multiplication (3).

III. MECHANISMS BY WHICH ACTIVATED MONOCYTES INHIBIT *L. PNEUMOPHILA* INTRACELLULAR MULTIPLICATION

Activated monocytes inhibit *L. pneumophila* intracellular multiplication in two ways. First, they phagocytize approximately 50% fewer *L. pneumophila* than nonactivated monocytes, thereby restricting access of the bacteria to the intracellular milieu they require to multiply (6). This may occur as a result of down regulation of complement receptors that mediate phagocytosis of *L. pneumophila* (8). Second, activated monocytes dramatically slow the multiplication rate of intracellular bacteria, prolonging the doubling time by more than threefold (6). This may occur as the result of restricted access of the bacteria to intracellular iron upon which the bacteria are dependent for growth (9).

Activation of monocytes appears to have little influence on several features of phagosome biology. In both activated and nonactivated monocytes (10–13), *L. pneumophila* forms a specialized ribosome-lined phagosome, inhibits phagosome-lysosome fusion (although to a somewhat lesser degree in activated than nonactivated monocytes), and inhibits phagosome acidification (10–13).

IV. INTERACTION BETWEEN IFNγ-ACTIVATED MONOCYTES AND *L. PNEUMOPHILA*

Human monocytes activated by recombinant human IFNγ inhibit *L. pneumophila* intracellular multiplication (13). The effect of IFNγ on monocytes is dose

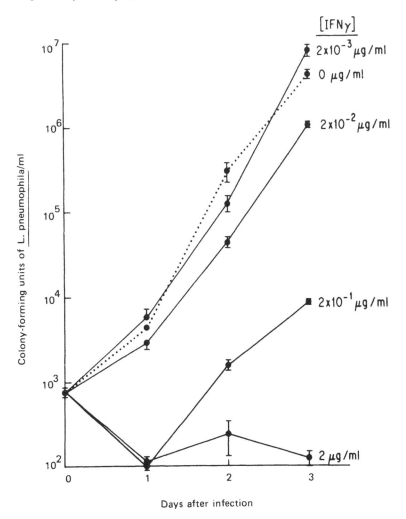

Figure 1 Monocytes activated with human recombinant IFNγ inhibit *L. pneumophila* intracellular multiplication. Freshly explanted monocytes in monolayer culture were incubated with various concentrations of IFNγ ranging from 0 to 2 μg/ml for 24 hr. *L. pneumophila* (10^3 CFU/ml) were then added to the cultures, and the cultures were assayed daily for colony-forming units of *L. pneumophila*. Each point represents the average of triplicate culture wells ± SE. (From Ref. 13.)

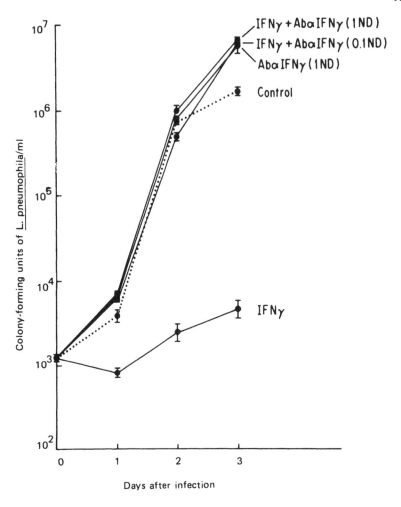

Figure 2 Monoclonal anti-IFNγ antibody neutralizes the capacity of IFNγ to activate monocytes such that they inhibit *L. pneumophila* intracellular multiplication. Monocytes were incubated for 24 hr with IFNγ alone, IFNγ plus anti-IFNγ antibody (2 concentrations), anti-IFNγ antibody alone, or control medium, as indicated, infected, and assayed for colony-forming units of *L. pneumophila* as in Figure 1. The concentration of IFNγ was 1 μg/ml (1.1 × 10⁴ antiviral units/ml). The concentrations of antibody were 1.1 × 10⁴ viral-neutralizing units/ml, defined as 1 neutralizing dose (1 ND), or 1.1 × 10³ viral neutralizing units/ml (0.1 ND). (From Ref. 13.)

dependent (Fig. 1); maximal inhibition of *L. pneumophila* multiplication consistently occurs with 2 μg/ml IFNγ (2.2 $\times$ 10^4 antiviral units/ml), although in some experiments maximal inhibition occurs at concentrations of IFNγ 10- or 100-fold less. The capacity of IFNγ to activate human monocytes in this way is completely neutralized by monoclonal antibody against IFNγ (Fig. 2).

The IFNγ-activated monocytes are larger, more spread out, and more aggregated than nonactivated monocytes (13), as are monocytes activated by mitogen-induced supernates (6).

The length of time monocytes are kept in culture is an important determinant of the kinetics of activation. In monocytes infected with *L. pneumophila* at 1 day after explantation, inhibition of bacterial multiplication is independent of the length of time monocytes are preincubated with IFNγ before infection; monocytes treated with IFNγ 0 or 24 hours before infection are comparably inhibitory. However, in monocytes infected with *L. pneumophila* two days after explantation, inhibition of bacterial multiplication is dependent on preincubation time with IFNγ; monocytes treated with IFNγ for 12–48 hr before infection are more inhibitory than monocytes treated 0 hr before infection.

Monocytes treated with IFNγ after infection with *L. pneumophila* also inhibit intracellular multiplication, provided the treatment is administered sufficiently soon. In monocytes infected one day after explantation, treatment at 0, 1, or 2 hr after infection results in maximal inhibition, treatment at 6 or 12 hr after infection results in partial inhibition, and treatment at 24 hr after infection results in little or no inhibition of *L. pneumophila* intracellular multiplication (13).

Interferon γ does not have to be present in the cultures for monocytes to inhibit *L. pneumophila* intracellular multiplication. Monocytes treated with IFNγ, washed, and infected with *L. pneumophila* in the absence of IFNγ, inhibit *L. pneumophila* multiplication as strongly as monocytes treated with IFNγ both before and after infection (13).

Monocytes require only a brief exposure to IFNγ to become maximally activated. Monocytes exposed to IFNγ for only 1 hr before infection appear morphologically activated within 24 hr, and these monocytes maximally inhibit *L. pneumophila* intracellular multiplication (13).

V. ACTIVATED MONOCYTES FAIL TO KILL INTRACELLULAR *L. PNEUMOPHILA*

Monocytes activated with IFNγ or with supernates of concanavalin A-stimulated mononuclear cell cultures inhibit *L. pneumophila* multiplication, but do not kill the intracellular bacteria (6,13). In the presence of anti-*L. pneumophila* antibody, both activated and nonactivated monocytes kill a small proportion of an inoculum of *L. pneumophila*, but activated monocytes do not kill more than nonactivated monocytes (6,13).

VI. INTERACTION BETWEEN IFNγ-ACTIVATED ALVEOLAR MACROPHAGES AND *L. PNEUMOPHILA*

Alveolar macrophages activated with IFNγ also strongly inhibit *L. pneumophila* intracellular multiplication (15). Inhibition is dose dependent (2 > 0.2 > 0.02 μg/ml IFNγ), although differences in inhibitory capacity among alveolar macrophages treated with 0.02 to 2 μg/ml IFNγ (220 to 22,000 antiviral units/ml) are small. Inhibition is also time dependent, that is the degree of alveolar macrophage inhibition of *L. pneumophila* intracellular multiplication is proportional to pretreatment time with IFNγ (96 > 72 > 48 > 24 hr for macrophages infected at 96 hr after explantation, and 48 > 24 hr for macrophages infected at 48 hr after explantation).

VII. COMBINED INFLUENCE OF IFNγ AND ANTIBIOTICS ON INTRACELLULAR *L. PNEUMOPHILA*

As noted above, IFNγ-activated monocytes inhibit *L. pneumophila* intracellular multiplication, but do not kill these bacteria. Similarly, the antibiotics erythromycin, rifampin, and clindamycin, when added to infected monocytes, inhibit multiplication of intracellular *L. pneumophila*, but do not kill the bacteria (14,16). In contrast to IFNγ, which exerts its antibacterial effect indirectly by activating the monocytes, the antibiotics act directly on the bacteria by inhibiting protein synthesis (erythromycin and clindamycin) or RNA synthesis (rifampin). Since neither IFNγ nor antibiotics alone result in intracellular killing of *L. pneumophila*, their combined effects were studied to determine if the two agents would act synergistically to kill intracellular *L. pneumophila*.

Interferon γ and antibiotics together do not kill intracellular *L. pneumophila* regardless of the sequence with which they are added to infected monocytes (16). That is IFNγ activation of monocytes containing antibiotic-inhibited *L. pneumophila* does not result in killing of the bacteria, nor does antibiotic treatment of IFNγ-activated monocytes containing *L. pneumophila*. These results suggest that the combination of IFNγ and antibiotics may offer no therapeutic advantage over antibiotics alone in the treatment of Legionnaires' disease and perhaps other diseases caused by intracellular pathogens, for example, leprosy and tuberculosis.

VIII. CONCLUSIONS

Cell-mediated immunity plays a major role in host defense against *L. pneumophila*. The importance of IFNγ to the immunobiology of *L. pneumophila* derives from its capacity to activate the mononuclear phagocyte, the effector arm of cell-mediated immunity. IFNγ-activated human monocytes and alveolar macrophages inhibit the intracellular multiplication of *L. pneumophila*. Activation is dose and time dependent. Although activated monocytes strongly inhibit multiplication, they do not kill intracellular *L. pneumophila*. Treatment of monocytes with IFNγ and antibiotics in combination also does not result in killing of intracellular *L. pneumophila*.

ACKNOWLEDGMENTS

Dr. Horwitz is Gordon MacDonald Scholar at UCLA and recipient of a Faculty Research Award from the American Cancer Society. This work is supported by Grant AI 22421 from the National Institutes of Health.

REFERENCES

1. Fraser, D.W., T.R. Tsai, W. Orenstein, W.E. Parkin, H.J. Beecham, R.G. Sharrar, J. Harris, G.F. Mallison, S.M. Martin, J.E. McDade, C.C. Shepard, P.S. Brachman, and the Field Investigation Team. 1977. Legionnaires' disease. Description of an epidemic of pneumonia. *N. Engl. J. Med. 297*: 1189–1197.

2. Horwitz, M.A., and S.C. Silverstein. 1980. The Legionnaires' disease bacterium (*Legionella pneumophila*) multiplies intracellulary in human monocytes. *J. Clin. Invest. 66*:441–450.

3. Nash, T.W., D.M. Libby, and M.A. Horwitz. 1984. Interaction between the Legionnaires' disease bacterium (*Legionella pneumophila*) and human alveolar macrophages. Influence of antibody, lymphokines, and hydrocortisone. *J. Clin. Invest. 74*:771–782.

4. Horwitz, M.A., and S.C. Silverstein. 1981. Interaction of the Legionnaires' disease bacterium (*Legionella pneumophila*) with human phagocytes. I. *L. pneumophila* resists killing by polymorphonuclear leukocytes, antibody, and complement. *J. Exp. Med. 153*:386–397.

5. Horwitz, M.A., and S.C. Silverstein. 1981. Interaction of the Legionnaires' disease bacterium (*Legionella pneumophila*) with human phagocytes. II. Antibody promotes binding of *L. pneumophila* to monocytes but does not inhibit intracellular multiplication. *J. Exp. Med. 153*:398–406.

6. Horwitz, M.A., and S.C. Silverstein. 1981. Activated human monocytes inhibit the intracellular multiplication of Legionnaires' disease bacteria. *J. Exp. Med. 154*:1618–1635.

7. Horwitz, M.A. 1983. Cell-mediated immunity in Legionnaires' disease. *J. Clin. Invest. 71*:1686–1697.

8. Payne, N.R., and M.A. Horwitz. 1987. Phagocytosis of *Legionella pneumophila* is mediated by human monocyte complement receptors. *J. Exp. Med. 166*:1377–1389.

9. Byrd, T.F., and M.A. Horwitz. 1987. Intracellular multiplication of *Legionella pneumophila* in human monocytes is iron-dependent and the capacity of activated monocytes to inhibit intracellular multiplication is reversed by iron-transferrin. *Clin. Res. 35*:613A.

10. Horwitz, M.A. 1983. Formation of a novel phagosome by the Legionnaires' disease bacterium (*Legionella pneumophila*) in human monocytes. *J. Exp. Med. 158*:1319-1331.

11. Horwitz, M.A. 1983. The Legionnaires' disease bacterium (*Legionella pneumophila*) inhibits phagosome-lysosome fusion in human monocytes. *J. Exp. Med. 158*:2108-2126.

12. Horwitz, M.A., and F.R. Maxfield. 1984. *Legionella pneumophila* inhibits acidification of its phagosome in human monocytes. *J. Cell. Biol. 99*: 1936-1943.

13. Bhardwaj, N., T. Nash, and M.A. Horwitz. 1986. Gamma interferon-activated human monocytes inhibit the intracellular multiplication of *Legionella pneumophila. J. Immunol. 137*:2662-2669.

14. Horwitz, M.A., and S.C. Silverstein. 1983. The intracellular multiplication of Legionnaires' disease bacteria (*Legionella pneumophila*) in human monocytes is reverisbly inhibited by erythromycin and rifampin. *J. Clin. Invest. 71*:15-26.

15. Nash, T.W., D.M. Libby, and M.A. Horwitz. 1988. Interferon gamma-activated human alveolar macrophages inhibit the intracellular multiplication of *Legionella pneumophila. J. Immunol.* (in press).

16. Bhardwaj, N., and M.A. Horwitz. 1988. Gamma interferon and antibiotics fail to act synergistically to kill Legionella pneumophila in human monocytes. *J. Interferon Res.* (in press).

16

Interferon Effects on Infection with Enteroinvasive Bacteria

Miklos Degré and Geir Bukholm / The Wilhelmsens Institute of Bacteriology, University of Oslo, National Hospital, Oslo, Norway

I. INTRODUCTION

Interferons have been known for more than 30 years, especially for their antiviral activities. It is also generally accepted that interferons play an important role in the host defense against viral infections. It is now well established that in addition to their antiviral activities interferons exert a wide variety of effects on physiological functions of normal cells. These effects include general functions, for example, effect on cell multiplication and on functions of specialized cells such as production of enzymes. Several such cellular functions are involved in the complex host defense against infectious agents. Nevertheless it is still not generally accepted that interferons may play a role in the host defense against bacterial infections. We have been especially interested in interactions between different types of infectious agents, and the mechanism(s) of such interactions. We believe that there are clear indications that at least part of the interaction is mediated by interferons and that interferons do influence host defense mechanisms against bacterial infections. In addition, recent data indicate that interferons can influence the direct interaction between bacteria and cells and thereby influence the development of the bacterial infection in the host. In this chapter we present data on bacterial infections and interferons, with special emphasis on the direct interaction between cells and enterobacteria.

II. PATHOGENESIS OF INFECTION WITH ENTEROINVASIVE BACTERIA

The ability of bacteria to penetrate the epithelial lining of the intestinal mucosa by invading the epithelial cells is an important mechanism for enteric disease. Several bacteria have the ability of establishing a partly intracellular, partly extracellular state of infection. For these bacteria the intracellular phase is essential for their survival in the host. Bacteria that give rise to infections in humans and that enter their host via the human gut have developed different ways of competing with the existing bacterial colonization. Some of them, such as *Vibrio cholerae* and the enterotoxigenic *Escherichia coli* (ETEC) colonize the surface of the gut lining, others such as *Shigella*, enteroinvasive *E. coli*, *Salmonella*, and *Yersinia enterocolitica* enter the epithelial cells and thereby avoid an extracellular competition with the other bacteria. This process of entering the epithelial cells is referred to, in this chapter, as bacterial invasion.

We know that bacillary dysentery is caused by destruction of the epithelial lining of the mucosa of the large bowel after invasion by bacteria belonging to the *Shigella* genus (1). A similar disease is also caused by the enteroinvasive *E. coli* (EIEC) (2). *Salmonella* bacteria seem to have chosen their target somewhat more proximal in the bowel than the *Shigella*. *Salmonella* bacteria penetrate the epithelium of the small intestine and multiply intracellularly in the enteric mucosa. The destruction of the small intestinal epithelium is not so prominent as for *Shigella* and the result is often a watery diarrhea with little or no blood loss (2).

The more virulent salmonella types which produce typhoid or paratyphoid fever very often cause a less prominent enteric disease though it is clear that the bacteria are introduced through the intestines and penetrate the epithelial mucosa by invading the cells there (2). However, the bacteria do not multiply in the intestine, but pass into the mesenteric glands, to the bloodstream, and further on to tissues of the liver and spleen. After replication in these organs a secondary bacteremic phase results in a generalized spread of bacteria throughout the body and a secondary invasion of the intestines takes place.

Yersinia enterocolitica is also thought to produce a disease in humans by an enteroinvasive mechanism. The serotype 0:8 strains seem to be especially potent invaders of epithelial tissues (3).

Both bacterial and host cellular factors are of major importance for development of disease. This phenomenon has been clearly illustrated in several model systems where the different phases of bacterial invasion can be studied in detail. At least two models have been widely used for the study of bacterial invasiveness. An in vitro model was described by Serény (4). The ability of producing keratoconjunctivitis in the eyes of guinea pigs correlated well with the ability of *Shigella* bacteria and invasive *E. coli* to produce enteric disease in humans. The

Salmonella bacteria do not give a positive Serény test unless the eyes of the animal are pretreated with bile salts. Also *Y. enterocolitica* serotype 0:8 invades the guinea pig eye, but the serotypes 0:3 and 0:9 do not (3). Other in vivo models, such as infection in primates and the "starved guinea pig test" have been performed but none of them have received any wide application (5,6).

In 1961 Gerber and Watkins (7) established an in vitro model for the study of the invasive potential of dysenteria bacteria. A correlation between the ability to produce disease in man and the ability to invade and survive inside epithelial-derived cell monolayer was demonstrated. Later reports have confirmed this correlation and similar correlations have been found for *Salmonella* bacteria and for enteroinvasive *E. coli* as well. The most used cell lines to demonstrate this invasive potential of enteric pathogens are the HeLa and the HEp-2 cells. They are considered to be about equally sensitive to invasiveness by the intestinal bacteria (8).

A major problem in connection with in vitro invasiveness studies is to differentiate between intracellular and extracellular bacteria. It is often difficult if not impossible to differentiate between bacteria attached to the surface and those internalized into the cells with conventional light microscopy. Alternative techniques employ electron microscopy (9) and radioactive-labelled bacteria (10). These methods are time consuming and expensive. We have developed a light microscopical reading system (11) combining two microscopical methods on the same microscope: ultraviolet (UV) incident light microscopy and Nomarski differential interference contrast microscopy. The narrow focus which characterizes the Nomarski method allows determination of the bacterial location in the cells, intra- or extracellularly. When parallel samples were examined with the Nomarski method and scanning electron microscopy no significant differences were found in the distribution pattern of bacteria (12). Also other methods for discrimination between extra- and intracellular location have been applied: cell cultures have been treated with aminoglycosides (kanamycin or gentamicin) to kill extracellular bacteria and intracellular bacteria are quantitated by colony count from the cellular lysate (13). Also the ability of gamma globulins to remain in live cell cultures or glutaraldehyde-fixed cells extracellularly, but to penetrate the cell membrane of acetone-fixed cells has been used (14).

Hale and Bonventre (13) showed that heating of a *Shigella flexneri* culture to 56°C for short periods of time resulted in loss of infectivity for the cells. Infectivity was also reduced after UV irradiation and after treatment with kanamycin. Also Kaplan and collaborators (15) showed that UV and short-term gentamicin treatment of *S. typhimurium* abolished the infectivity of the bacterium for HeLa cells.

The mechanisms of bacterial invasiveness became further characterized when the dependence of extrachromosomal genetic information was discovered. The ability of *Shigella* bacteria and enteroinvasive *E. coli* to be invasive in cell cultures and in human gut mucosa seems to be dependent on the presence of a high-molecular weight plasmid. The phenotypic expression of this plasmid genetic information seems to be certain polypeptides, some of which is thought to be necessary for invasion. These polypeptides are probably binding to specific host cell receptors, inducing a receptor-mediated endocytosis of the bacterium. The genetic information on the plasmid seems to be sufficient to induce endocytosis in in vitro cell cultures, and invasive potential can be transferred from one bacterium to another by transferring the plasmid (16). However, chromosomal sequences are also necessary to express virulence in vivo. If only the plasmid is transferred the bacterium is still avirulent and therefore unable to produce a positive Serény test. If, however, a sequence of the chromosomal DNA is transferred to the new bacterium, Serény test positivity and in vivo virulence are expressed (16).

Virulence-associated plasmids are described among *Salmonella* bacteria. For *S. typhimurium* the plasmid size is about 62,000 dalton (17), for *S. enteritidis* the size is 36,000 dalton (18). The exact role of these plasmids is not certain. Some reports claim that the presence of these plasmids increases invasiveness and adhesiveness of the bacteria in cell cultures; also virulence for mice seem to be enhanced (17). On the other hand, the bacteria are still invasive after loss of the plasmid, so invasiveness genes are probably located to the chromosomes.

Both clinical observations and experimental studies show that the state of the host and the host cell is essential for the result of an infection with an invading organism. Most studies of host cell-dependent factors for enteroinvasive bacteria have been performed on cell culture models. Several groups have shown that bacteria enter the cells by means of endocytosis. If cell cultures are pretreated with cytochalasin or dihydrocytochalasin B, the endocytosis apparatus of the cells is blocked (19,20). In such cell cultures no invasiveness of *Shigella* or *Salmonella* bacteria can be demonstrated. On the other hand it seems as *Y. enterocolitica* is able to invade cells that are cytochalasin B treated (19).

Many experiments have shown that preinfections with other infectious agents, especially with virus, can enhance the effect of a later bacterial infection. Not many of these have been concerned with enteric organisms, but Tzipori and collaborators reported in 1982 that foals infected with rotavirus and enterotoxigenic *E. coli* developed a more serious disease than foals that were given either of the two agents alone (21). These results have been reproduced in an in vitro cell culture model in our laboratory. Human rotavirus-infected cells were more susceptible to infection by *S. typhimurium* than control cells. In in vitro models it was shown that infection of cell cultures with coxsackie B-1 virus or measles virus enhanced the invasion of *S. typhimurium* or *S. flexneri* (22–24).

III. INTERFERON EFFECT ON INFECTIONS WITH ENTEROINVASIVE BACTERIA IN ANIMAL MODELS

Salmonella typhimurium induces a disease in mice that is very similar to the disease caused by *S. typhi* in humans. When administered orally, the bacterium penetrates the intestinal mucosa, and spreads via blood to liver and spleen. Both the potential of invasiveness in enteric mucosa and the ability of the bacterium to hide inside cells of the body to be protected against antibacterial defense factors are important for development of disease.

Izadkhah and collaborators reported in 1980 (25) that when *S. typhimurium* were injected intraperitoneally into mice, the lethality was much lower in the mice that had been treated with interferons than in control mice. In 1984 Bukholm and collaborators (26) showed that *S. typhimurium* given by oral route via a gastric tube were capable of killing normal infant mice. Ninety minutes after administration of bacteria they could be demonstrated inside cells of the intestinal mucosa. As early as 8 hours after administration some of the animals succumbed from the infection, and after 24 hours most of the animals were dead. Most of the animals that survived for 24 hours recovered. When mice were treated with interferons, the number of bacteria per intestinal cell was significantly reduced and development of disease was delayed. In a group of 17 mice, 4 survived the infection in control groups, but as many as 15 survived after interferon treatment (Table 1). Also low concentrations of IFNγ protected mice against subsequent challenge with *S. typhimurium* (69).

The mechanisms behind this interferon effect is not clear. Several interferon effects probably work in concert in an animal, both general activities such as effect on the immune apparatus and effect on phagocytosis, and local activities at the place of invasion, namely effect on internalization of bacteria. In the following sections we will discuss interferon activities that may alter the course of bacterial infections.

Table 1 Effect of Mouse Fibroblast Interferon on Mortality of Infant Mice After Challenge With *Salmonella typhimurium*

Interferon dose (U/animal)	Time after bacterial inoculation (no. of live mice/total no. of mice)		
	12 hr	18 hr	22 hr
0	13/17	7/17	4/17
500	16/17	16/17	15/17

Source: From Ref. 26.

IV. PRESENCE OF INTERFERONS DURING BACTERIAL INFECTION

A. Interferon Induction by Preceding Viral Infections

It is an old clinical observation that viral infections, especially in the respiratory tract, are frequently followed by bacterial superinfections. The first description of *Hemophilus influenzae* by Pfeiffer as a possible cause of influenza, and the large numbers of bacterial complications during the "Spanish flu" are just two representative examples for this phenomenon. The viral agents often directly affect important antibacterial host defense factors, for example ciliary activity in the trachea and bronchi (27), or the phagocytic activity of alveolar macrophages (28,29). In addition most of these viral infections induce production of interferons. The interferons are usually present at the time of the secondary bacterial invasion. The main bulk of virally induced interferons are of type I, α and β.

B. Interferon Induction by Bacteria

Interferon can be present in the body fluids during natural infection with bacterial agents. Thus several authors demonstrated interferon activity in the cerebrospinal fluid during bacterial meningitis (30,31). Howie and co-workers (32) reported interferon activity in the middle ear fluids obtained from patients with acute bacterial otitis media. The activities were characterized as leukocyte-type interferons. The bacteria isolated simultaneously included both gram-negative rods as *H. influenzae* and Enterobacteriaceae and gram-positive cocci, as *Streptococcus pneumoniae, S. pyogenes*, and *Staphylococcus aureus.*

A wide range of bacteria, both gram positive and negative, can induce interferon production in several experimental animals. Both viable and inactivated bacteria induced significant titers of interferons when injected into mice (33–36). The bacteria include gram-positive cocci as streptococci and staphylococci, gram-positive rods as *Corynebacterium sp.* and *Bacillus subtilis*, and gram-negative rods as *E. coli, Klebsiella sp. Salmonella sp., H. influenzae, Brucella sp.*, and *Listeria monocytogenes*, and various extracts from the same bacteria (37–41). The type of interferons were to a large extent interferon α/β but also IFNγ was present, especially in the spleens (33).

Bacteria or extracts of bacteria can also induce interferon production in vitro in cultures of different types of cells. The majority of these experiments were performed with splenocytes, that produce mostly IFNγ, but other cell types, such as fibroblasts can also produce IFNβ when stimulated with bacterial products, such as extracts from *H. influenzae* (42).

C. Immune Interferon during Bacterial Infection

Various types of bacteria, including gram-negative rods, or extracts from them can induce production of IFN when injected into mouse or rabbits, or if spleen cells from sensitized animals are stimulated in vitro (42–47). Also, addition of bacteria can enhance the production of immune interferon induced by mitogens (48).

V. INTERFERON EFFECT ON GENERAL DEFENSE FACTORS AGAINST BACTERIA

We know that interferons may influence the activities of several nonspecific defense factors. Phagocytosis, the major single factor in the antibacterial host defense is altered by interferons. Pretreatment of both mononuclear or polymorphonuclear phagocytic cells with type I, α or β, interferons generally enhance the ingestion of bacteria, whether the phagocytosis is mediated by nonspecific receptors, Fc-, or C3b-receptors (49,50). This activity is time and dose dependent, and neutralized by specific anti-interferon sera (50). Also, IFNγ enhances phagocytosis, although this effect seems to be more complicated than that of interferon α/β (51). It might be especially interesting to mention that viable bacteria are ingested more actively by interferon-treated phagocytes than by untreated control cells (49–51). Several model studies indicate that this effect is not restricted to in vitro systems. Thus, mice treated with interferons eliminated intravenously injected particles more efficiently than untreated animals (52). Also cells obtained from mice pretreated with interferon phagocytosed more actively than cells from control mice (53). On the other hand, monocytes obtained from patients treated with IFNα for extended periods, had a reduced phagocytic capacity (54).

Interferons generally enhance, or in some instances depress, immunological responses. It is not known whether such alterations have any significance on antibacterial defense. The immune defense is especially involved during repeated infections, and it seems reasonable to believe that interferons might influence these activities.

VI. INTERFERON EFFECT ON INVASIVENESS OF ENTEROBACTERIA IN CELL CULTURES

The effect of interferons on interaction between invasive bacteria and cultivated cells has been studied extensively in our laboratory (11,12,19,22–24,26,55,56). Invasiveness of *S. typhimurium* and *S. paratyphi-B* was inhibited in HEp-2 cell cultures pretreated with human leukocyte and human recombinant alpha-2 interferons. The effect was dose dependent and followed a biphasic curve, and it was neutralized with anti-interferon globulin. Both the number of cells with intracellular bacteria and the number of bacteria per infected cell were significantly reduced. Maximum inhibitory effect was obtained with 100 units/ml of human leukocyte interferon and 500 units of recombinant alpha-2 interferon. Reduction of invasiveness was around 90% at these concentrations of interferon. At lower and higher concentrations of interferon the inhibitory effect was less pronounced (Fig. 1). This effect was not restricted to this particular system. A comparable effect was observed using A-549 cells, an interferon-sensitive line of human laryngeal carcinoma origin. A similar effect was obtained also in mouse fibroblast L-929 cells treated with murine fibroblast interferon (26). The effect was species specific, so murine interferon did not influence invasiveness in human cells and vice versa.

We have examined the effect of human γ IFN on invasiveness of different enterobacteria (56). In HEp-2 cells the effect was very similar to that of type I interferons: invasiveness of *S. typhimurium* was inhibited in a dose-dependent manner. The inhibition potential was high in this cell system, comparable to that of IFNα. Maximum inhibitory effect was seen at 10 units/ml. At this dose, a tenfold reduction of invading bacteria was observed. Compared to its antiviral activity, IFNγ showed a higher anti-invasive effect than IFNα/β. This is in accordance with other nonantiviral effects of interferons.

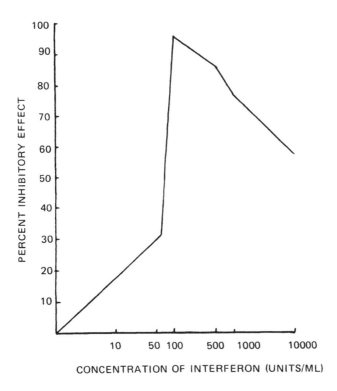

Figure 1 Inhibition of invasiveness of *S. typhimurium* in HEp-2 cells by various concentrations of interferon.

The mechanisms for the anti-invasive effect of interferon is not clear. The invasion process includes several phases that may be influenced by interferon treatment: A phase of reversible adhesion, a phase of irreversible adhesion, a phase of endocytosis, and a phase of intracellular multiplication. Several earlier reports indicate that interferon alters the functional properties of the cell membrane. Thus interferon impaired cell motility, increased plasma membrane rigidity, and reduced lateral movement of some surface receptors (57). Following interferon treatment the cell membrane seems to be less permeable to some low molecular substances (58,59).

To further investigate this direct plasma membrane effect of interferon, the effect on bacterial adhesiveness was studied. The product of virulence-associated plasmid genes are thought to be certain polypeptides essential to endocytosis or adherence. However, in order to study adhesiveness as an isolated phenomenon in an invasive species such as *Salmonella*, endocytosis had to be blocked. This was achieved by treating cells with dihydrocytochalasin B. Cell cultures were challenged with *S. typhimurium*, and the number of adherent bacteria was recorded. Only a slight, but significant, effect of interferon on bacterial adherence was demonstrated: after 3 hr of incubation bacterial adhesiveness was reduced from 59% to 52%. This effect was too small to account for the relatively more marked effect on bacterial invasiveness. Thus, an interferon effect on bacterial adhesiveness does not seem to explain why invasiveness of *S. typhimurium* and *S. flexneri* is inhibited (Table 2).

The other explanation would be that interferon had an effect on the endocytosis process. Several data indicate that interferon treatment alters the penetration of cell membranes by viral particles, namely endocytosis of vesicular stomatitis virus (60), budding of retroviruses (61), and penetrations of soluble substances (58,59). Endocytosis and exocytosis are energy-consuming processes demanding an active cell metabolism. When incubations of bacteria in cell cultures were performed at 4°C, no interferon effect was recorded. Was this

Table 2 Effect of Interferon on Adhesiveness of *Salmonella typhimurium* to Cytochalasin B-Treated HEp-2 Cells

Interferon per (U/ml)	Cells with adhesive bacteria	Adhesive bacteria per cell	Bacteria per 100 cells
0	59	2.9	183
10	58	2.7	166
100	52	2.5	139
1000	55	2.3	133
10000	57	2.7	166

Source: From Ref. 62

energy-demanding process involving protein synthesis? The anti-invasive effect of interferon was fully blocked by substances interfering with ribosomal action. Such substances are cycloheximide, shiga toxin, and abrin, which all attack the 60 S subunit of the ribosome and all completely abolish interferon action on *S. typhimurium* invasiveness in HEp-2 cells. This indicates that the anti-invasive effect of interferon is dependent on a continuous protein synthesis, probably of interferon-induced proteins. Which protein or proteins involved remains to be shown (Table 3).

Interferon effect on invasiveness of *Shigella* bacteria has been studied by Bukholm et al. (62) and by Niesel et al. (63). In HEp-2 cells the effect of IFNα had no significant effect on invasiveness when bacterial inoculum was high. However, Niesel and collaborators were able to demonstrate an effect in this cell system when they kept the bacterial inoculum very low and only incubated the bacteria with cell culture for 20 min (63). Earlier experiments have shown that treatment interfering with cellular protein synthesis altered the anti-invasive effect of interferons (19). It was therefore logical to test whether shiga toxin had any effect on this model. All *Shigella* species secrete certain amounts of this potent toxin to the medium (64), and in cell cultures susceptible to the toxin at extremely low concentrations. For example, 20 pg/ml is sufficient to inhibit protein synthesis. HEp-2 cells are usually susceptible to shiga toxin, they provide the specific receptor, and are able to produce endocytosis of the toxin (65). Also in our system we could demonstrate a specific toxin effect of purified shiga toxin as well as of filtrates from our *S. flexneri* cultures. Experiments were therefore repeated in a cell culture system not susceptible to the toxin, the A-549 cells. In these cells purified toxin had no effect on protein synthesis and produced no cytotoxic effect. When the cells were treated with interferon, invasiveness of *S. flexneri* was inhibited in a dose-dependent manner, very similar to that observed on HEp-2 cells infected with *S. typhimurium*. The percentage of infected cells was reduced from 40% in the controls to 9% at maximum IFN

Table 3 Effect of Ribosomal 60S Subunit Inactivators on Bacterial Invasiveness in HEp-2 Cell Cultures Pretreated with Interferon

Interferon (U/ml)	Toxin type	Toxin conc. (μg/ml)	Percent inhibition of invasiveness
	–	–	0
500	–	–	82
500	cycloheximide	1	0
500	shiga toxin	0.0001	0
500	abrin	10	0

Source: From Ref. 62

inhibition. The maximum inhibitory effect was observed at 500 u/ml with human leukocyte interferon and 1000 u/ml of recombinant alpha-2 interferon. Similar findings were recorded for enteroinvasive strains of *E. coli* and other members of the *Shigella* genus. However, our strain of *S. sonnei* was also partially inhibited in the HEp-2 cells by interferon treatment probably because *S. sonnei* is a weak toxin producer.

Gamma interferon had no effect on invasiveness of *S. flexneri* or enteroinvasive *E. coli* in HEp-2 cells, but was easily demonstrated in A-549 cells similar to that on *Salmonella*. The effect was dose dependent and maximal inhibitory effect was recorded at 10 u/ml (62, and unpublished data).

We also infected HEp-2 cells with measles virus before we inoculated them with *Shigella* and observed an increased invasiveness shortly after virus infection (24). When these virus-infected cell cultures were pretreated with interferons, bacterial invasiveness was reduced. This reduction also followed a biphasic dose-response curve and not a linear one as expected if the interferon effect was due to an effect on virus multiplication only. In fact, the dose-response curve was very similar to that observed when using A-549 cells. We have at present no evidence of any transformation of A-549 cells after measles infection that would turn them less susceptible to shiga toxin. However, the measles virus-infected HEp-2 cells had acquired new properties regarding response to interferon treatment compared to the noninfected mother cells.

Interferon effect was also tested on invasiveness of *Y. enterocolitica*. No effect was observed when interferon-treated HEp-2 cells were infected with *Y. enterocolitica* 0:3 or 0:8 with intact virulence-associated plasmids. Both type I and type II interferons were tested. However, it seems as plasmid-cured variants are more susceptible to interferon treatment (data not published).

The last part of the invasive process that can be influenced by interferon is the phase of intracellular multiplication. Although most studies, including our own, conclude that interferon has a significant effect on cellular endocytosis of bacteria, most studies do not exclude the possibility of effect on intracellular multiplication. Indeed, some earlier studies concluded that interferons can influence this phase. Green reported in 1973 (66) that lysates from cells treated with interferons inhibited extracellular growth of *S. flexneri*. This report could indicate that the intracellular environment is altered in interferon-treated cells. Also Gober and collaborators (67) showed that intracellular environment could be significantly altered by interferon treatment when they demonstrated that interferons and interferon inducers could inhibit the intracellular growth of *S. flexneri*. In addition to these experiments, Whitaker-Dowling and co-workers (68) have shown more recently that intracellular growth of *Legionella* species was also inhibited. We know that none of these intestinal pathogen bacteria are affected directly by interferon in in vitro cultures, thus this effect on intracellular multiplication must be mediated by host cell mechanisms presently not known.

VII. CONCLUSIONS

Both in vitro experiments and animal experiments indicate that interferons can influence the development of infection with intestinal pathogen gram-negative rods. These bacteria do induce interferon production in the organism and interferon also can be present following preceding viral infections. Both general host mechanisms and local interaction between target cells and bacteria might be affected. The mechanism of interferon effect on bacterial interaction of cell is only partly known, but several phases of intracellular infection are involved.

REFERENCES

1. LaBrec, E.H., H. Schneider, T.J. Magnani, and S.B. Formal. 1964. Epithelial cell penetration as an essential step in the pathogenesis of bacillary dysentery. *J. Bacteriol. 88*:1503–1518.

2. Parker, M.T. 1984. Enteric infections: typhoid and paratyphoid fever. In *Topley and Wilson's Principles of Bacteriology, Virology and Immunology.* Edward Arnold, London, Vol. 3, pp. 407–433.

3. Une, T. 1977. Studies on the pathogenicity of *Yershinia enterocolitica.* II. Interaction with cultured cells in vitro. *Microbiol. Immunol. 21*:365–377.

4. Serény, B. 1957. Experimental keratoconjunctivitis shigellosa. *Acta Microbiol. Hung. 4*:367–376.

5. Formal, S.B., H.L. DuPont, R. Hornick, M.J. Snyder, J. Libonati, and E.H. LaBrec. 1971. Experimental models in the investigation of virulence of dysentery bacilli and *Escherichia coli. Ann. NY Acad. Sci. 176*:190–196.

6. O'Brien, A., D.L. Rosenstreich, I. Scher, G. Campbell, R.P. MacDermott, and S. Formal. 1980. Genetic control of susceptibility to *Salmonella typhimurium* in mice: Role of the LPS gene. *J. Immunol. 124*:20–24.

7. Gerber, D.F., and H.M.S. Watkins. 1961. Growth of *Shigella* in monolayer tissue cultures. *J. Bacteriol. 82*:815–822.

8. Day, N.P., S.M. Scotland, and B. Rowe. 1981. Comparison of an HEp-2 tissue culture test with the Sereny test for detection of enteroinvasiveness in *Shigella* spp. and *Escherichia coli. J. Clin. Microbiol. 13*:596–597.

9. Rollag, H., and T. Hovig. 1984. Phagocytosis of non-opsonized Escherichia coli by mouse peritoneal macrophages. An electron microscopic study. *Zbl. Bakt. Hyg. A 257*:93–107.

10. Rollag, H. 1979. Uptake of non-opsonized *E. coli* by unstimulated mouse peritoneal macrophages. *Acta Path. Microbiol. Immunol. Scand.* Sect. C *87*:99–105.

11. Bukholm, G., B.V. Johansen, and E. Namork. 1984. A method differentiating between bacterial adhesiveness and invasiveness in cell culture monolayer. *J. Microscopy 133*:79–81.

12. Bukholm, G., B.V. Johansen, E. Namork, and J. Lassen. 1982. Bacterial adhesiveness and invasiveness in cell culture monolayer. 1. A new light optical method evaluated by scanning electron microscopy. *Acta Path. Microbiol. Immunol. Scand.* Sect. B *90*:403–408.

13. Hale, T.L., and P.F. Bonventre. 1979. *Shigella* infection of Henle intestinal epithelial cells: Role of the bacterium. *Infect. Immun. 24*:879–886.

14. Hale, T.L., R.E. Morris, and P.F. Bonventre. 1979. *Shigella* infection of Henle intestinal epithelial cells: Role of host cells. *Infect. Immun. 24*: 887–894.

15. Kaplan, P., C.E. Benson, and J.S. Gots. 1978. Abstr. Ann. Meet. *Am. Soc. Microbiol. 41*:20.

16. Sansonetti, P.J., T.L. Hale, G.J. Dammin, C. Kapfer, H.H. Collins Jr., and S.B. Formal. 1983. Alterations in the pathogenicity of *Escherichia coli* K-12 after transfer of plasmid and chromosomal genes from *Shigella flexneri. Infect. Immun. 39*:1392–1402.

17. Jones, G.W., D.K. Rabert, D.M. Svinarich, and H.J. Whitfield. 1982. Association of adhesive, invasive and virulent phenotypes of *Salmonella typhimurium* with autonomous 60-megadalton plasmids. *Infect. Immun. 38*:476–486.

18. Helmuth, R., R. Stephan, C. Bunge, B. Hoog, A. Steinbeck, and E. Bulling. 1985. Epidemiology of virulence-associated plasmids and other membrane protein patterns within seven common *Salmonella* serotypes. *Infect. Immun. 48*:175–182.

19. Bukholm, G. 1984. Effect of cytochalasin B and dihydrocytochalasin B on invasiveness of entero-invasive bacteria in HEp-2 cell cultures. *Acta Path. Microbiol. Immunol. Scand.* Sect. B *92*:145–149.

20. Kihlström, E., and L. Nilssson. 1977. Endocytosis of *Salmonella typhimurium* 395 MS and MR 10 by HeLa cells. *Acta Path. Microbiol. Scand.* Sect. B *85*:322–328.

21. Tzipori, S., T. Makin, M. Smith, and F. Krautil. 1982. Enteritis in foals induced by rotavirus and enterotoxigenic *Escherichia coli. Aus. Vet. J. 51*:20–23.

22. Bukholm, G, and M. Degré. 1984. Invasiveness of *Salmonella typhimurium* in HEp-2 cell cultures preinfected with coxsackie B 1 virus. *Acta. Path. Microbiol. Immunol. Scand.* Sect. B *92*:45–51.

23. Bukholm, G., M. Holberg-Petersen, and M. Degré. 1985. Invasiveness of *Salmonella typhimurium* in HEp-2 cell cultures preinfected with UV-inactivated coxsackie virus. *Acta Path. Microbiol. Immunol. Scand.* Sect. B *93*:61–65.

24. Bukholm, G., K. Modalsli, and M. Degré. 1986. Effect of measles-virus infection and interferon treatment on invasiveness of *Shigella flexneri* in HEp-2-cell cultures. *J. Med. Microbiol. 22*:335–242.

25. Izadkhah, Z., A.D. Mandel, and G. Sonnenfeld. 1980. Effects of treatment of mice with sera containing gamma interferon on the course of infection with *Salmonella typhimurium. J. IFN Res. 1*:137–145.

26. Bukholm, G., B.P. Berdal, C. Haug, and M. Degré. 1984. Mouse fibroblast interferon modifies *Salmonella typhimurium* infection in infant mice. *Infect. Immun. 45*:62–66.

27. Degré, M. 1971. Synergistic effect in viral-bacterial infection. 5. Functional studies on the role of the ciliary activity in the mouse trachea. *Acta Path. Microbiol. Scand.* Sect. B *79*:137–141.

28. Degré, M. 1970. Synergistic effect in viral-bacterial infection. 2. Influence of viral infection on the phagocytic ability of alveolar macrophages. *Acta Path. Microbiol. Scand.* Sect. B *78*:41-50.

29. Jakab, G.J., and G.M. Green. 1976. Defect in intracellular killing of *Staphylococcus aureus* within alveolar macrophages in Sendai virus-infected murine lungs. *J. Clin. Invest. 57*:1533-1539.

30. Michaels, R.H., M.M. Wienberger, and M. Ho. 1965. Circulating interferon-like viral inhibitor in patients with meningitis due to *Haemophilus influenzae. New Engl. J. Med. 272*:1148-1162.

31. Haahr, S. 1968. The occurrence of interferon in the cerebrospinal fluid in patients with bacterial meningitis. *Acta Path. Microbiol. Scand. 73*:264-274.

32. Howie, V., R.B. Pollard, K. Kleyn, B. Lawrence, T. Peskuric, K. Paucker, and S. Baron. 1982. Presence of interferon during bacterial otitis media. *J. Infect. Dis. 145*:811-814.

33. Baron, S., V. Howie, M. Langford, E.M. Macdonald, G.J. Stanton, J. Reitmeyer, and D.A. Weigent. 1982. Induction of interferon by bacteria, protozoa, and viruses: defensive role. *Tex. Rep. Biol. Med. 41*:150-157.

34. Nakane, A., T. Minigawa, and I. Yasuda. 1985. Induction of alpha/beta interferon in mice infected with *Listeria monocytogenes* during pregnancy. *Infect. Immun. 50*:877-880.

35. Degré, M., and H. Dahl. 1974. Interferon production and prevention of viral infections in mice by components of a mixed bacterial vaccine. *Acta Path. Microbiol. Scand.* Sect. B *82*:904-910.

36. Singer, S.H., and M.C. Hardegree. 1971. Induction of interferon by bacterial vaccines and allergenic extracts. *J. Allergy 47*:332-340.

37. Ho, M. 1964. Interferon-like viral inhibitor in rabbits after intravenous administration of endotoxin. *Science 145*:1472-1474.

38. Grossberg, S.E., G. Burleson, P. Morahan, and P. Jameson. 1972. A bacterial protein-inducing antiviral resistance and high titers of interferon. *Progr. Immunobiol. Standard. 5*:274-278.

39. Galabov, A.S., and S.M. Galabov. 1973. Interferon induction by detoxicated bacterial endotoxins. *Acta Virol. 17*:493-500.

40. Youngner, J.S., D.S. Feingold, and J.K. Chen. 1973. Involvement of a chemical moiety of bacterial lipopolysaccharide in production of interferons in animals. *J. Infect. Dis. 128*:227-231.

41. Feingold, D.S., G. Keleti, and J.S. Youngner. 1976. Antiviral activity of *Brucella abortus* preparations: separation of active components. *Infect. Immun. 13*:763-767.

42. De Clercq, E., and T.C. Merigan. 1969. An active interferon inducer obtained from *Hemophilus influenzae* type B. *J. Immunol. 103*:899-906.

43. Kojima, Y., F. Yoshida, and Y. Nakase. 1973. Interferon production by *Bordetella pertussis* components in rabbits and rabbit cell cultures. *Jap. J. Microbiol. 17*:160-161.

44. Tanamoto, K., C. Abe, J.Y. Homma, and Y. Kojima. 1979. Regions of the lipopolysaccharide of *Pseudomonas aeruginosa* essential for antitumor and interferon-inducing activities. *Eur. J. Biochem. 97*:623-629.

45. Cavaillon, J.-M., Y. Riviere, J. Svab, L. Montagnier, and J.E. Alouf. 1982. Induction of interferon by *Streptococcus pyogenes* extracellular products. *Immunol. Letters.* 5:323-326.

46. Heremans, H., M. De Ley, A. Billiau, and P. De Somer. 1982. Interferon induced in mouse spleens by *Staphylococcus aureus. Cell. Immunol.* 71: 353-364.

47. Blanchard, D.K., T.W. Klein, H. Friedman, and W.E. Stewart II. 1985. Kinetics and characterization of interferon production by murine spleen cells stimulated with *Legionella pneumophilia* antigens. *Infect. Immun.* 49:719-723.

48. Havell, E.A. 1982-1983. Augmented capacity for alpha and beta interferon production during bacterial infection. *Ann. Rep. Trudeau Inst.* 38-40.

49. Rollag, H., and M. Degré. 1981. Effect of interferon preparations on the uptake of non-opsonized *Escherichia coli* by mouse peritoneal macrophages. *Acta Path. Microbiol. Scand.* Sect. B 89:153-159.

50. Melby, K., M. Degré, and T. Midtvedt. 1982. Effect of leukocyte interferon on phagocytic activity of polymorphonuclear leukocytes. *Acta Path. Microbiol. Scand.* Sect. B 90:181-184.

51. Rollag, H., M. Degré, and G. Sonnenfeld. 1984. Effects of interferon-α/β and interferon-γ preparations on phagocytosis by mouse peritoneal macrophages. *Scand. J. Immunol.* 20:149-155.

52. Degré, M., and H. Rollag. 1979. Influence of interferon on the in vivo phagocytic activity of RES cells. *J. Reticuloendothel. Soc.* 25:489-493.

53. Donahoe, R.M., and K.Y. Huang. 1976. Interferon preparations enhance phagocytosis in vivo. *Infect. Immun.* 13:1250-1257.

54. Einhorn, S., H. Blomgren, C. Jarstrand, H. Strander, and J. Wasserman. 1983. Influence of interferon-therapy on functions of the human immune system. In *The Biology of the Interferon System.* Edited by E. De Maeyer and H. Schellekens. Elsevier, Amsterdam, pp. 347-351.

55. Bukholm, G., and M. Degré. 1983. Effect of human leukocyte interferon on invasiveness of *Salmonella* species in HEp-2 cell cultures. *Infect. Immun.* 42:1198-1202.

56. Bukholm, G., and M. Degré. 1985. Effect of human gamma interferon on invasiveness of *Salmonella typhimurium* in HEp-2 cell cultures. *J. IFN Res.* 5:45-53.

57. Pfeffer, L.M., E. Wang, and I. Tamm. 1980. Interferon effects on microfilament organization, cellular fibronectin distribution and cell motility in human fibroblasts. *J. Cell Biol.* 85:9-17.

58. Brouty-Boyé, D., and M.G. Tovey. 1978. Inhibition by interferon of thymidine uptake in chemostatic cultures of L1210 cells. *Intervirology* 9:243-252.

59. Degré, M. 1978. Effect of human leukocyte interferon on the permeability of cytoplasmamembrane of cultured cells. *Acta Path. Microbiol. Scand.* Sect. B 86:303-307.

60. Whitaker-Dowling, P.A., D.K. Wilcox, C.C. Widnell, and J.S. Youngner. 1983. Interferon mediated inhibition of virus penetration. *Proc. Natl. Acad. Sci.* 80:1083-1086.

61. Billiau, A., V.G. Edy, H. Sobis, and P. DeSomer. 1974. Influence of interferon on virus particle synthesis in oncornavirus-carrier lines. II. Evidence for a direct effect on particle release. *Int. J. Cancer 14*:335–340.

62. Bukholm, G., M. Bergh, and M. Degré. 1987. Mechanisms for anti-invasive effect of interferon. In *The Biology of the Interferon System*. Edited by K. Cantell and H. Schellekens. Martinius Nijhoff Publishers, Dordrecht, Boston, Lancaster, pp. 183–190.

63. Niesel, D.W., C.B. Hess, Y.C. Cho, K.D. Klimpel, and G.R. Klimpel. 1986. Natural and recombinant interferons inhibit epithelial cell invasion by *Shigella* spp. *Infect. Immun. 52*:828–833.

64. Bartlett, A.V. III, D. Prado, T.C. Cleary, and L. Pickering. 1986. Production of shiga toxin and other cytotoxins by serogroups of shigella. *J. Infect. Dis. 154*:996–1002.

65. Eiklid, K., and S. Olsnes. 1980. Interaction of *Shigella shigae* cytotoxin with receptors on sensitive and insensitive cells. *J. Receptor Res. 1*:199–213.

66. Green, J.A. 1973. Inhibition of *Shigella flexneri* by lysates of cells treated with interferon-containing preparations. *Infect. Immun. 7*:1006–1008.

67. Gober, L.L., A.E. Friedman-Kien, E.A. Havell, and J. Vilcek. 1972. Suppression of the intracellular growth of *Shigella flexneri* in cell cultures by interferon preparations and polyinosinic-polycytidylic acid. *Infect. Immun. 5*:370–376.

68. Whitaker-Dowling, P., J.D. Dowling, L. Liu, and J.S. Youngner. 1986. Interferon inhibits the growth of *Legionella micdadei* in mouse L cells. *J. IFN Res. 6*:107–114.

69. Gould, C.L., and G. Sonnenfeld. 1987. Effect of treatment with interferon-γ and concavalin A on the course of infection of mice with *Salmonella typhimurium* strain LT-2. *J. IFN Res. 7*:255–260.

Index

Printed and bound by CPI Group (UK) Ltd, Croydon, CR0 4YY

17/10/2024

01775703-0005